MW01534026

FINITE MATHEMATICS AND CALCULUS WITH APPLICATIONS

The IEP Series in Mathematics
under the consulting editorship of
Richard D. Anderson, Louisiana State University
Alex Rosenberg, Cornell University

FINITE MATHEMATICS AND CALCULUS WITH APPLICATIONS

G.M. STRATOPOULOS
United States International University, San Diego

IEP, A Dun-Donnelley Publisher, New York

Published simultaneously in Canada by Fitzhenry & Whiteside, Ltd., Toronto.

Library of Congress Cataloging in Publication Data
Stratopoulos, G M 1931–
Finite mathematics and calculus with applications.

(The IEP series in mathematics)
Bibliography: p.
Includes index.
1. Mathematics—1961– I. Title: Finite mathematics and calculus . . .
QA39.2.S77 1976 510 75-34204
ISBN 0-7002-2486-6

IEP A Dun-Donnelley Publisher
666 Fifth Avenue
New York, New York 10019

Manufactured in the United States of America

CONTENTS

PREFACE

I began this book with the conviction that most students need mathematics. Whether they will be professionally concerned with the behavioral, management, or social sciences, or will spend their working lives in administrative or corporate offices, they will inevitably be involved with mathematical applications and will, ideally, realize the pleasurable gains in productivity that come with the mastery of certain techniques, and the awareness of others.

Mathematics is intellectually stimulating, aesthetically pleasing, and certainly useful. Some of the newer applied areas as well as many of those developed and refined over many years are applicable in an amazing range of circumstances. So I have brought together, and tried to order logically, elements of optimization theory, the quantitative underpinnings of management science, operations research, econometrics, linear algebra, and linear programming. I have presented game, sampling, and probability theories as essential tools in the decision-making process. The content of the book became, inevitably, an introduction to topics and applications in finite mathematics and calculus for the behavioral, management, and social sciences. As it turns out, this content is also very suitable for those taking mathematics as a liberal arts requirement. Given the backgrounds and interests of my students, the level of presentation is quite elementary but always related to real world situations, and dependent only on some familiarity with intermediate algebra.

Proofs of theorems are in general omitted. Instead, examples are extensively used to demonstrate the validity of theorems and corollaries. Frequently, illustrations from real situations serve to introduce and clarify certain mathematical notions. In addition, the substantial number of examples and exercises in each section demonstrate both mathematical principles and their relevancy. Answers to selected problems are provided within each exercise. For further discussions of the topics in the book and for proofs of certain theorems and corollaries, we provide references for supplementary readings at the end of each chapter. Throughout the text, "footnote numbering" is used to identify the appropriate reference in the listing of supplementary readings at the end of each chapter.

The table of contents describes the material that can adequately be covered in two semesters or in three quarters either as a prerequisite course or as a liberal arts requirement in mathematics. To provide flexibility in the selection and sequence of content coverage many chapters are written independently of others. For instance, one instructor may choose to introduce the calculus section after presenting Chapters 1 and 2. Another may elect to study only the chapters on determinants and matrices and their applications on (a) linear systems of equations and inequalities, (b) linear transformations and the characteristic value problem, (c) Markov chains, and (d) linear programming and game theory without any loss of continuity.

I would like to express my appreciation to the following people who assisted me in the selection and organization of the materials in this manuscript: Dr. Phillip Beukema, professor of business and economics; Dr. David Jacobs, professor of psychology; Dr. James Gehrmann, professor of mathematics; my secretary, Phyllis Smart, for being such a beautiful typist; and my wife, Dr. Irene C. Stratopoulos, professor of English, for her very substantial contributions to the book.

PART ONE

Finite Mathematics

CHAPTER 1

Theory of Sets

1.1 Introduction

The construction of mathematical models essential to the analysis and solution of problems in the behavioral, social, management, and physical sciences requires a certain amount of technical vocabulary, some understanding of the laws of logic,[1] and certainly the assumption of certain statements referred to as either axioms or postulates. Thus we shall begin the development of finite mathematics and calculus with the assumption that the student has been introduced to the basic characteristics of mathematics in a high school curriculum and that he has taken at least an intermediate algebra course. Moreover, since the language and techniques of set theory provide not only the necessary technical vocabulary but often the basic tools for the construction of appropriate mathematical models, we begin with an introduction to the fundamentals of set theory. Its extensive and convenient application will become increasingly apparent as we consider the study of combinatorial analysis, sampling theory, and theory of probability as well as other fundamental topics in finite mathematics and calculus. The theory of sets was created and first used by the great mathematician Georj Cantor (1845–1918), who was born in Russia, educated in Germany, and taught there during the years 1874–1895 when his works on this subject were published.

1.2 Definitions and Symbols

A set is simply a collection, class, family, or aggregate of distinct objects that are called "elements" or "members of the set." The elements of a set may be people, points, lines, numbers, or mental concepts. So we may be speaking of the set of counting (natural, positive integers) numbers 1, 2, 3, . . . , or the set of people with IQ's less than 100, or the set of all persons under welfare in the state of California, or even the set of all banks that pay 7% interest on all savings accounts. From the definition of a set it is clear that there exist two types of sets: namely, those with a finite number of elements, on the one hand; and those with an infinite number of elements, on the other.

DEFINITION 1.1 We say that a set is finite if the number of its elements can be represented by a certain natural number n. Otherwise we say the set is infinite.

For example, the set of all natural numbers is infinite, and the set of all people with IQ's less than 100 is finite.

It is customary to denote sets by means of capital letters, and the elements of sets by small letters. We write $a \in A$ to denote that "a is a member of the set A," or to say that "a belongs to A." Similarly, we write $a \notin A$ to denote that "a does not belong to A," or "a is not an element of set A."

A set may be specified in the following two ways:

i. By listing all its elements within braces.
ii. By stating a characteristic property or properties that determine whether or not a given object is an element of that set.

The notation usually adopted for the second method comprises two parts, separated by a vertical line, within braces: The first part tells us what type of element is being considered and the second part specifies the characteristic property. For example, if N is the set of all natural numbers and if S is the set of all those natural numbers whose square is less than 12, then S can be specified as follows:

$$S = \{1, 2, 3\}$$

or

$$S = \{x \in N \mid x^2 < 12\}$$

The latter is read as "S is the set of all natural numbers x for which x^2 is less than 12."

The empty or void set denoted by $\emptyset$ is the set that contains no elements at all. Although the empty set may seem at first to be an artificial notion, its usefulness and acceptance as a bona fide set become increasingly apparent in the study of the algebra of sets.[2] In such an algebra the empty set plays the role (as well as other roles) comparable to the zero number in the algebra of real or complex numbers.

DEFINITION 1.2 Two sets A and B are said to be equal, written $A = B$, if and only if they contain the same elements.

In terms of the membership relation the above definition may be expressed as $A = B$ means "$x \in A$ if and only if $x \in B$." For example, if

$$A = \{1, 2, 3\}$$

if

$$B = \{1, 2, 3, 4\}$$

and if

$$C = \{x \in N \mid x^2 < 13\}$$

then the following are true:

a. $A \neq B$ since $4 \in B$ but $4 \notin A$.
b. $A = C$ since $C = \{x \in N \mid x^2 < 13\}$ implies that

$$C = \{1, 2, 3\}$$

and

$$A = \{1, 2, 3\}$$

The previous example indicates that every element of the set A is an element of the set B, but A is not the same as the set B. The idea illustrated in this example is generalized in Definition 1.3.

DEFINITION 1.3 A set A is said to be a subset of a set B, written $A \subseteq B$, if and only if every element of A is an element of B; A is said to be a proper subset of B, written $A \subset B$, if and only if $A \subseteq B$ but $A \neq B$. Thus $A \subseteq B$ means "if $x \in A$, then $x \in B$." For example, if

$$A = \{x \in N \mid x^3 \leq 27\} \quad \text{and} \quad B = \{x \in N \mid x^3 \leq 64\}$$

then

$A \subseteq B$ and $A \neq B$

since

$$A = \{1, 2, 3\}$$
$$B = \{1, 2, 3, 4\}$$

and every element of A is in B, whereas the number 4 is in B but not in A. Thus A is a proper subset of B. The subset relationship $A \subseteq B$ is often written as $B \supseteq A$.

Theorem 1.1

a. Every set is a subset of itself.
b. The empty set is a subset of every set.
c. If A and B are any two sets, then $A = B$ if and only if $A \subseteq B$ and $B \subseteq A$.

In examining different relationships among sets, let us suppose we have the following three sets:

1. The set B of all independent banks in the state of California.
2. The set P of all presidents of the independent California banks.
3. The set E of all employees of the independent California banks.

It is clear that to each element b in B there corresponds one and only one element p in P, and conversely to each element of P there corresponds one and only one element of B. However, this type of correspondence does not hold true between the sets B and E since the Bank of America in California has more than one employee. This type of relation between two sets is generalized in Definition 1.4.

DEFINITION 1.4 Two sets A and B are said to be in a one-to-one correspondence if and only if their elements can be paired in such a way that each element of A is paired with one and only one element of B, and conversely. Such sets are said to be equivalent, and the equivalence between them is specified by writing $A \sim B$.

In view of this definition we say that the sets B and P defined above are equivalent or that there can be established a one-to-one correspondence between them. On the other hand the sets B and E are obviously not equivalent. Another example of set equivalence is provided by the set M of all married males and the set F of all married females in a society that allows no polygamy for either sex. The necessary one-to-one correspondence for M and F is provided by the marriage license. Yet another example of such a one-to-one correspondence between two infinite sets, in this case, can be established between the set E of all even numbers, and the set O of all odd numbers. (How?)

DEFINITION 1.5 We say that a set of numbers S is bounded if and only if there exist numbers m and M such that $m \leq x \leq M$ for all $x \in S$. The numbers m and M are called "lower" and "upper" bounds of S, respectively. Otherwise we say the set is unbounded. Moreover, we say that S is bounded above (or below) if and only if it has an upper (or lower) bound exclusively. The different bounds of a set may or may not be members of the set.

For example, let

$$A = \{x, \text{real numbers} \mid x > 20\}$$

$$B = \{x, \text{real numbers} \mid x < 0\}$$

and

$$C = \{x \in N \mid 1 \leq x \leq 10\}$$

Then A is bounded below, and a lower bound is $m = 20$; B is bounded above, and upper bound is $M = 0$; and C is bounded with a lower bound $m = 1$ and an upper bound $M = 10$. On the other hand the set of all real numbers is unbounded. (Why?)

DEFINITION 1.6 By a variable we mean a symbol representing an unspecified element of a given set containing more than one element. A symbol representing the element of a set that contains only one member is called a "constant."

For example, if

$$A = \{1, 2, 7\} \qquad \text{and} \qquad B = \{5\}$$

then $x \in A$ implies x is a variable, and $y \in B$ implies y is a constant.

Exercises

Whenever possible, designate each of the sets in Exercises 1 through 8 in two ways.

1. The natural numbers between 3 and 9.
2. The even numbers between 2 and 21.
3. The set of stocks listed on the New York Stock Exchange.
4. The set of all real numbers satisfying each of the equations

   a. $6x^2 + x - 2 = 0$.
   b. $5x^2 + 3x + 7 = 0$.

5. The set of all possible outcomes in the game of tossing and matching pennies by two players.

   Ans. Four in number

6. The set of possible stimulus configurations available in an experiment on concept formation in small children in which the experimenter wishes to vary the dimensions of color, size, and number of petals in a flower sketch whenever there are

   a. Two colors, two sizes, and two petal configurations.
   b. Three colors, three sizes, and three petal configurations.

   Ans. (b) 27

7. The set of possible scores of a person required to take three different tests T_1, T_2, and T_3 in each of the two different areas E_1 and E_2.

   Ans. Six in number

8. The set of all numbers x such that

   a. $2^x > 4$.
   b. $2^x < 4$.
   c. $1/16 \leq 2^x \leq 64$.

   Ans. $-4 \leq x \leq 6$

9. About the previous eight exercises determine the following.

   a. All the finite sets.
   b. All the infinite sets.
   c. All the bounded sets.
   d. All the sets bounded above or below only.
   e. All the unbounded sets.

10. If $A = \{a, b, c\}$, then list all the subsets of A.

    Ans. Eight

11. If a psychiatrist wishes to meet individually and in all possible groups of twos and threes a group of four of his patients, how many different meetings must he schedule?

    Ans. 14

12. Let the sets U, A, B, and C be defined as follows:

    $U = \{1, 2, 3, 4, 5, 6\}$.
    $A = \{1, 2, 3, 4\}$.
    $B = \{4, 5, 6\}$.
    $C = \{4, 5\}$.

Then, replace the comma with $\subseteq$ or $\nsubseteq$ in each of the following:

a. A, U. b. C, A. c. A, B. d. C, B.

13. About the sets:

$N = \{$all natural numbers$\}$.
$A = \{$all even natural numbers$\}$.
$B = \{$all odd natural numbers$\}$.
$C = \{x \mid 1 < x < 5\}$.
$D = \{x \mid x < 9\}$.

Determine which of the following are true?

a. $A \sim D$ b. $C = D$ c. $C \subset D$
d. $A \sim B$ e. $A = B$ f. $D \subset C$
g. $A \sim N$ h. $A \subset B$ i. $A \subset N$
j. $C \sim D$ k. $C \subset A$ l. $C \subset B$.

14. State all the possible one-to-one correspondences between the sets

$A = \{1, 2, 3, 4\}$ and $B = \{a, b, c, d\}$

Ans. 24 in number

1.3 The Algebra of Sets

In practice it is often useful as well as very convenient to adopt geometric language for sets. For instance, we may choose to call "point" an "element of a set" even though the element has no obvious geometric character. When we adopt such descriptive language, we must bear in mind that we do so only for convenience and must carefully refrain from assuming any properties not legitimately established.

Such a geometric interpretation of sets suggests the use of sketches, called "Venn diagrams," to represent sets and relations between sets. Thus the subset relation $A \subset B$ can be shown graphically as in Figure 1.1.

John Venn (1834–1923), the great English logician and ordained priest, who resigned his order in 1883 to devote all his time to the study and teaching of logic, first employed these diagrams in his work *Symbolic Logic*, published in 1881.

These diagrams provide insight concerning sets, and often suggest methods by which statements about sets can be proved or disproved. However, diagrams are not valid substitutes for formal proofs.

With these preliminaries in mind, let us now consider a set U, called the "universal set," together with the set K consisting of all the subsets of U, and let us define the following important operations on the set K, that is, on the subsets of U.

Figure 1-1

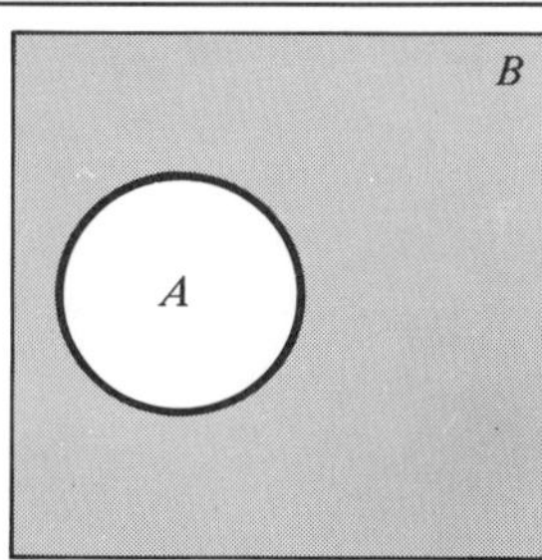

DEFINITION 1.7 Let A and B be any two subsets of U, and let us define the following sets:

i. The set $C = \{x \in U \mid x \in A \text{ or } x \in B\}$. This set is called the "union" of A and B, and it is denoted by $C = A \cup B$.
ii. The set $D = \{x \in U \mid x \in A \text{ and } x \in B\}$. This set is called the "intersection" of A and B, and it is denoted by $D = A \cap B$.
iii. The set $A' = \{x \in U \mid x \notin A\}$. This set is called the "complement" of A with respect to U.

Therefore if A and B are any elements of K, then the set "A union B," $(A \cup B)$, consists of all elements of the universal set U that are elements of either A or B or both. The set "A intersection B," $(A \cap B)$, consists of all elements of U that are elements of both A and B. Finally, the "complement of A in U" consists of all the elements of U that are not elements of A. The complement of A may be denoted by either A' or $\bar{A}$.

DEFINITION 1.8 We say that two sets A and B are disjoint if and only if $A \cap B = \emptyset$.

The above two definitions can be nicely illustrated by the Venn diagrams shown in Figure 1.2.

Thus in Figure 1.2 the shaded regions of the plane in (a) and (b) represent the union and intersection of the sets A and B, respectively, whereas the shaded region in (c) represents the complement of A with respect to U.

Based on the above definitions, the collection K of all the subsets of a universal set U forms an interesting algebraic system (in many ways similar to that of real and complex numbers). Such algebraic systems, commonly known as Boolean algebras after the name of the great British mathematician George Boole (1815–1864), were first

Figure 1-2

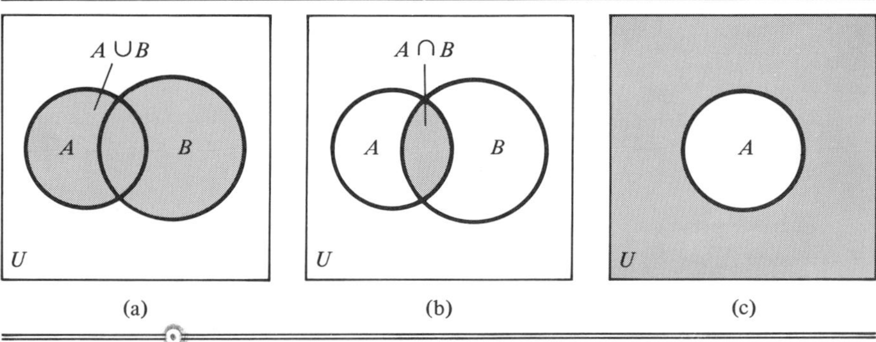

(a) (b) (c)

introduced by him in his book *An Investigation into the Laws of Thought* published in 1854. One of the most prominent applications of Boolean algebras is found in the design of digital computers.[3] Thus K forms a Boolean algebra about which we could prove the following theorems. However, since mathematical rigor is beyond the scope of this book, we shall state these theorems and illustrate only the last two, leaving the demonstration of the remaining ones as an exercise for the student.

Theorem 1.2 *Identity laws* If $A \in K$, then

$$A \cup \emptyset = A$$
$$A \cap \emptyset = \emptyset$$
$$A \cup U = U$$

and

$$A \cap U = A$$

Theorem 1.3 *Complement laws* If $A \in K$, then

$$A \cup A' = U$$
$$A \cap A' = \emptyset$$

and

$$(A')' = A$$

Theorem 1.4 *Idempotent laws* If $A \in K$, then

$$A \cup A = A \quad \text{and} \quad A \cap A = A$$

Theorem 1.5 *Commutative laws* If A and B are elements of K, then

$$A \cup B = B \cup A \quad \text{and} \quad A \cap B = B \cap A$$

Theorem 1.6 *Associative laws* If A, B, and C are elements of K, then

$$A \cup (B \cup C) = (A \cup B) \cup C$$

and

$$A \cap (B \cap C) = (A \cap B) \cap C$$

Theorem 1.7 *Distributive laws* If A, B, and C are elements of K, then

$$A \cup (B \cap C) = (A \cup B) \cap (A \cup C)$$

and

$$A \cap (B \cup C) = (A \cap B) \cup (A \cap C)$$

Theorem 1.8 *De Morgan's laws* If A, B, and C are elements of K, then

$$(A \cup B)' = A' \cap B'$$

and

$$(A \cap B)' = A' \cup B'$$

Now for the purpose of illustrating Theorems 1.7 and 1.8, let

$$U = \{0, 1, 2, 3, 4, 5, 6\}$$
$$A = \{0, 1, 2, 3, 4\}$$
$$B = \{2, 3, 5\}$$

and

$$C = \{2, 3, 6\}$$

Then, U is the universal set, and obviously A, B, and C are subsets of U.

Illustrations

THEOREM 1.7

Since $B \cap C = \{2, 3\}$,

$$A \cup (B \cap C) = \{0, 1, 2, 3, 4\} \cup \{2, 3\}$$

or

$$A \cup (B \cap C) = \{0, 1, 2, 3, 4\}$$

Similarly,

$$A \cup B = \{0, 1, 2, 3, 4, 5\}$$

and

$$A \cup C = \{0, 1, 2, 3, 4, 6\}$$

imply

$$(A \cup B) \cap (A \cup C) = \{0, 1, 2, 3, 4\}$$

That is,

$$A \cup (B \cap C) = (A \cup B) \cap (A \cup C)$$

Now, to show that

$$A \cap (B \cup C) = (A \cap B) \cup (A \cap C)$$

we determine that

$$B \cup C = \{2, 3, 5, 6\}$$

and thus

$$A \cap (B \cup C) = \{2, 3\}$$

Likewise,

$$A \cap B = \{2, 3\} \quad \text{and} \quad (A \cap C) = \{2, 3\}$$

Therefore,

$$(A \cap B) \cup (A \cap C) = \{2, 3\}$$

and

$$A \cap (B \cup C) = (A \cap B) \cup (A \cap C)$$

THEOREM 1.8

$$A \cup B = \{0, 1, 2, 3, 4, 5\}$$

thus

$$(A \cup B)' = \{6\}$$

Now,

$$A' = \{5, 6\} \quad \text{and} \quad B' = \{0, 1, 4, 6\}$$

Therefore

$$A' \cap B' = \{6\} \quad \text{or} \quad (A \cup B)' = A' \cap B'$$

Similarly, since

$$A \cap B = \{2, 3\}$$

we have

$$(A \cap B)' = \{0, 1, 4, 5, 6\}$$

However,

$$A' = \{5, 6\} \qquad \text{and} \qquad B' = \{0, 1, 4, 6\}$$

Therefore

$$(A \cap B)' = A' \cup B'$$

Exercises

1. Using your own examples of sets, illustrate Theorems 1.2, 1.3, 1.4, 1.5, and 1.6.
2. If

$$U = \{-3, -2, -1, 0, 1, 2, 3\}$$
$$A = \{-1, 0, 2, 3\}$$

and

$$B = \{-2, -1, 0, 2\}$$

then illustrate the following:

a. $A \subseteq A \cup B$. b. $A \cap B \subseteq A$. c. $(A \cup B)' = A' \cap B'$.
d. $(A \cap B)' = A' \cup B'$. e. $A \sim B$.

3. Using your own examples of sets, demonstrate the following theorems:

a. $A \subseteq B$ if and only if $A \cup B = B$.
b. $B \subseteq A$ if and only if $A \cap B = B$.
c. If $A \subseteq B$, then $A' \supseteq B'$.

4. Draw Venn diagrams that illustrate each of the following relations:

a. $A \subset B$. b. $A \cap B \neq \emptyset$. c. $A \cap B \supset C$.
d. $A \cap B \subset C$. e. $A \cup B = C$. f. $A \cap B \cap C \neq \emptyset$.

5. If U is the set of all congressmen, F is the set of all congressmen who vote in favor of a bill sponsored by the majority party, and M is the set of minority congressmen, then draw Venn diagrams to illustrate the following: (a) The set of all defectors from the minority party. (b) The set of loyal minority party members.

1.4 Applications

In analyzing the previous discussion of the general theory of sets, we find that the operations union, intersection, and complementation, together with the geometric description of sets, provide a very convenient approach to the building and studying of workable mathematical models. Thus by way of the following examples we consider some models in the behavioral, social, and management sciences.

Example 1.1 In a survey of 100 students by the registrar of a school it was found that the number of students studying various subjects was as follows:

English, 56	English and History, 14
History, 38	English and Mathematics, 12
Mathematics, 30	History and Mathematics, 9
	All three subjects, 5

About these students, the registrar's office would like to know (a) how many students studied none of these subjects; (b) how many students enrolled in english only; and (c) how many studied english and history but not mathematics?

Solution Let E, H, and M be the sets of students studying english, history, and mathematics, respectively. Let Figure 1.3 be a Venn diagram of this set theory problem. Finally, let us write the number of students in each of the sets of the eight basic regions in Figure 1.3. Then we begin with the intersection of the sets E, H, and M; that is, we begin with the five students who are studying all three subjects. Now, by subtracting this number, 5, from the numbers in the intersections of two sets we obtain the numbers for three more regions of Figure 1.3. Continuing in this fashion we obtain the numbers in each of the eight regions of Figure 1.3. Therefore

a. The set $(E \cup H) \cup M$ of students studying at least one of these subjects has $35 + 9 + 20 + 4 + 5 + 7 + 14 = 94$ students, and the set $[(E \cup H) \cup M]'$ of students studying none of these subjects has $100 - 94 = 6$ students.

Figure 1-3

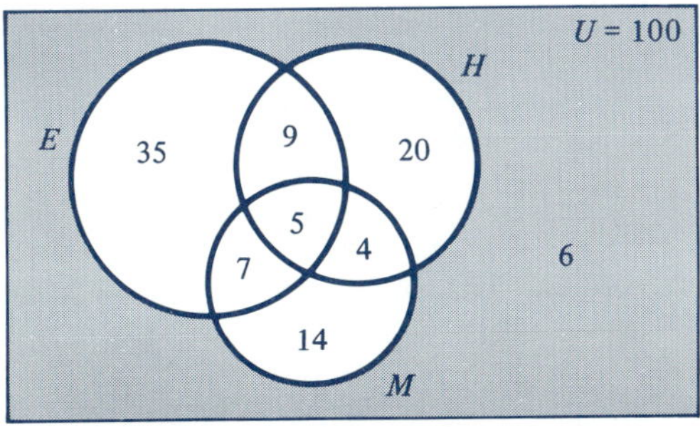

b. The number of students enrolled in english only is 35.
c. The number of students studying english and history but not mathematics is 9.

Example 1.2 In a labor force model let each person be classified according to Table 1.1. Let U and L be the sets of all persons in the labor force and all the employed ones, respectively. Now, if no person with 4 or more years of college is unemployed, then find the relationship of the set of unemployed persons other than those who have neither 4 or more years of college nor only 5 or 6 years of grade school, and the set of unemployed persons having only 5 or 6 years of grade school.

Table 1.1

Place of Residence	Age	Education
R_1: rural, farm	A_1: 20–24	E_1: 5–6 years of grade school
R_2: rural, nonfarm	A_2: 25–29	E_2: 7–8 years of grade school
R_3: city, 2500–25,000	A_3: 30–34	E_3: 1–3 years of high school
R_4: city, 25,000–100,000	A_4: 35–39	E_4: 4 years of high school
	A_5: 40–49	E_5: 1–3 years of college
	A_6: 50–64	E_6: 4 or more years of college

Solution We know that L is the set of all employed persons. Thus let R_1, R_2, R_3, R_4, and E_1, E_2, E_3, E_4, E_5, E_6 be the sets of persons listed in the categories of Table 1.1. Then L', the complement of L, is the set of all unemployed persons. Similarly E_6' is the set of all persons who do not have 4 or more years of college, and E_1' is the set of those who do not have 5–6 years of grade school. The problem requires that the sets

$$L' \cap (E_6' \cap E_1')'$$

and

$$L' \cap E_1$$

be compared. Thus

$$\begin{aligned} L' \cap (E_6' \cap E_1')' &= L' \cap [(E_6')' \cup (E_1')'] && \text{by De Morgan's law} \\ &= L' \cap (E_6 \cup E_1) && \text{by the complement law} \\ &= (L' \cap E_6) \cup (L' \cap E_1) && \text{by the distributive law} \\ &= \emptyset \cup (L' \cap E_1) && \text{since } L' \cap E_6 = \emptyset \end{aligned}$$

Therefore,

$$L' \cap (E_6' \cap E_1')' = L' \cap E_1$$

or the set of unemployed persons other than those who have neither

4 or more years of college nor 5–6 years of grade school, and the set of unemployed persons having only 5 or 6 years of grade school are equal.

Example 1.3 A school psychologist, in studying membership structure of high school cliques, uses the following sets:

U = the set of all students.
M = the set of all male students.
F = the set of all female students.
D = the set of all students who date.
C = the set of all students who are college bound.
O = the set of all students in the group of the "out group."
I = the set of all students in the "in group."

Translate each of the following observations into either one equation or inequality of sets using the symbols,

$$=, \quad \neq, \quad \emptyset, \quad \cap, \quad \cup, \quad ', \quad \subseteq$$

a. Some male students do not plan to attend college.
b. Not all female students date.
c. Those girls who date or plan to attend college are in the "in group."
d. No student is a member of both the "out group" and the "in group."
e. Neither the male students who are noncollege bound nor the girls who are nondaters are members of the "in group."
f. Some boys who do not date but who plan to attend college are members of the "in group."

Solution This problem constitutes little more than a good exercise of definitions and symbols on the operations of sets. Thus the answers to these questions are as follows:

a. $M \cap C' \neq \emptyset$.
b. $F \cap D' \neq \emptyset$.
c. $F \cap (D \cup C) \subseteq I$.
d. $O \cap I = \emptyset$.
e. $(M \cap C') \cup (F \cap D') \subseteq I'$.
f. $M \cap D' \cap C \cap I \neq \emptyset$. (Why?)

Exercises

1. In a survey of dining habits of people, the following information was established:

50 ordered salad.
65 ordered dessert.
50 ordered soup.
20 ordered salad and dessert.
15 ordered salad and soup.
25 ordered dessert and soup.
10 ordered all three.
20 ordered none of the three.

a. How many people were involved in the survey?
b. How many people ordered salad and dessert but not soup?
c. How many people ordered salad but not dessert?
d. How many people ordered salad only?

Ans. (a) 135; (b) 10; (c) 30; (d) 25

2. A survey of female college students revealed the following:

44 were beautiful.
30 were intelligent.
51 were personable.
8 were beautiful and intelligent.
11 were beautiful and personable.
9 were intelligent and personable.
3 were beautiful, intelligent, and personable.

a. How many girls were in the survey?
b. How many girls were beautiful but not personable?
c. How many girls were intelligent or personable but not beautiful?

Ans. (a) 100; (b) 33; (c) 56

3. In a survey attempting to determine the reading habits of people on three magazines X, Y, and Z it was found that

33% read magazine Y.
29% read magazine X.
22% read magazine Z.
13% read magazines Y and Z.
6% read magazines X and Y.
14% read magazines X and Z.
6% read all three magazines.

a. What percent read none of the three magazines?
b. What percent read magazines X and Y but not Z?
c. What percent read magazine X if and only if they read magazine Y?

Ans. (a) 43%; (b) 0%; (c) 6%

4. Seven employees of a company are to meet with the management of the company in groups of three in such a manner that each employee appears with each other employee in one and only one meeting during the sequence of meetings. How many three-employee meetings must be held in order to meet this requirement?

5. A company has a top management committee consisting of four vice-presidents and the president. The committee reaches its decision by simple majority vote.

a. List the set of all possible outcomes of the vote on a given motion, assuming that no member abstains from voting.

b. List the subset of winning outcomes, that is, those outcomes that result in the motion being passed.

In addition list the subset of winning outcomes if

c. The president has veto power.
d. The president has two votes.
e. The president votes only on tied votes.

6. Four patients A, B, C, and D are waiting to consult a psychiatrist. The psychiatrist elects to see them singly or in groups of two, three, or four. List the collection of possible sets of patients with whom the psychiatrist might confer.

Ans. 15

7. The registration of an insurance agents' convention showed that the following types of agents were in attendance:

a. 180 life insurance.
b. 230 fire and casualty.
c. 240 auto insurance.

Of these agents, 140 were selling both auto insurance and fire and casualty insurance, 110 were selling both life insurance and auto insurance, 90 were selling both life and fire and casualty insurance, and 60 were selling all three. If all agents in attendance were selling at least one of the above mentioned types of insurance, then determine the following:

a. How many agents attended the convention?
b. How many agents were selling auto insurance but not fire and casualty?
c. How many agents were specializing in only one type of insurance?
d. How many agents were selling only fire insurance?

Ans. (a) 370; (b) 100; (c) 150; (d) 60

8. In a safety-estimates' prediction, let us assume that 59,000 motor vehicle fatalities will take place in 1978. Of these, say, 17,000 will occur in urban places, 25,000 will occur during daylight hours, and 18,500 will occur at night in rural places. Furthermore, let us predict that 4100 pedestrian fatalities will occur in rural places; 17,900 nonpedestrian fatalities will occur in rural places at night; 260 pedestrian fatalities will occur in urban places during the daylight hours; and 1800 pedestrian fatalities will occur at night in rural places.

a. What portion of daylight accident victims in rural places will be pedestrians?

b. What portion of the total pedestrian fatalities in rural areas will occur at night?
c. What portion of the total fatalities in rural areas will occur during daylight hours?
d. What portion of daylight accident victims in urban places will be nonpedestrians?

9. In a survey of voting habits of 1000 registered voters, it was found that 605 voted in the 1956 presidential election, 595 voted in the 1960 presidential election, and 675 voted in the 1964 presidential election. Of these, 415 voted in both the 1964 and 1960 elections, 385 voted in both the 1956 and 1960 elections, 395 voted in both the 1956 and 1964 elections, and 65 did not vote in any of the three elections.

a. How many voted in all three elections?
b. How many voted in exactly one election?
c. How many voted in two or more elections?

Ans. (a) 255; (b) 250; (c) 685

References for Supplementary Readings

1. A. Tarski, *Introduction to Logic*, 2nd rev. ed. (New York: Oxford, 1946); A. Church, *Introduction to Mathematical Logic*. (Princeton, New Jersey: Princeton University Press, 1956).
2. Joseph Breuer, *Introduction to the Theory of Sets*. (Englewood Cliffs, N.J.: Prentice-Hall, 1958).
3. C. E. Shannon, "A Symbolic Analysis of Relay and Switching Circuits," *Transactions of the American Institute of Electrical Engineers*, 57 (1938): 713–723.

   G. Hoernes, and M. F. Heilweil, *Introduction to Boolean Algebra and Logic Design*. (New York: McGraw-Hill, 1964).

CHAPTER 2

Elementary Function Theory

2.1 Relations

The statements that "9 is greater than 5," "Barbara is married to Park," and "Los Angeles is larger than San Diego" demonstrate what is commonly called a relationship. Expressions of the type "is married to," "is greater than," and "is larger than" are generally classified as relations. A relation is a correspondence or an association between the elements of two sets. For example, the elements of the sets $X = \{1, 2, 3, \ldots\}$ and $Y = \{1^2, 2^2, 3^2, \ldots\}$ are in a relation or correspondence determined by $y = x^2$ where $x \in X$ and $y \in Y$. Another example of a relation is provided by a salesman's commission of $50 received on each completed sale as the following table indicates.

Number of sales	0	1	2	3	4	...
Dollars of commission	0	50	100	150	200	...

A more convenient way of presenting this relation is its symbolic representation; that is, if n represents the number of sales and y the dollar's commission, then the relation or rule determining the salesman's commission is written as

$$y = 50n$$

The equations

$$y = x, \qquad y = x^2 - 1, \qquad \text{and} \qquad y = \frac{2}{x+1}$$

provide further examples of relations.

DEFINITION 2.1 A relation is a set of ordered pairs. The *domain* of a relation is the set of all first members of the ordered pairs, and the *range* is the set of all second members of the ordered pairs.

Hence, the set $\{(1, 2), (-3, 4), (1, 5), (3, -3)\}$ is a relation with domain $D = \{1, -3, 3\}$ and range $R = \{2, 4, 5, -3\}$. Similarly, the set $\{(x, y) \mid y = x\}$ is also a relation with domain the set of all real numbers x, and range the set of all real numbers y. Now, since we shall consider only those relations that are formed from real numbers or are represented only by real numbers, we can use the *Cartesian coordinate system* to represent relations as points in a plane. This representation of a relation shall be referred to as the *graph* of the relation. Thus the graph of a relation is its geometric representation.

The idea of coordinates and graphs can be traced to the ancient Egyptians and later to the Roman surveyors and the Greek map makers. The Greek Apollonius investigated thoroughly the geometry and graphs of conic sections which he named ellipses, parabolas, and hyperbolas. Although some earlier works relating algebra and geometry by way of coordinates and graphs exist, the French mathematicians Pierre de Fermat (1601–1665) and Rene Descartes (1596–1650) are credited with the invention of analytic geometry and its applications on graphs.

Example 2.1 If $U = \{1, 2, 3\}$ is the universal set for each of the following relations, then list the members of the relation, and indicate its domain, range, and graph.

a. $R_1 = \{(x, y) \mid y = x, x \in U\}$.
b. $R_2 = \{(x, y) \mid y > x, x \in U, y \in U\}$.

Solution The members of the relation R_1 are the ordered pairs $(1, 1)$, $(2, 2)$, and $(3, 3)$. Clearly, both its domain and range consist of the numbers 1, 2, and 3. The graph of R_1 consists of the three points shown in Figure 2.1.

Figure 2-1

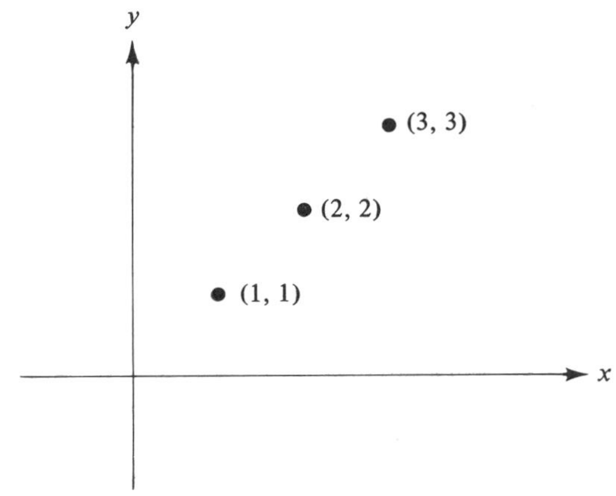

Similarly, the members of R_2 are the ordered pairs $(1, 2)$, $(1, 3)$, and $(2, 3)$. The domain of R_2 is the set $D = \{1, 2\}$, and its range is $R = \{2, 3\}$. The points of Figure 2.2 constitute the graph of the relation R_2.

Example 2.2 If R is the set of all real numbers, then graph the following relations:

a. $R_1 = \{(x, y) \mid y = x, x \in R\}$.
b. $R_2 = \{(x, y) \mid y = 1/(x + 1), x \in R\}$.
c. $R_3 = \{(x, y) \mid y > x, x \in R\}$.

Figure 2-2

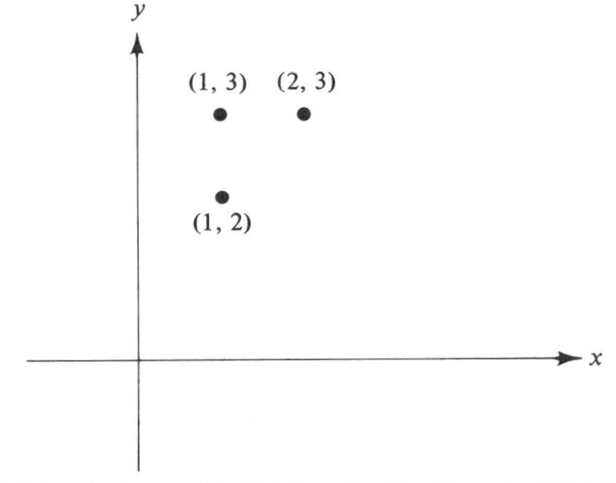

Graphs: a. Since x can be any real number, we may give arbitrary values to x and obtain corresponding values for y as the following table indicates.

x	y = x
0	0
1	1
2	2
—2	—2
—3	—3
.	.
.	.
.	.

Thus the set of ordered pairs

$$\{(0, 0), (1, 1), (2, 2), (-2, -2), \ldots\}$$

constitutes the relation R_1 whose graph, in fact, is the set of all points on the line l indicated in Figure 2.3.

b. Again, since $y = 1/(x + 1)$ for all real numbers x, the following table yields a representative number of ordered pairs in the relation R_2.

x	y = 1/(x + 1)
0	1
1	$\frac{1}{2}$
2	$\frac{1}{3}$
—1	∞ (undefined)
—$\frac{1}{2}$	2
—2	—1
—3	—$\frac{1}{2}$
.	.
.	.
.	.

Thus the graph of R_2 is the curve connecting the points $(0, 1)$, $(1, \frac{1}{2})$, $(2, \frac{1}{3})$, $(-2, -1)$, $(-3, -\frac{1}{2})$ . . . as shown in Figure 2.4.

c. Finally, since $y > x$ for all x, the graph of the relation R_3 must be the set of all points in the xy plane whose y coordinate is greater than the x coordinate. This graph is the shaded part of the xy plane as indicated in Figure 2.5 (the line $y = x$ is not included).

Whenever we have two sets A and B, we may form ordered pairs by choosing the first element of the pair from the set A and the second from the set B. The set of all possible ordered pairs thus formed

Figure 2-3

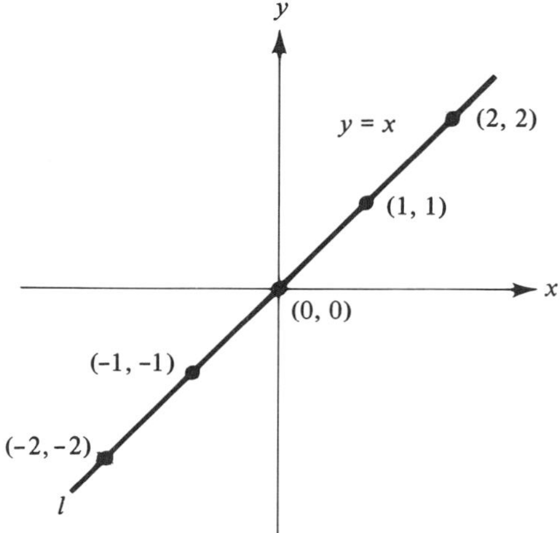

is called the *Cartesian product* of the sets A and B, and it is denoted by $A \times B$. Consequently, a relation is the Cartesian product $D \times R$ of some two sets D and R that are called the "domain" and "range" of the relation, respectively. Thus

$$A \times B = \{(a, b) \mid a \in A \text{ and } b \in B\}$$

$$D \times R = \{(d, r) \mid d \in D \text{ and } r \in R\}$$

Figure 2-4

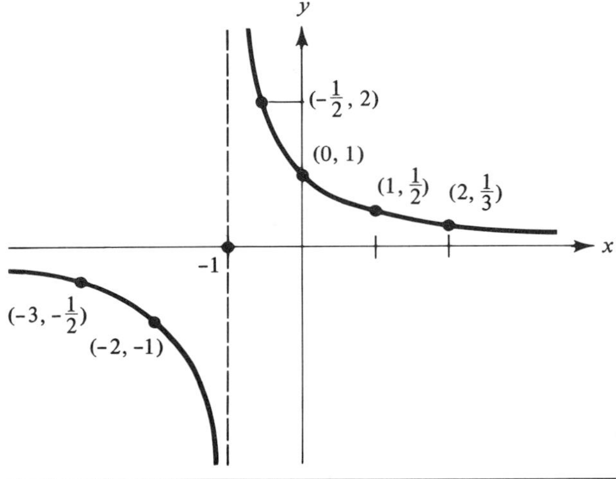

Figure 2-5

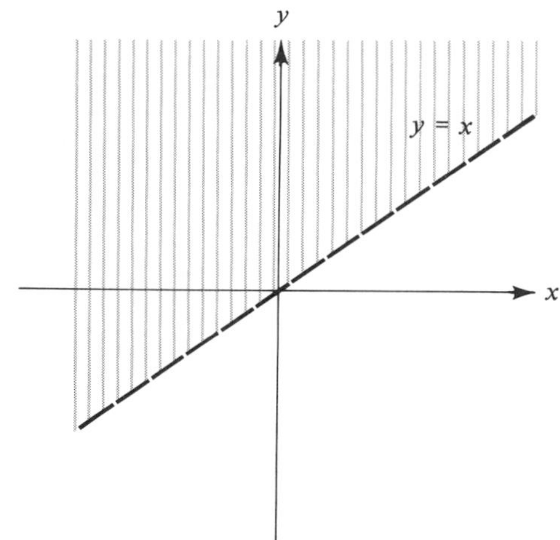

Example 2.3 If $A = \{1, 2, 3\}$ and $B = \{1, 4, 9\}$, then find the Cartesian products $A \times B$, $B \times A$, and graph the relations determined by them.

Solution: Since

$$A \times B = \{(a, b) \mid a \in A \text{ and } b \in B\}$$
$$A = \{1, 2, 3\}$$

and

$$B = \{1, 4, 9\}$$

we obtain

$$A \times B = \{(1, 1), (1, 4), (1, 9), (2, 1), (2, 4), (2, 9), (3, 1), (3, 4), (3, 9)\}$$

for the Cartesian product of A and B. The graph of the relation $A \times B$ consists of the nine points shown in Figure 2.6.

Similarly,

$$B \times A = \{(b, a) \mid b \in B, a \in A\}$$

implies that

$$B \times A = \{(1, 1), (1, 2), (1, 3), (4, 1), (4, 2), (4, 3), (9, 1), (9, 2), (9, 3)\}$$

for the given sets A and B. The graph of the relation $B \times A$ consists of the nine points shown in Figure 2.7.

Figure 2-6

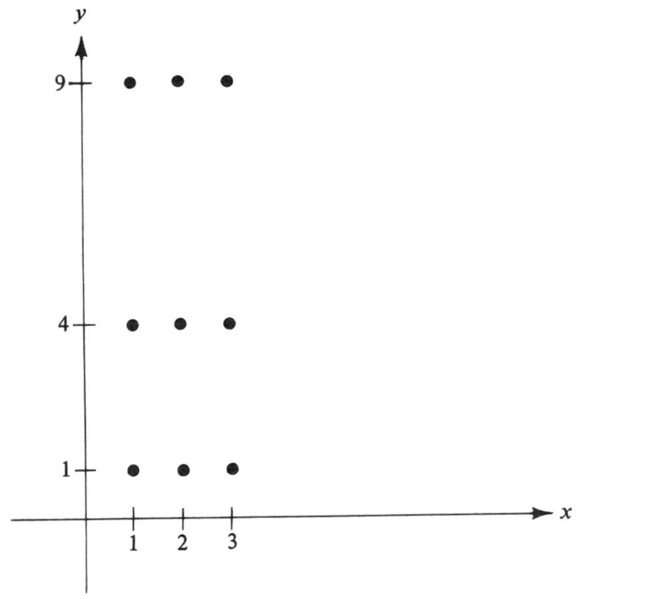

Figure 2-7

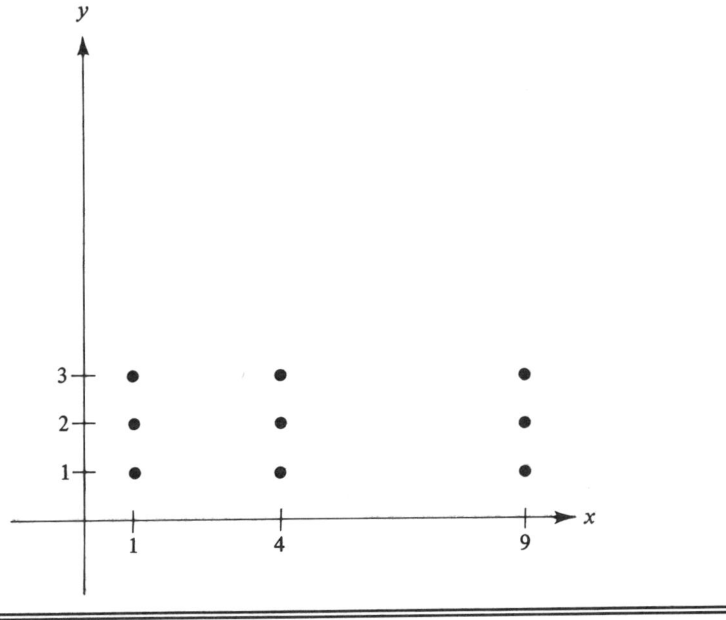

Clearly, the Cartesian products $A \times B$ and $B \times A$ are not equal since the points (a, b) and (b, a) are different unless $a = b$. For instance, the Cartesian products $A \times B$ and $B \times A$ of Example 2.3 are not equal since $(1, 4) \in A \times B$, but $(1, 4) \notin B \times A$. The concept of Cartesian product generalizes as Definition 2.2 indicates.

DEFINITION 2.2 By the Cartesian product of the sets $A_1, A_2, A_3, \ldots, A_n$ we mean the set

$$A_1 \times A_2 \times \cdots \times A_n = \{(a_1, a_2, \ldots, a_n) \mid a_i \in A_i, \; i = 1, 2, \ldots, n\}$$

where $(a_1, a_2, \ldots, a_n)$ is called an ordered n-tuple. The n-tuples $(a_1, a_2, \ldots, a_n)$ and $(b_1, b_2, \ldots, b_n)$ are said to be equal whenever $a_1 = b_1, a_2 = b_2, \ldots, a_n = b_n$.

Example 2.4 If

$$A = \{a, b\}, \qquad B = \{1, 2\}, \qquad \text{and} \qquad C = \{x, y\}$$

Figure 2-8

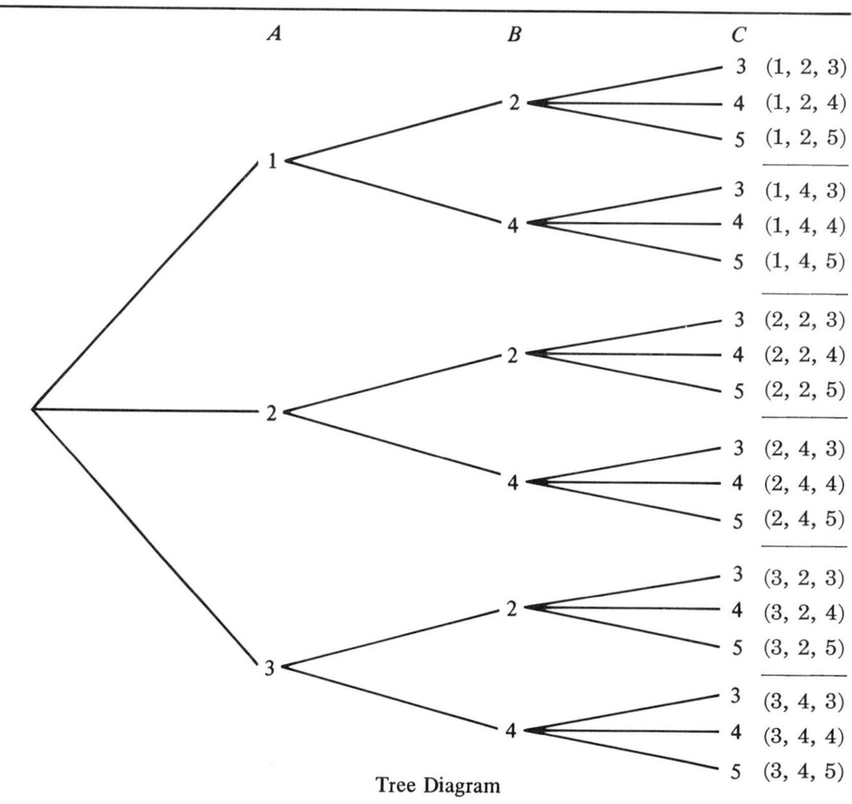

Tree Diagram

then the Cartesian product

$$A \times B \times C = \{(a, 1, x), (a, 1, y), (a, 2, x), (a, 2, y), (b, 1, x), (b, 1, y), (b, 2, x), (b, 2, y)\}$$

Example 2.5 If

$$A = \{1, 2, 3\}, \qquad B = \{2, 4\}, \qquad \text{and} \qquad C = \{3, 4, 5\}$$

then find the Cartesian product $A \times B \times C$.

Solution: The so-called "tree" diagram shown in Figure 2.8 provides a very convenient method of determining $A \times B \times C$. The "tree" is constructed from left to right, and $A \times B \times C$ consists of the ordered triples listed to the right of the tree. Thus $A \times B \times C$ contains the 18 triples indicated at the right of the tree diagram. Clearly, tree diagrams can be constructed for any type of Cartesian product.

Exercises

1. Which of the following sets are relations?

   a. $\{(1, 2), (2, 3), (3, 4)\}$, b. $\{(1, 3), (1, 4), (3, 1), (4, 1)\}$.
   c. $\{1, 2, 0, 5\}$. d. $\{(a, b), (b, a)\}$.

   Ans. (a), (b), (d)

2. Let $U \times U$ be the universal set for each of the following relations. If $U = \{0, 1, 2, 3, 4\}$, then tabulate each of the relations, state its domain and range, and graph it.

   a. $R = \{(x, y) \mid y = -x\}$. b. $R = \{(x, y) \mid y < x\}$.
   c. $R = \{(x, y) \mid y = x^3\}$. d. $R = \{(x, y) \mid x + y = 1\}$.
   e. $R = \{(x, y) \mid x^2 + y^2 = 4\}$.

3. Shade in the region of the plane that contains the graph of the relation with domain D and range R as indicated below.

   a. $D = \{x \mid -1 \leq x \leq 1\}$ and $R = \{y \mid 3 \leq y \leq 5\}$.
   b. $D = \{x \mid x \geq 0\}$ and $R = \{y \mid y = 5\}$.
   c. $D = \{x \mid x \leq 0\}$ and $R = \{y \mid y \geq 0\}$.
   d. $D = \{2, 3\}$ and $R = \{y \mid -2 \leq y \leq 3\}$.
   e. $D = \{x \mid -1 \leq x < 0\}$ and $R = \{y \mid 1 < y < 4\}$.

4. Use set notation to describe each of the following implied relations. State the domain and range, and graph each relation.

   a. $y = 2x + 1$. b. $y = -2$. c. $y^2 = x$.
   d. $y < 2x + 1$. e. $y = x^2$. f. $x = 7$.
   g. $x^2 + y^2 = 9$. h. $3x + 4y = 2$.

2.2 Functions

A function is a correspondence that assigns to each member in a certain set, called the *domain* of the function, exactly one member in a second set, called the *range* of the function; that is, a function is a relation in which no two distinct ordered pairs have the same first elements. For example, the amount of income tax withheld from one's paycheck is a function of the income tax rate. The amount of sales tax charged for a certain purchase is a function of the sales tax rate, and so on. Tables such as the one below, which lists the number of hypothetical car thefts in selected states for the year 1972, constitute useful examples of functions.

Table 2.1

State	Illinois	California	Utah	New York	Kansas
Car thefts	50	75	10	250	15

Clearly, the Table 2.1 contains ordered pairs (x, y) where x represents a state and y a number. Thus it defines a relation that is a function since each state (first element) appears only once. Since functions are special relations, they can be represented graphically. Hence, by the graph of a function we shall mean the graph of the relation defining it.

Functions such as the cost production, marginal product, demand, marginal revenue, utility, and marginal utility, as well as the function of the learning curve, and habit strength, among many others, constitute the major concern of economists and behavioral scientists. Graphs of these functions constitute the so-called profiles of economic systems and populations.[1]

Example 2.6 Graph the total cost function $C = 6 + 4x + 3x^2$ and the marginal cost function $M = 4 + 6x$ where x is the number of units of a certain commodity produced.

Graphs

a. Cost function: If we give arbitrary values to x, we obtain corresponding values of C as the following table indicates. The numbers in the first column of the table are the values of the independent variable x, and those in the second column are the values of the dependent variable C or the values of the function.

x	$C = 6 + 4x + 3x^2$
0	6
1	13
—1	5
2	26
—2	10
.	.
.	.
.	.

Now, the smooth curve connecting the points (x, C) obtained from this table and shown in Figure 2.9 is the graph of the cost function.

b. Similarly, we obtain the following table and the graph shown in Figure 2.10 for the marginal cost function $M = 4 + 6x$

Example 2.7 Graph the function $y = 1/(2x - 1)$. How does this function differ from those of Example 2.6?

Solution: If we give arbitrary values to x, we obtain the corresponding values of y as shown in the following table. Now, plotting the points (x, y) obtained in the table and "appropriately" connecting them, we obtain the graph of this function as shown in Figure 2.11.

This function differs from those of Example 2.6 in that the number $\frac{1}{2}$ is not in its domain, whereas the domain of the functions in Example 2.6 includes all the real numbers. Functions that are not defined for certain values of x are said to be discontinuous at those values of x.

Figure 2-9

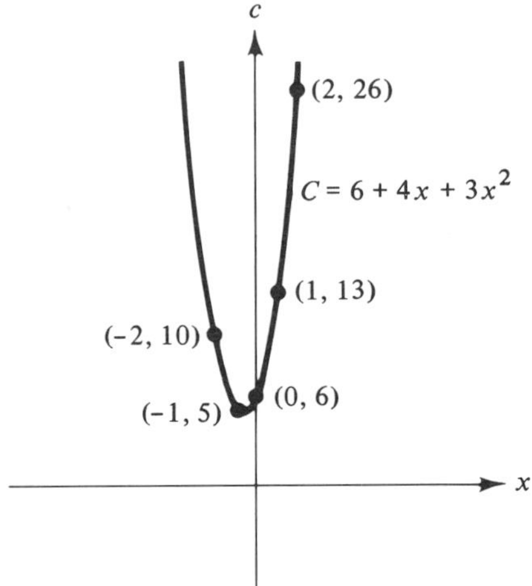

Figure 2-10

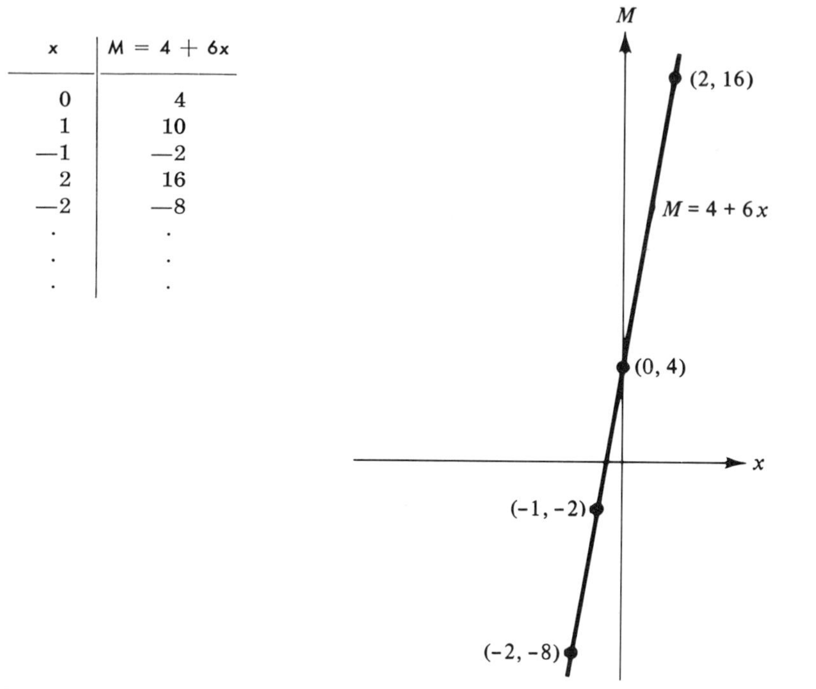

x	$M = 4 + 6x$
0	4
1	10
—1	—2
2	16
—2	—8
.	.
.	.
.	.

Figure 2-11

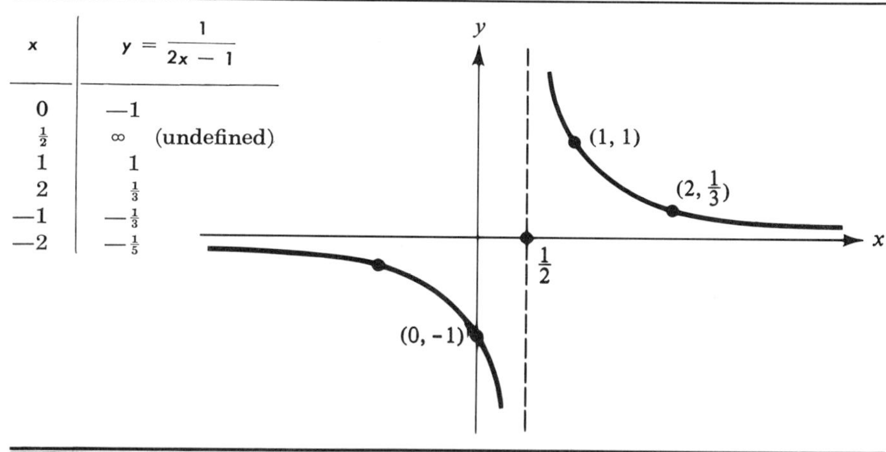

x	$y = \frac{1}{2x - 1}$
0	—1
$\frac{1}{2}$	∞ (undefined)
1	1
2	$\frac{1}{3}$
—1	$-\frac{1}{3}$
—2	$-\frac{1}{5}$

The definition of a function implies that it can be represented or described by a set of ordered pairs. This representation gives a complete picture of the function in the sense that the ordered pairs indicate which member of the domain is associated with each member of the range. For example, if a salesman receives a commission of \$5 on each completed sale, then his commission income is a function of how much he sells, and it may be represented by the following set of ordered pairs

$$\{(0, 0), (1, 5), (2, 10), (3, 15), \ldots\}$$

where the first number of a pair represents the number of sales and the second, the commission income. However, a more common and convenient convention is the symbolic representation of the function. In our example, if x stands for the number of sales and y for the dollars of commission, then the function or formula (rule) determining the salesman's commission of \$5 per sale can be expressed as

$$y = 5x$$

Moreover, this particular function can be thought of as a "machine" or "system" that accepts the numbers $x = 0, 1, 2, \ldots$ as inputs and produces the corresponding number $y = 0, 5, 10, 15, \ldots$ as outputs (Figure 2.12).

In general, a function f may be described in any one of the following ways:

a. A set of ordered pairs (x, y) where the set of x's is the domain of the function and the set of y's is its range.
b. A "machine" or "system" that accepts elements x of the domain as inputs and produces corresponding elements y of the range as outputs.
c. A "mapping" of members of the domain to corresponding mem-

Figure 2-12

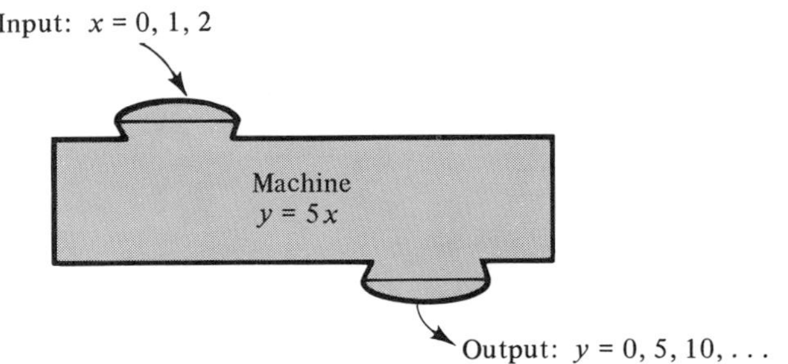

bers of the range; that is, if (x, y) is a member of the function f, then f associates x of the domain with y of the range, and we say f maps x into y or x is mapped into y by f or, equivalently, y is the image of x under f. This we write as

$$f\colon x \to y$$

or, if a function is to be described by an equation (formula or rule), we write

$$y = f(x)$$

and read it as "y is a function of x" which means that y is the image of x under f and not f times x. When the notation $y = f(x)$ is used, we refer to x as the independent variable and y as the dependent variable. The number $f(x) = y$ is called the "value of the function" or the "functional value" at x.

Example 2.8 Interpret each of the following functions in as many ways as possible

a. $f = \{(1, 2), (3, 4), (4, 4)\}$.
b. $y = x^2$.

Interpretations

a. The function $f = \{(1, 2), (3, 4), (4, 4)\}$ can be interpreted as

   i. A mapping whereby f maps 1 into 2, 3 into 4, and 4 into 4.
   ii. An equation in which $2 = f(1)$, $4 = f(3)$, and $4 = f(4)$.
   iii. A machine or system with inputs 1, 3, 4 and outputs 2, 4, 4, respectively (Figure 2.13).

b. Similarly, the function $y = x^2$ can be interpreted as

   i. A mapping whereby the function maps x of the domain set

Figure 2-13

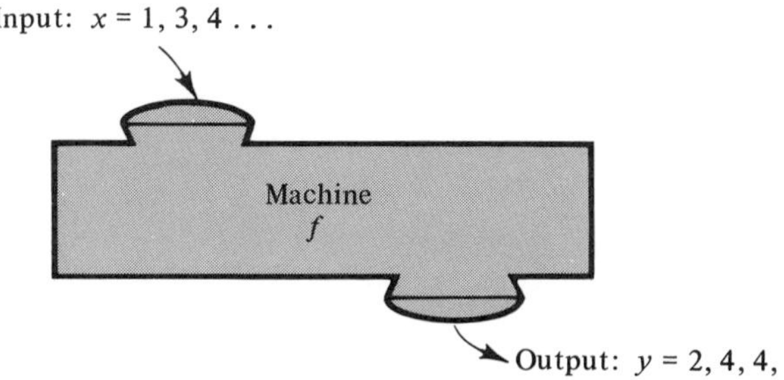

Figure 2-14

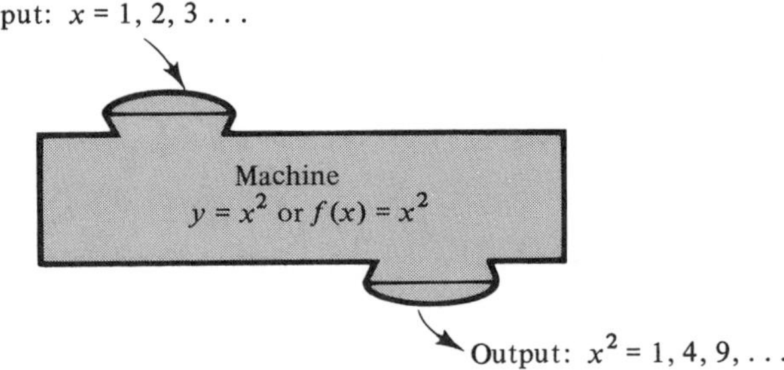

into x^2 of the range set; that is, the number 1 of the domain is mapped into 1 of the range set, 2 into 4, 3 into 9, ...

ii. An equation $f(x) = x^2$, that is, $f(1) = 1$, $f(2) = 4$, $f(3) = 9, \ldots$.

iii. A machine with input x and output x^2 for each x (Figure 2.14).

Example 2.9 If $f(x) = 2x - 1$, then evaluate

$$f(0), f(1), f(a), f(a+b), \qquad \text{and} \qquad \frac{f(a+h) - f(h)}{h}$$

Solution: Since $f(x) = y$ means that x is mapped into y, or that for input x the output is y, we have

$f(0) = 2(0) - 1 = -1$ or that $f(x) = 2x - 1$ maps 0 into -1

$f(1) = 2(1) - 1 = 1$ or that $f(x) = 2x - 1$ maps 1 into 1

$f(a) = 2a - 1$ or that $f(x) = 2x - 1$ maps a into $2a - 1$

$f(a+b) = 2(a+b) - 1$ or that $f(x) = 2x - 1$ maps $a + b$ into $2(a+b) - 1$

Thus

$$f(a+h) = 2(a+h) - 1 \qquad \text{and} \qquad f(h) = 2h - 1$$

Consequently,

$$\frac{f(a+h) - f(h)}{h} = \frac{[2(a+h) - 1] - (2h - 1)}{h}$$

$$= \frac{2a + 2h - 1 - 2h + 1}{h} = \frac{2a}{h}$$

Exercises

1. Identify which of the following relations is or is not a function. Explain each case, state the domain and range, and graph each relation.

   a. $\{(-8, 0), (-7, 1), (-2, 2), (-8, 3), (2, -2)\}$.
   b. $\{(x, y) \mid -3x^2 = y\}$.
   c. $\{(1, 1), (2, 1), (4, 1)\}$.
   d. $f(x) = 1/\sqrt{x - 1}$.
   e. $\{(x, y) \mid x^2 + y^2 = 4\}$.
   f. $y = \sqrt{x}$.
   g. $\{(x, y) \mid y = 5x + 2\}$.
   h. $y = 1/x^2$.
   i. $\{(x, y) \mid y \leq x + 3\}$.
   j. $\{(x, y) \mid xy < 1\}$.

   Ans. (a) No; (d) Yes, for all $x > 1$; (e) Yes, for all $-2 \leq x \leq 2$; (h) Yes, for all $x \neq 0$; (i) No

2. If $f(x) = x^2 - 1$, then find the following:

   a. $f(4)$.
   b. $f(3/a)$.
   c. The image of -3 under f.
   d. The output set, if the input set is $\{-2, 0, 3\}$, and the function f is acting as a machine.
   e. $f(a + h) - f(a)$.

   Ans. (c) 8; (d) 3, −1, 8

3. Evaluate $\dfrac{f(x + h) - f(x)}{h}$, $h \neq 0$, for each of the following functions:

   a. $f(x) = 4$.
   b. $f(x) = 4x - 3$.
   c. $f(x) = x^2/(x + 1)$.
   d. $f(x) = x^2 + 2$.

   Ans. (b) 4; (d) $2x + h$

4. If f and g are any two functions and if D_f and D_g denote the domain of f and g, respectively, then the sum, difference, product, and quotient of f and g are defined as follows:

$$f + g = \{(x, y) \mid (f + g)(x) = f(x) + g(x),\ x \in D_f \cap D_g\}$$
$$f - g = \{(x, y) \mid (f - g)(x) = f(x) - g(x),\ x \in D_f \cap D_g\}$$
$$f \cdot g = \{(x, y) \mid (f \cdot g)(x) = f(x)g(x),\ x \in D_f \cap D_g\}$$
$$f/g = \left\{(x, y) \mid (f/g)(x) = \frac{f(x)}{g(x)},\ g(x) \neq 0,\ x \in D_f \cap D_g\right\}$$

Now, evaluate $f + g$, $f - g$, $f \cdot g$, and f/g for each of the following pairs of functions.

a. $f = \{(1, 1), (2, 4), (-1, 1), (-2, 4)\}$ and $g = \{(1, 4), (2, 1), (-2, 0)\}$.
b. $f(x) = x - 2$ and $g(x) = x^2 + 7$.

5. Graph the learning curve

$$f(x) = \frac{Lx + Lc}{x + c + a}$$

where L is the limit of practice, x is the amount of formal practice, c is equivalent previous practice, and a is the rate of learning, if $L = 80$, $a = 20$, and $c = 5$.

6. Graph the total cost function

$$c(x) = 8 + 5x + 4x^2$$

and the marginal cost function

$$M(x) = 5 + 8x$$

for a firm, if x is the amount produced.

7. The total cost C of a certain commodity is given in terms of the quantity Q as

$$C = f(Q) = 3 - 2Q + Q^2$$

Calculate the values of C for $Q = 1, 2, 3, 4$, and plot the corresponding points (Q, C).

Ans. 2, 3, 6, 11

8. If a salesman is making a \$20 commission on each completed sale and if x represents the number of sales and y the dollars of commission income, then write a formula for the function determining the salesman's commission, and graph this function for the first five sales.

Ans. $y = 20x$

9. A worker's daily pay is based on the number of potato sacks he packs and loads. During a certain five-day period, his output and pay were as follows:

Day	1	2	3	4	5
Output(x)	125	184	173	112	150
Pay(y)	\$8.50	\$12.75	\$9.80	\$6.25	\$9.00

Graph y as a function of x.

10. If an amount of A dollars is deposited in a bank that pays compound interest at a rate i per year, the amount on deposit at the

end of x years is given by

$$y = A(1 + i)^x$$

Graph y as a function of x if $A = \$100$ and the rate is at 6% interest compounded annually.

2.3 Types of Functions

Advanced knowledge of certain characteristics of a function describing some economic, social, or physical behavior can often determine the action to be taken by the specialist in charge. For example, if the cost function of a certain firm is constantly "increasing," if social crime is always on the "rise," if the temperature inside a space craft is described by a "dangerously increasing function," if the chart (or function) describing a mental patient under treatment shows "no change," or if the heartbeat of a patient is not "normally periodic," then the person in charge must consider a new line of action. Thus, besides learning how to identify a function and determine its domain, range, and graph, there will be occasions when we need to determine whether or not a function is "linear," "nonlinear," "increasing," "decreasing," "even," or "odd." Furthermore it will be necessary to establish the "periodicity" as well as the "maximum" and "minimum" of many functions whenever they exist. Thus we now consider a sequence of examples each of which demonstrates some of the above characteristics of functions. These examples also provide some obvious generalizations described in subsequent definitions.

LINEAR FUNCTIONS AND EQUATIONS OF LINES

In this category we find the equation of the unique line l passing through two distinct points $A(x_1, y_1)$ and $B(x_2, y_2)$ in the xy plane; that is, we find an algebraic expression or equation such that the line l is identical with the set of all points satisfying that equation.

If the x coordinates x_1 and x_2 of all points A and B are equal, then the line l (or AB) must be vertical and its equation is $x = x_1$; that is, line l (or AB) must be vertical and its equation is $x = x_1$; that is,

$$l = \overleftrightarrow{AB} = \{(x, y) \mid x = x_1\}.$$

On the other hand if $x_1 \neq x_2$, then the equation of the line is

$$\frac{y - y_1}{x - x_1} = \frac{y_2 - y_1}{x_2 - x_1} \quad \text{or} \quad y - y_1 = \left(\frac{y_2 - y_1}{x_2 - x_1}\right)(x - x_1)$$

or

$$y - y_1 = m(x - x_1) \tag{2.1}$$

where the number

$$m = \left(\frac{y_2 - y_1}{x_2 - x_1}\right)$$

is called the slope of the line l, and Equation (2.1) is referred to as the point-slope equation of the line. Since the slope of a line remains the same no matter which two points of the line are used, any two points uniquely determine the slope and the equation of the line containing them. For a geometric interpretation of l and its slope see Figure 2.15.

The geometric interpretation of the slope of a line indicates that it serves to measure the extent to which the line is inclined.

DEFINITION 2.3 The equation

$$ax + by = c \qquad \text{or} \qquad y = -\frac{a}{b}x + \frac{c}{b}$$

where a, b, and c are fixed real numbers is called a "linear" function; it represents a straight line with slope $-a/b$, and y intercept the point $(0, c/b)$ or simply the number c/b, and x intercept the point $(c/a, 0)$ or simply the number c/a.

The x and y intercepts of a line are thus its intersections with the x and y axes, respectively.

To demonstrate further the linear function we consider the special cases given in the following examples.

Figure 2-15

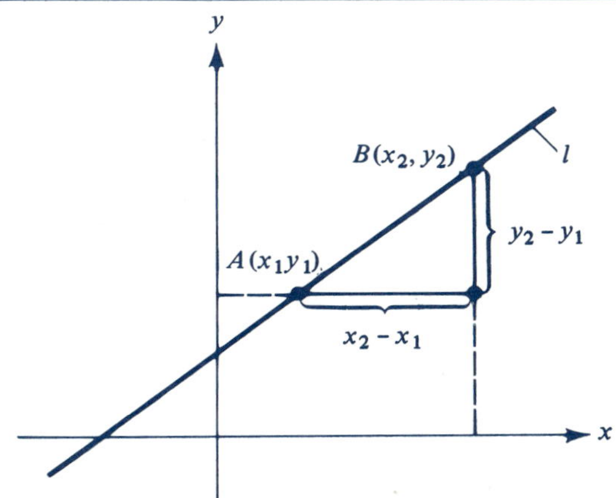

Figure 2-16

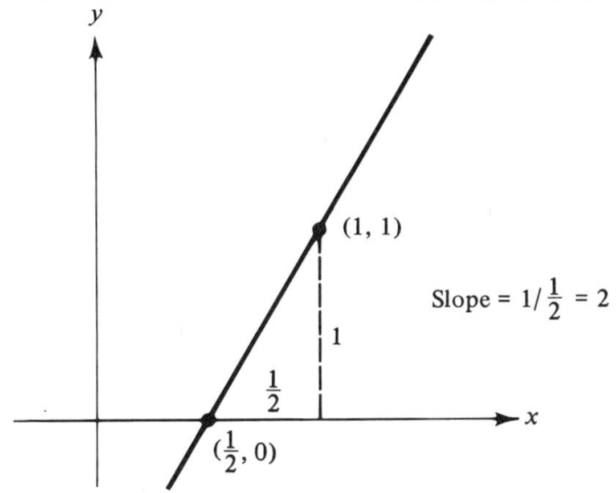

Example 2.10 Graph the linear function $y = 2x - 1$.

Solution: Since a linear function represents a straight line, $y = 2x - 1$ is the line containing the points $(0, -1)$ and $(\frac{1}{2}, 0)$ and having slope $m = 2$. The points $(0, -1)$ and $(\frac{1}{2}, 0)$, obtained from the equation $y = 2x - 1$ by letting $x = 0$ and $y = 0$ and solving for y and x, respectively, are the x and y intercepts of the line, respectively (Figure 2.16). The same terminology applies, of course, to the intersections of any graph (or curve) and the x and y axes.

Example 2.11 Find the equations of the following straight lines:

a. The line determined by the points $(-2, -3)$ and $(2, 4)$.
b. The line whose slope is $m = \frac{1}{2}$ and contains the point $(-1, -2)$.

Solution

a. Using Equation (2.1) we obtain

$$y - 4 = \frac{4 - (-3)}{2 - (-2)}(x - 2) \qquad \text{or} \qquad y - 4 = \frac{7}{4}(x - 2)$$

or

$$7x - 4y = -2$$

for the required equation.

b. Again, using Equation (2.1) we obtain

$$y + 2 = \tfrac{1}{2}[x - (-1)] \qquad \text{or} \qquad 2y - 4 = x + 1$$

or

$$x - 2y = -5$$

Example 2.12 Find the equation and graph the line whose points have coordinates x and y satisfying the following equations:

$$x = 2t - 1 \tag{2.2}$$

$$y = 3t + 4 \tag{2.3}$$

Solution: If we solve the first equation, $x = 2t - 1$, for t, we find

$$t = \frac{x + 1}{2}$$

Thus if we substitute this t in the second equation, $y = 3t + 4$, we obtain

$$y = 3\left(\frac{x + 1}{2}\right) + 4 \qquad \text{or} \qquad 2y = 3x + 11$$

for the equation of the line whose graph is shown in Figure 2.17. Equations such as (2.2) and (2.3) are called parametric equations of a line with parameter t.

DEFINITION 2.4 If (x, y) is any point of a curve C, then the equations

$$x = f(t) \qquad \text{and} \qquad y = g(t)$$

are called the parametric equations of C with parameter t.

Figure 2-17

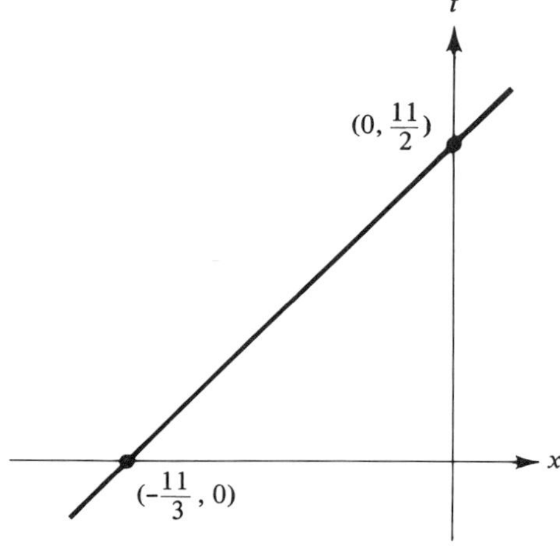

In business and economics if the linear equation (or line)

$$y = mx + c$$

represents the total cost of manufacturing x items, then the economists' term for the slope m of the line $y = mx + c$ is *marginal cost*[2] per item, and the y intercept c is called the *fixed cost.*[2] For example, if

$$y = 3x + 2$$

represents the total cost in dollars of manufacturing x items, then the total cost for 100 items is

$$y = 3(100) + 2 \qquad \text{or} \qquad y = \$302$$

On the other hand the marginal cost is \$3 per item and the fixed cost is \$2. For further explanations of the terms "marginal and fixed costs" we refer the reader to selected readings in business and economics.

DEFINITION 2.5 If A and B are any two points on a line, then by the segment AB we shall mean the set of all points between A and B together with the points A and B.

In terms of inequalities if A and B are the points (x_1, y_1) and (x_2, y_2), respectively, then the segment AB determined by them is the set

$$\overline{AB} = \{(x, y) \mid x_1 \leq x \leq x_2 \quad \text{and} \quad y_1 \leq y \leq y_2\}$$

shown in Figure 2.18.

Concerning lines and segments the following theorems can now be established.

Figure 2–18

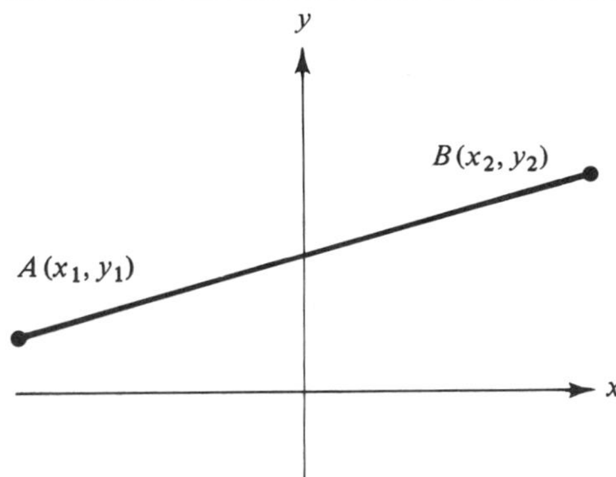

Theorem 2.1 If A and B are the points (x_0, y_0) and (x_1, y_1), respectively, then the following are true.

a. The midpoint of $\overline{AB}$, that is, the point M that divides $\overline{AB}$ into two equal segments, has coordinates $(x_0 + x_1)/2$ and $(y_0 + y_1)/2$.
b. The distance between A and B, denoted by AB, is given by

$$|AB| = \sqrt{(x_1 - x_0)^2 + (y_1 - y_0)^2}$$

Theorem 2.2 Two lines with slopes m_1 and m_2 are parallel if and only if $m_1 = m_2$, and perpendicular if and only if $m_1 m_2 = -1$.

Example 2.13 Given the points $A(6, 1)$ and $B(3, -3)$, find the midpoint of the segment AB and its length $|AB|$.

Solution: If we apply Theorem 2.1, we obtain the midpoint

$$M\left(\frac{6+3}{2}, \frac{1+(-3)}{2}\right) \quad \text{or} \quad M\left(\frac{9}{2}, -1\right),$$

and the distance between A and B or the length of segment AB to be

$$|AB| = \sqrt{(6-3)^2 + (1-(-3))^2}$$

or

$$|AB| = \sqrt{9+16} \quad \text{or} \quad |AB| = 5$$

Example 2.14 Determine whether or not the following pairs of lines are parallel.

a. l_1: $2x - y = 1$ and l_2: $x + 2y = 4$.
b. l_1: $x + y = 1$ and l_2: $x - 2y = 0$.
c. l_1: $x + y = 1$ and l_2: $3x + 3y = 5$.

Solution

a. Since the lines l_1 and l_2 can be written as $y = 2x - 1$ and $y = \frac{1}{2}x + 2$, respectively, their slopes are $m_1 = 2$ and $m_2 = -\frac{1}{2}$. Hence $m_1 m_2 = -1$, and the lines are perpendicular and not parallel.
b. Similarly, the slopes of $x + y = 1$ and $x - 2y = 0$ are $m_1 = -1$ and $m_2 = \frac{1}{2}$. Thus the lines are neither parallel nor perpendicular.
c. Finally, since the slopes of the lines $x + y = 1$ and $3x + 3y = 5$ are $m_1 = -1$ and $m_2 = -1$, the lines must be parallel.

INCREASING AND DECREASING FUNCTIONS

Example 2.14 introduces the very important class of increasing and decreasing functions.

Example 2.15 Graph the following functions:

a. $y = 2x + 4$. b. $y = -2x + 2$.
c. $y = 2^x$. d. $y = (\frac{1}{2})^x$.

Solution: The graphs and the associated tables of values for each function are shown in Figures 2.19 and 2.20.

A close observation of these functions and their graphs reveals that the values in Figures 2.19(a) and 2.20(a) increase as x increases. On the other hand the values of the functions in Figures 2.19(b) and

Figure 2–19

x	y = 2x + 4
0	4
−2	0
1	6
.	.
.	.
.	.

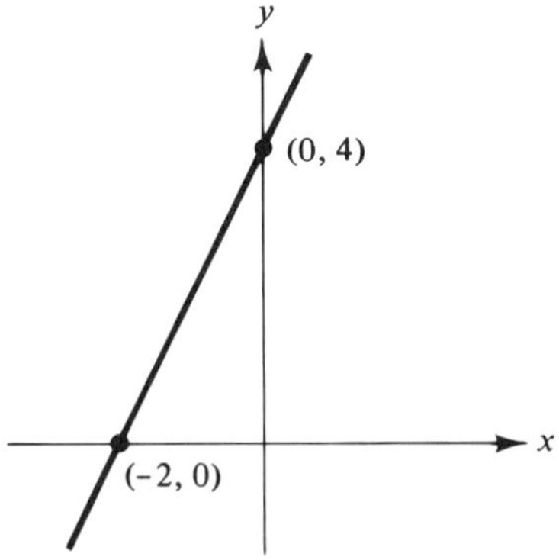

x	y = −2x + 2
0	2
1	0
2	−2
.	.
.	.
.	.

(a)

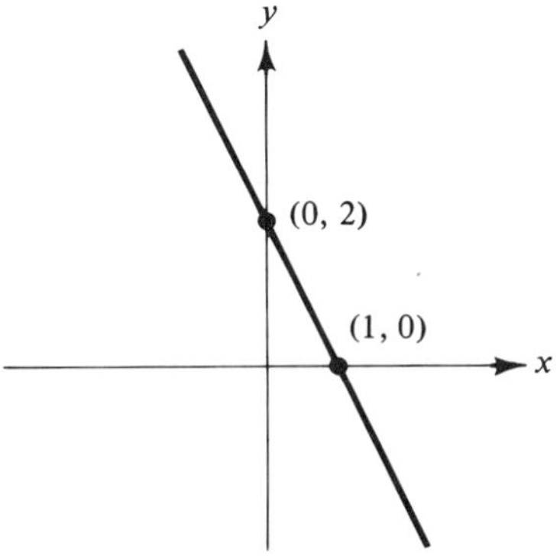

(b)

Figure 2-20

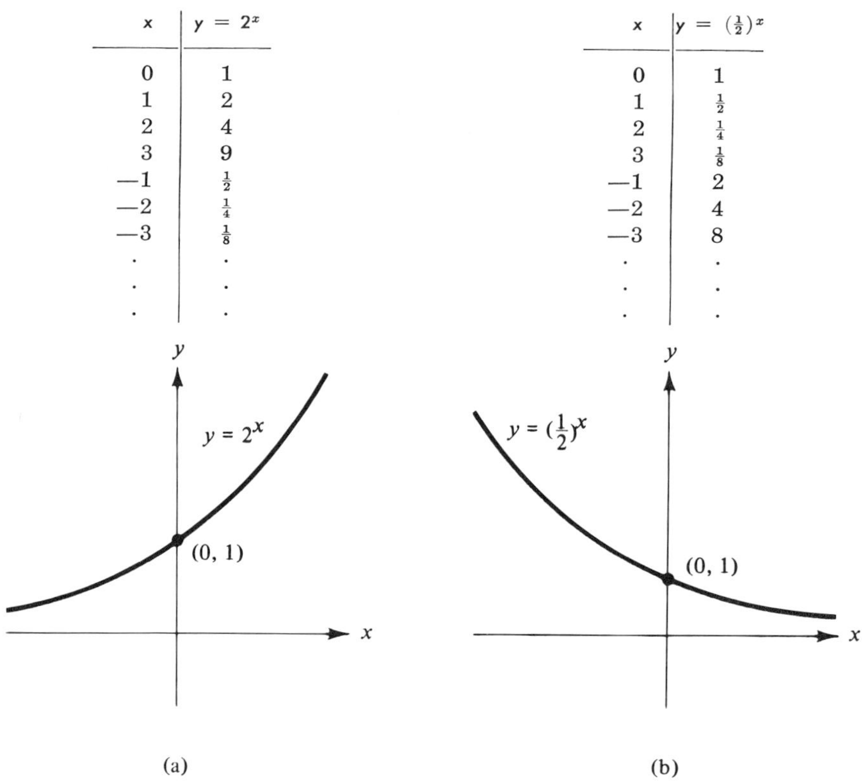

x	$y = 2^x$
0	1
1	2
2	4
3	9
—1	$\frac{1}{2}$
—2	$\frac{1}{4}$
—3	$\frac{1}{8}$
.	.
.	.
.	.

x	$y = (\frac{1}{2})^x$
0	1
1	$\frac{1}{2}$
2	$\frac{1}{4}$
3	$\frac{1}{8}$
—1	2
—2	4
—3	8
.	.
.	.
.	.

2.20(b) decrease as x increases. These observations lead to the following important definition.

DEFINITION 2.6 A function $f(x)$ is said to be increasing if and only if its values increase as x increases. Similarly, a function is said to be decreasing if and only if the value of the function decreases as x increases.

In terms of inequalities, the above definition implies that $f(x)$ is increasing if and only if $f(x_1) < f(x_2)$ for all x_1 and x_2 for which $x_1 < x_2$. Analogously, the function $g(x)$ is decreasing if and only if $g(x_1) < g(x_2)$ for all x_1 and x_2 for which $x_1 > x_2$. Thus the function $y = x^2$ shown in Figure 2.21 increases for $x > 0$ and decreases for $x < 0$.

However, the function $y = x^3$ (Figure 2.22) is increasing for all values of x.

Figure 2–21

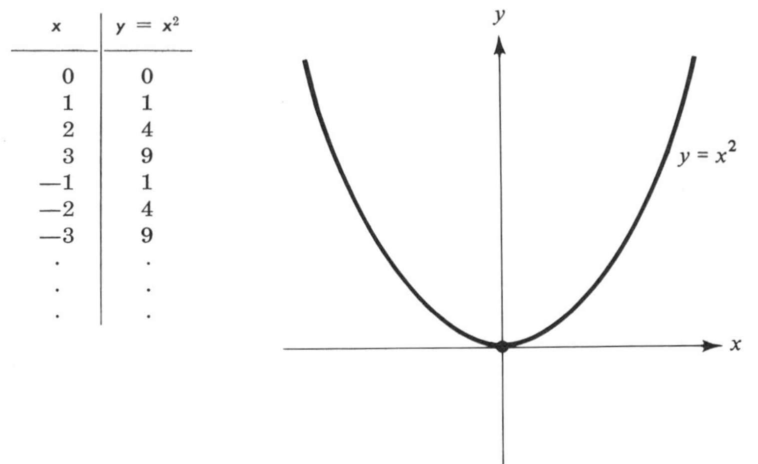

x	$y = x^2$
0	0
1	1
2	4
3	9
—1	1
—2	4
—3	9
.	.
.	.
.	.

Figure 2–22

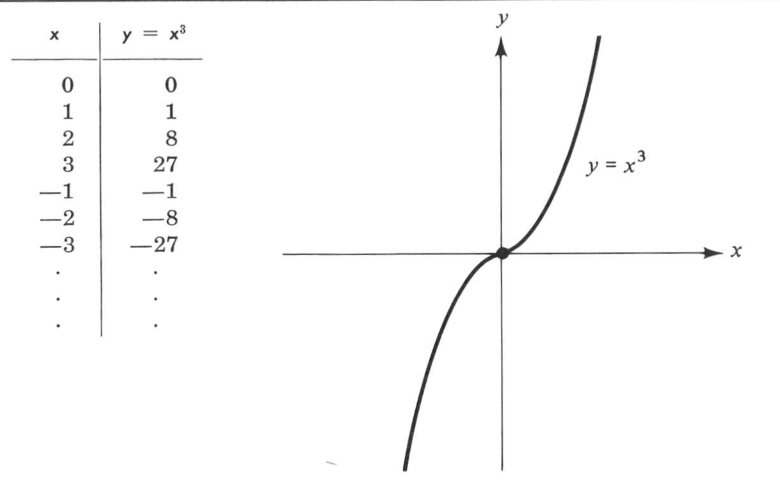

x	$y = x^3$
0	0
1	1
2	8
3	27
—1	—1
—2	—8
—3	—27
.	.
.	.
.	.

EVEN AND ODD FUNCTIONS (THE ABSOLUTE VALUE)

Additional observations of the last two functions $f(x) = x^2$ and $g(x) = x^3$ indicate that

$$f(-x) = (-x)^2 = x^2 \qquad \text{or} \qquad f(-x) = f(x)$$

and

$$g(-x) = (-x^3) = -x^3 \qquad \text{or} \qquad g(-x) = -g(x)$$

that is, the value of $f(x)$ remains the same even though x takes on opposite values. On the other hand, the values of $g(x)$ become opposite to those of $g(-x)$. This observation leads to the following generalizations.

DEFINITION 2.7 A function $f(x)$ is said to be even if and only if $f(-x) = f(x)$. Similarly, we say that the function $g(x)$ is odd if and only if $g(-x) = -g(x)$.

Geometrically, the graph of an even function is symmetric with respect to the y axis whereas that of an odd function is symmetric with respect to the origin (see Figure 2.22). Another useful example of an even function is provided by the so-called absolute value function defined as follows.

$$f(x) = |x| = \begin{cases} x & \text{if } x \geq 0 \\ -x & \text{if } x < 0 \end{cases}$$

Thus $|5| = 5$ and $|-5| = 5$. Generally,

$$f(-x) = |-x| = x = f(x)$$

for all values of x. Hence the absolute value function is an even function whose values are always positive or zero. Moreover, the table of values and the graph of $f(x) = |x|$ (Figure 2.23) show that the function is "linearly" increasing for all $x > 0$ and "linearly" decreasing for all $x < 0$.

Figure 2-23

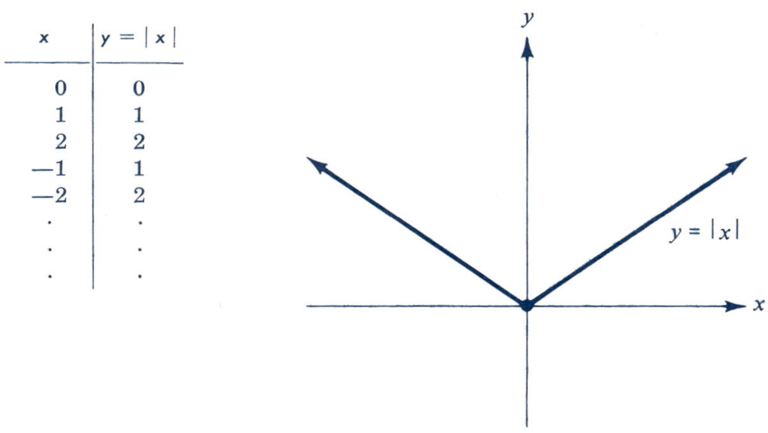

x	y = |x|
0	0
1	1
2	2
—1	1
—2	2
.	.
.	.
.	.

PERIODIC FUNCTIONS

A very important class of functions is that in which a function repeats itself at regular intervals. For example, the function of the human heart repeats itself; the sunrise and sunset are regulated by a repeating function; and the motion of vibrating strings in musical instruments, as well as the engine of an automobile, are based on repeating functions.

A geometric representation of a "periodic" function is presented in Figure 2.24. As the graph of the function $f(t)$ indicates, the function repeats itself every two units of t. This value of two units' interval is referred to as the "period" of the function $f(t)$. Generally, the concept of a periodic function is that given in Definition 2.8.

DEFINITION 2.8 We say that a function $f(t)$ is periodic if and only if there exists a number p such that

$$f(t+p) = f(t)$$

The smallest such number p is called the period of the function, and its reciprocal $1/p$ is referred to as its frequency.

SEQUENCES and the FACTORIAL FUNCTION

A very special type of function is one in which the domain consists of natural numbers and perhaps zero. Such a function is commonly known as a sequence, and is denoted by $f(n)$ for $n = 0, 1, 2, 3, \ldots$. The graph of a sequence is the set of all points of the form

Figure 2–24

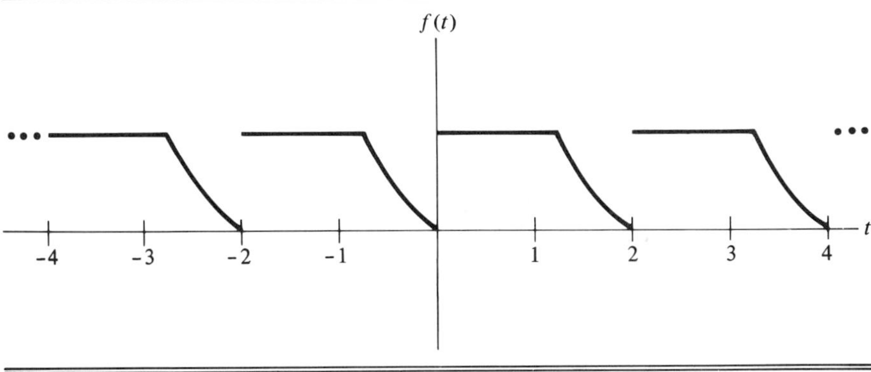

Figure 2-25

n	$f(n) = 1/n$
1	1
2	$\frac{1}{2}$
3	$\frac{1}{3}$
4	$\frac{1}{4}$
5	$\frac{1}{5}$
.	.
.	.
.	.

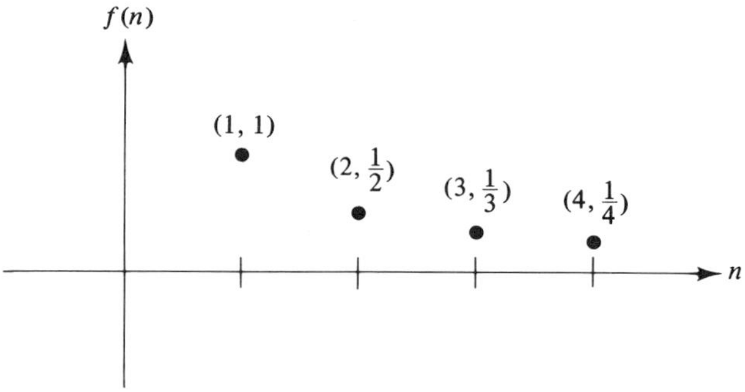

$f(n)$

$(1, 1)$ $(2, \frac{1}{2})$ $(3, \frac{1}{3})$ $(4, \frac{1}{4})$

n

$(n, f(n))$. The values (or terms) of the sequence $f(n) = 1/n$ and part of its graph are shown in Figure 2.25.

A still more specialized sequence, which is very useful in the development of combinatorial analysis and the binomial theorem, determines the so-called factorial function defined as follows:

$$f(n) = n! = 1 \cdot 2 \cdot 3 \cdots (n-1) \cdot n \qquad \text{and} \qquad 0! = 1$$

Thus

$$\frac{3!}{0!} = \frac{\cancel{1} \cdot 2 \cdot 3}{\cancel{1}} = 6, \qquad 4! = 1 \cdot 2 \cdot 3 \cdot 4 = 24,$$

$$6! = 1 \cdot 2 \cdot 3 \cdot 4 \cdot 5 \cdot 6 = 720,$$

and

$$\frac{5!}{3!} = \frac{\cancel{1} \cdot \cancel{2} \cdot \cancel{3} \cdot 4 \cdot 5}{\cancel{1} \cdot \cancel{2} \cdot \cancel{3}} = 20$$

We shall discuss the concept and the development of logarithmic and exponential functions in Section 2.5 immediately following the discussion of the inverse of a function.

Exercises

1. Graph the lines whose equations are given by $y = mx + 3$ for $m = 1, -1, 2, 0$.
2. For each of the following find the equation of the line having the given slope m and passing through the given point. In each case find another point on the line.

   a. $m = 0$ and $(4, -2)$. b. $m = -3$ and $(-1, 3)$.
   c. $m = \frac{1}{2}$ and $(2, -4)$. d. $m = 3$ and $(3, 3)$.

   Ans. (b) $3x + y = 0$, $(0, 0)$; (d) $3x - y = 6$, $(0, -6)$

3. In each of the following, find an equation and the slope, if any, of each of the lines through the given pairs of points.

   a. (1, 1) and (1, 2). b. (−1, 3) and (5, −2).
   c. (−2, 0) and (4, 0). d. (1, 1) and (3, 4).

   Ans. (a) $x = 1$, $m = \infty$; (b) $5x + 6y = 13$, $m = -5/6$; (d) $3x - 2y = 1$, $m = 3/2$

4. In each of the following equations, y represents the cost in dollars of manufacturing x items. The number of items produced is given with each equation. Find the following for each equation.

   a. The total cost.
   b. The average cost per item (total cost divided by the number of items produced).
   c. The fixed cost.
   d. The marginal cost per item.

   i. $y = 6x + 4$, 100 items produced.
   ii. $y = 7x + 2$, 50 items produced.
   iii. $y = 50x + 1000$, 200 items produced.

   Ans. (a) 604; (b) 6.04; (c) 4; (d) 6

5. An investment of \$10,000 yields an income of \$1000, whereas an investment of \$50,000 yields an income of \$4000. Assuming that the income is a linear function of investment and the investment is not less than \$10,000, determine the function and its equation. What is the domain, the slope, and the interpretation of the slope?

   Ans. $3x - 40y = -10{,}000$

6. A corporation pays taxes on profits according to the following rates: 30% up to \$40,000 and when the profit exceeds \$40,000, \$9000 plus 50% of the excess over \$40,000. Express the tax T in terms of the profit P, and sketch the graph of T.

7. In each of the following, find the equations of the perpendicular and parallel lines to the given line through the given point. Also sketch the corresponding graphs.

   a. $y = 2x - 1$ and (1, 3). b. $2x + y = 3$ and (−1, −2).
   c. $3x + 2y = 5$ and (1, 1). d. $x - y = 0$ and (−1, 2).

   Ans. (b) $2x + y = -4$, $x - 2y = 3$; (d) $x - y = -2$, $x + y = 0$

8. Graph the lines whose parametric equations are as follows:

   a. $x = 2t - 3$ and $y = t - 1$.

b. $x = 3t - 5$ and $y = 2t - 7$.

What is the slope of each line?

9. Graph the curves whose parametric equations are

a. $x = t$ and $y = 1/2t$. b. $x = t$ and $y = t - 1$.

10. Find the midpoint and the length of the segments determined by the following pairs of points:

a. $(1, 2)$ and $(5, 6)$. b. $(2, 5)$ and $(6, 2)$.
c. $(-2, -3)$ and $(-2, 5)$.

Ans. (a) $(3, 4)$, $4\sqrt{2}$; (c) $(-2, 1)$; 8

11. If $A(0, 0)$, $B(2, 2)$, and $C(2, -2)$ are the vertices of the triangle ABC, then find the midpoints of its sides and show that the segments determined by them are parallel and equal to half the corresponding sides of the triangle.
12. For what values of x are the following functions increasing or decreasing. Sketch the graph of each function.

a. $y = 5x + 2$. b. $2x + y = 1$.
c. $y = x^2 - 3$. d. $y = 2x^2 + 1$.
e. $y = x(x - 1)$. f. $y = |x - 1|$.

Ans. (b) Decreases for all x; (d) Increases for $x > 0$ and Decreases for $x < 0$; (f) Increases for $x > 1$ and Decreases for $x < 1$

13. Determine whether or not the following functions are even or odd. Sketch the graph of each function.

a. $y = x^3$. b. $y = 3x$.
c. $y = x^2 + 3$. d. $y = 2x + 1$.
e. $y = \sqrt{9 - x^2}$. f. $y = \frac{1}{4}\sqrt{36 - 9x^2}$.

Ans. (b) Odd; (f) Even

14. Plot five points of the graph of the following sequences.

a. $f(n) = 1/(n + 1)$. b. $f(n) = 2/(n^2 + 1)$.
c. $f(n) = 1 + (-1)^n$. d. $f(n) = 1/[n(n + 1)]$.

15. Graph $y = |2x - 1|$ and $y = |2x + 1|$.
16. Evaluate each of the following:

a. $10!/8!2!$. b. $6!/3!3!$.
c. $0!/4!$. d. $5!/3!2!$.
e. $n!/[(n - k)!k!]$ if $k \leq n$.

Ans. (a) 45; (b) 20; (d) 10

2.4 The Inverse of a Function

Functions are often manipulated in essentially the same way as numbers; that is, they are added, subtracted, multiplied, and divided. Thus it is possible to have functions that are combinations of other functions. It is therefore meaningful to write

$$f + g \qquad f - g \qquad f/g \qquad \text{and} \qquad f \cdot g$$

for any two functions f and g. For instance, if

$$f(x) = 2x - 1$$

and

$$g(x) = x^2 + 2x^3$$

then

a. The function $h = f + g$ is defined as

$$h(x) = f(x) + g(x) \qquad \text{or} \qquad h(x) = (2x - 1) + (x^2 + 2x^3).$$

That is,

$$h(x) = 2x^3 + x^2 + 2x - 1$$

b. The function $H = f - g$ is defined to be

$$H(x) = f(x) - g(x) \qquad \text{or} \qquad H(x) = (2x - 1) - (x^2 + 2x^3)$$

Thus

$$H(x) = -2x^3 - x^2 + 2x - 1$$

c. The function $q = f/g$ is defined as

$$q(x) = \frac{f(x)}{g(x)}, \qquad g(x) \neq 0$$

or

$$q(x) = \frac{2x - 1}{x^2 + 2x^3}, \qquad x^2 + 2x^3 \neq 0$$

d. The function $p = f \cdot g$ is defined to be

$$p(x) = f(x)g(x) \qquad \text{or} \qquad p(x) = (2x - 1)(x^2 + 2x^3)$$

Hence,

$$p(x) = 4x^4 - x^2$$

However, the operation of expressing a "function of a function" is somewhat more complicated. This operation on the functions f and

g is called the composition of f by g or g by f, and it is denoted by

$$y = f(g(x)) \qquad \text{or} \qquad Y = g(f(x))$$

respectively.

Frequently the functions

$$y = f(g(x)) \qquad \text{and} \qquad Y = g(f(x))$$

are called the composite function of f by g and g by f, respectively. For example, if

$$f(x) = x^2 \qquad \text{and} \qquad g(x) = 2x - 3$$

then the composite function of f by g is

$$y = f(g(x)) = (g(x))^2 = (2x - 3)^2 = 4x^2 - 12x + 9$$

and the composite function of g by f is

$$Y = g(f(x)) = 2f(x) - 3 = 2x^2 - 3$$

Clearly, the composition of functions is not a commutative operation; that is, $f(g) \neq g(f)$ as the previous example demonstrates. However, in many instances

$$f(g(x)) = g(f(x)) \text{ for all } x$$

In fact, sometimes

$$f(g(x)) = g(f(x)) = I(x)$$

where I is the identity function, that is, the function that maps each x into itself $(I(x) = x)$. The latter property of composition of functions motivates Definition 2.9.

DEFINITION 2.9 We say that two functions f and g are inverses of each other if and only if

$$f(g(x)) = g(f(x)) = I(x) = x$$

for all x, where I is the identity function.

In general the inverse of a function f is denoted by f^{-1} which is read "the inverse of f" and not "f to the -1 power." Therefore the inverse of $f(x) = x + 5$ is the function $g(x) = x - 5$ or conventionally $f^{-1}(x) = x - 5$ since

$$f(g(x)) = g(x) + 5 = (x - 5) + 5 = x$$

and

$$g(f(x)) = f(x) - 5 = (x + 5) - 5 = x$$

Similarly, the functions

$$f(x) = 3x + 7 \qquad \text{and} \qquad g(x) = \frac{x - 7}{3}$$

are inverses of each other since

$$f(g(x)) = 3g(x) + 7 = 3\left(\frac{x - 7}{3}\right) + 7 = x$$

and

$$g(f(x)) = \frac{f(x) - 7}{3} = \frac{(3x + 7) - 7}{3} = x$$

that is,

$$f(g(x)) = g(f(x)) = I(x) = x.$$

Whenever the inverse of a function $y = f(x)$ exists, it can be obtained by application of the following two steps:

Step 1. Solve the equation $y = f(x)$ for x in terms of y.
Step 2. Interchange the variables x and y; that is, substitute x by y and y by x in the resulting equation of Step 1.

This method of finding the inverse of a function we illustrate in the following examples.

Example 2.16 Find the inverse of the function $y = (3x - 1)/2$, and graph both the function and its inverse. Is the inverse a function?

Solution If we solve $y = (3x - 1)/2$ for x in terms of y, as suggested by Step 1, we obtain $x = (2y + 1)/3$. Furthermore, if we interchange x and y in the last equation (Step 2), we obtain

$$y = f^{-1}(x) = \frac{2x + 1}{3}$$

for the inverse of the function

$$y = f(x) = \frac{3x - 1}{2}$$

Since both the function and its inverse are linear, their graphs are those of the straight lines shown in Figure 2.26.

The inverse

$$f^{-1}(x) = \frac{2x + 1}{3}$$

Figure 2-26

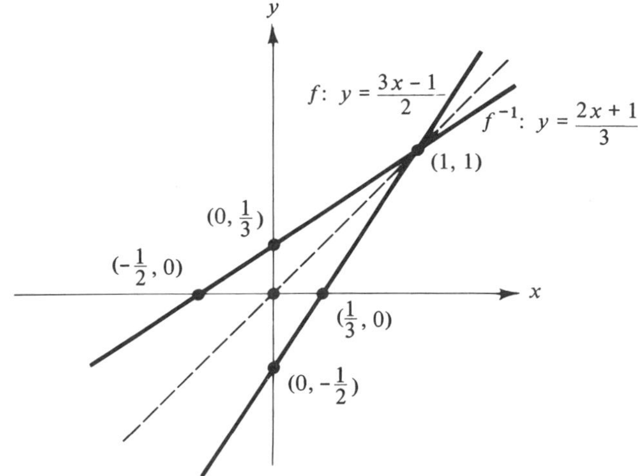

is indeed a function, since for each value of x we obtain one and only one value for y; that is, $f^{-1}(x)$ is single valued.

Example 2.17 Find the inverse of $f(x) = x^3$, and draw the graphs of both the function and its inverse $f^{-1}(x)$.

Solution: An application of Step 1 yields $x = \sqrt[3]{y}$. Furthermore, the

Figure 2-27

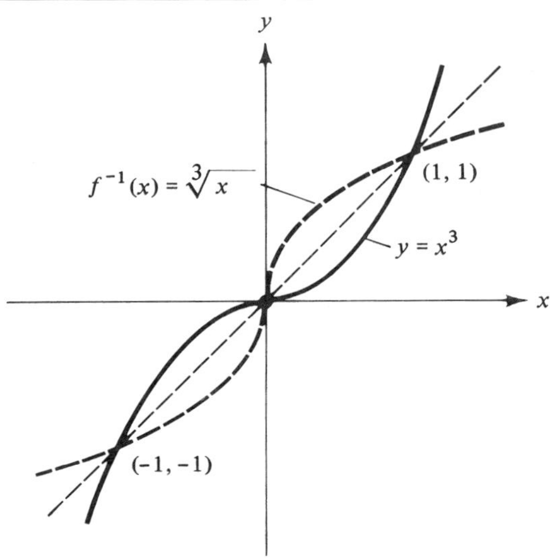

application of Step 2 implies that

$$y = \sqrt[3]{x} \qquad \text{or} \qquad f^{-1}(x) = \sqrt[3]{x}$$

is the required inverse of $f(x) = x^3$.

The graphs of $f(x) = x^3$ and its inverse $f^{-1}(x) = \sqrt[3]{x}$ are shown in Figure 2.27.

The last two examples suggest that the graphs of a function $f(x)$ and its inverse $f(x)$ are symmetric relative to the line (dotted line) dividing the first quadrant into 45° angles. It can be shown that this is indeed the case for all functions and their inverses.

Exercises

In each of the following find the inverse, and graph the function and its inverse. State whether or not the inverse, in each case, is a function.

a. $f(x) = 5x + 2$.
b. $f(x) = x^{1/3}$.
c. $f(x) = x - 1$.
d. $f(x) = 1/(x^2 - 1)$.
e. $f(x) = x^2 + 3$.
f. $g(t) = t/(t - 2)$.
g. $h(t) = 1/t$.
h. $s(t) = 16t^2$.

Ans. (a) $f^{-1}(x) = (x - 2)/5$; (e) $f^{-1}(x) = \sqrt{x - 3}$; (f) $f^{-1}(t) = 2t/(t - 1)$

2.5 Logarithmic and Exponential Functions

In investment problems an amount of dollars deposited in a bank that pays compound interest at the rate i per year becomes

$$T = A(1 + i)^x$$

at the end of x years. For instance, an amount of \$1000 deposited in a bank that pays compound interest at 6% annually will become

$$T = 1000(1 + 0.06)^{10} \qquad \text{or} \qquad T = 1000(1.06)^{10}$$

at the end of 10 years.

In bacteriology it is known that bacteria multiply at a rate proportional to the number of bacteria present at any time t and that this number is given by the formula

$$N = N_0(2.718)^{kt}$$

where N_0 is the number of bacteria present at the beginning (at $t = 0$) and k is a constant of proportionality. For example, if we begin (at $t = 0$), with 100 bacteria, then at $t = 10$ the number of bacteria present will be

$$N = 100(2.718)^{10k}$$

Finally, in learning theory the habit strength $H(n)$ is given in terms of the number n of repetitions by the formula

$$H(n) = 100(1 - 2.718^{-an})$$

where a is a positive constant indicating the rate at which habit strength is acquired.

In all the above instances, as well as in many other areas of mathematical applications, the function

$$f(x) = b^x, \quad b > 0 \quad \text{and} \quad b \neq 1$$

is the main source of describing and modeling a particular situation. The values $b = 2$ and $b = \frac{1}{2}$ yield the functions $f(x) = 2^x$ and $g(x) = (\frac{1}{2})^x$ for which tables of functional values are given below and graphs are shown in Figure 2.28, (a) and (b).

Clearly, the function $f(x) = 2^x$ is increasing and $g(x) = (\frac{1}{2})^x$ is decreasing for all values of x. Moreover, in both cases the functional values are obtained as powers of 2 and $\frac{1}{2}$, respectively.

Figure 2-28

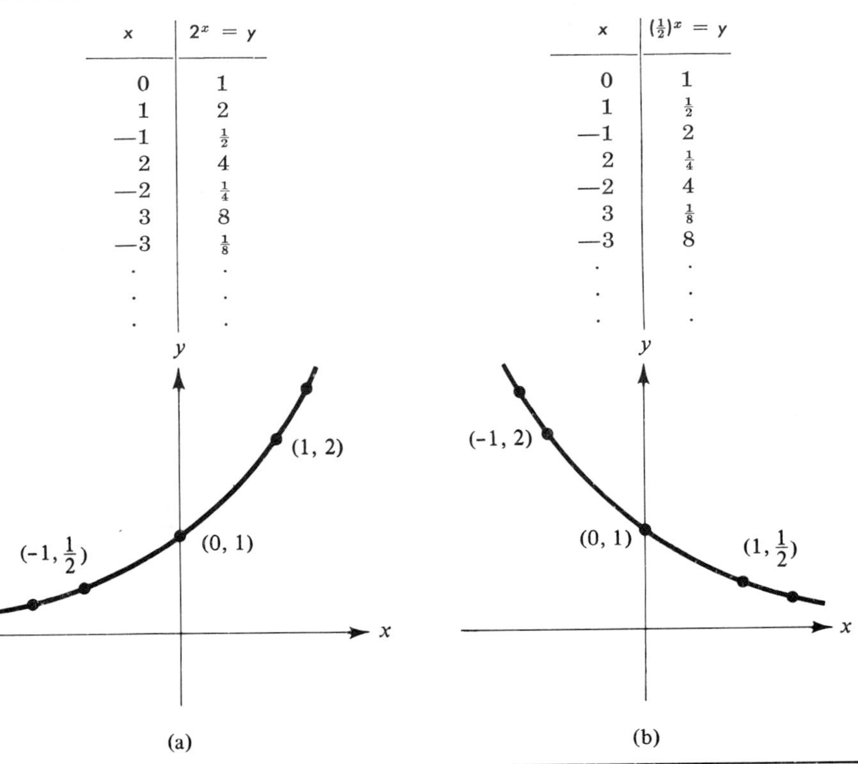

x	$2^x = y$
0	1
1	2
—1	$\frac{1}{2}$
2	4
—2	$\frac{1}{4}$
3	8
—3	$\frac{1}{8}$
⋮	⋮

x	$(\frac{1}{2})^x = y$
0	1
1	$\frac{1}{2}$
—1	2
2	$\frac{1}{4}$
—2	4
3	$\frac{1}{8}$
—3	8
⋮	⋮

(a) (b)

DEFINITION 2.10 The function $f(x) = b^x$, $b > 0$ and $b \neq 1$, is called the exponential function with base b or to the base b.

Obviously, the study of exponential functions requires familiarity with the laws of exponents which, for the sake of review, we state here in the form given in Theorem 2.3.

Theorem 2.3 For all real numbers a, b, m, and n we have

a. $a^m \cdot a^n = a^{m+n}$ product rule.
b. $a^m / a^n = a^{m-n}$ quotient rule.
c. $(a^m)^n = a^{mn}$ power to a power rule.
d. $(ab)^n = a^n b^n$ product to a power rule.

When the substitutions $m = n$ and $m = 0$ are made in part (b) of Theorem 2.3, we obtain $a^0 = 1$ and $a^{-n} = 1/a^n$, (zero and negative power rules), respectively.

Essentially, the different parts of Theorem 2.3 may be interpreted as follows:

a. The product of powers, with the same base, is a power of the same base and exponent the sum of their exponents. For example

$$3^4 \cdot 3^5 = 3^{4+5} = 3^9 \quad \text{and} \quad 5^{1/2} \cdot 5^{2/3} = 5^{1/2+2/3} = 5^{7/6}$$

b. The quotient of two powers, with the same base, is a power of the same base and exponent the difference of denominator exponent from that of the numerator. For example

$$\frac{5^6}{5^2} = 5^{6-2} = 5^4 \quad \text{and} \quad 3^{5/2}/3^{2/3} = 3^{5/2-2/3} = 3^{11/6}$$

c. The power of a number to another power is equal to a new exponential with unchanged base and exponent the product of the two (or more) powers. Thus

$$(4^5)^8 = 4^{40} \quad \text{and} \quad (5^{7/2})^{3/5} = 5^{21/10}$$

d. The power of a product is equal to the product of the respective powers. For instance,

$$(3 \cdot 5)^6 = 3^6 \cdot 5^6$$

e. As a consequence of Theorem 2.3, the zero power of any nonzero number (or quantity) is equal to 1, whereas the negative power of any nonzero number (or quantity) equals a fraction whose numerator is 1 and whose denominator is the same power with a positive exponent. For example,

$$(a + 2b)^0 = 1, \quad 5^0 = 1, \quad \text{and} \quad (x + y)^{-4} = \frac{1}{(x + y)^4}$$

if $x \neq -y$ and $a \neq -2b$.

Example 2.18 Using the laws of exponents, simplify the following:

a. $\dfrac{26x^3y^3}{13x^2y^5}$ b. $\dfrac{(x^2y^{-1})^3}{x^5y^2}$ c. $(x + 2y)^0\left(\dfrac{x^2x^{-5}y^4}{x^3y^2}\right)$

Solution

a. Applying the quotient rule of exponentials, we obtain

$$\frac{26x^3y^3}{13x^2y^5} = 2x^{3-2}y^{3-5} = 2xy^{-2} \quad \text{or} \quad \frac{26x^3y^3}{13x^2y^5} = \frac{2x}{y^2}$$

since

$$y^{-2} = \frac{1}{y^2}$$

b. The rules of raising a product to a power and a power to another power yield

$$\frac{(x^2y^{-1})^3}{x^5y^2} = \frac{x^6y^{-3}}{x^5y^2} = x^{6-5}y^{-3-2} = xy^{-5} = \frac{x}{y^5}$$

again, since

$$y^{-5} = \frac{1}{y^5}$$

c. The zero power rule, together with the other laws of exponents, results in

$$(x + 2y)^0 \cdot \left(\frac{x^2x^{-5}y^4}{x^3y^2}\right) = 1 \cdot \frac{x^{-3}y^4}{x^3y^2} = x^{-6}y^2 = \frac{y^2}{x^6}$$

if

$$x \neq -2y, \quad x \neq 0, \quad \text{and } y \neq 0$$

Returning now to the exponential function $f(x) = b^x$, we find that the inverse of $f(x) = 2^x$ contains the ordered pairs

$$(1, 0), (2, 1), (4, 2), (\tfrac{1}{2} - 1), (\tfrac{1}{4}, -2), \ldots$$

since the ordered pairs

$$(0, 1), (1, 2), (-1, \tfrac{1}{2}), (-2, \tfrac{1}{4}), \ldots$$

determine points on the graph of the function $f(x) = 2^x$. Similarly, the ordered pairs

$$(1, 0),\ (\tfrac{1}{2}, 1),\ (\tfrac{1}{4}, 2),\ (2, -1),\ (4, -2), \ldots$$

are contained in the inverse of the function $g(x) = (\frac{1}{2})^x$. (Why?) Generally, the inverse of $f(x) = b^x$, $b > 0$ and $b \neq 1$ is the set

$$f = \{(b^x, x) \mid f(x) = b^x \quad \text{and } x \text{ any real number}\}$$

Since the graph of any function and the graph of its inverse are symmetric to each other with respect to the 45° line, the graphs of the functions $f(x) = 2^x$ and $g(x) = (\frac{1}{2})^x$, together with the graphs of their inverses, are presented in Figure 2.29.

DEFINITION 2.11 The inverse of the exponential function $f(x) = b^x$, $b > 0$ and $b \neq 1$, is called the logarithmic function with base b or to the base b, and it is denoted by

$$y = \log_b x \qquad \text{or equivalently } b^y = x$$

Thus for $b = 10$

and

$$\begin{array}{lllll} x = & 1 & 10 & 100 & 1000 \ldots \\ y = & \log_{10} 1 & \log_{10} 10 & \log_{10} 100 & \log_{10} 1000 \ldots \end{array}$$

Figure 2-29

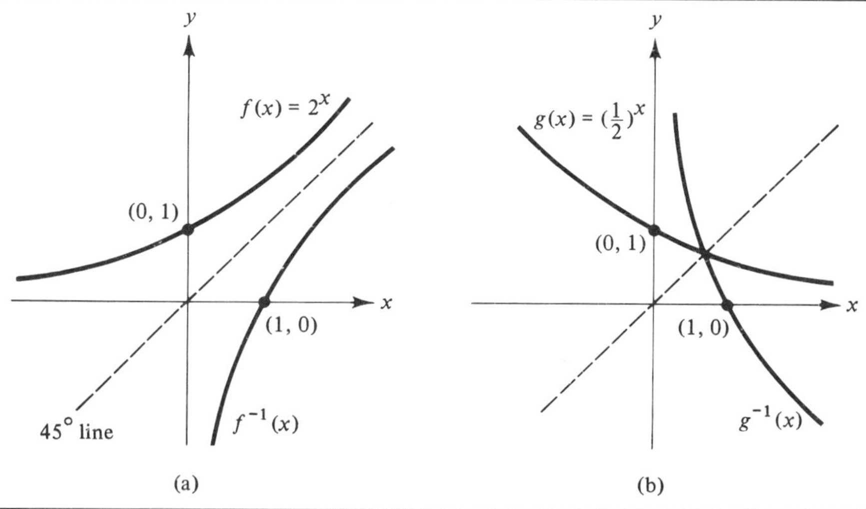

or

$$y = \quad 0 \qquad 1 \qquad 2 \qquad 3 \quad \ldots$$

since

$$10^0 = 1 \quad 10^1 = 10 \quad 10^2 = 100 \quad 10^3 = 1000 \ldots$$

Once again, for different values of $b > 1$ and $b < 1$, the graphs of the exponential functions $y = b^x$ and their inverses $y = \log_b x$ (or $b^y = x$) are shown in Figure 2.30. Evidently the logarithmic function $y = \log_b x$ (or $b^y = x$) is increasing for all x if $b > 1$ and decreasing if $b < 1$.

Example 2.19 Evaluate each of the following:

a. $\log_5 (25)$.
b. $\log_2 (1/16)$.
c. $\log_{1/2} (32)$.
d. $\log_3 (81)$.

Solution: Since $y = \log_b x$ if and only if $b^y = x$, we have

a. $\log_5 (25) = 2$, since $5^2 = 25$.

b. $\log (1/16) = -4$, since $2^{-4} = \dfrac{1}{2^4} = 1/16$.

c. $\log_{1/2} (32) = -5$, since $(\frac{1}{2})^{-5} = 32$.
d. $\log_3 (81) = 4$, since $3^4 = 81$.

Figure 2-30

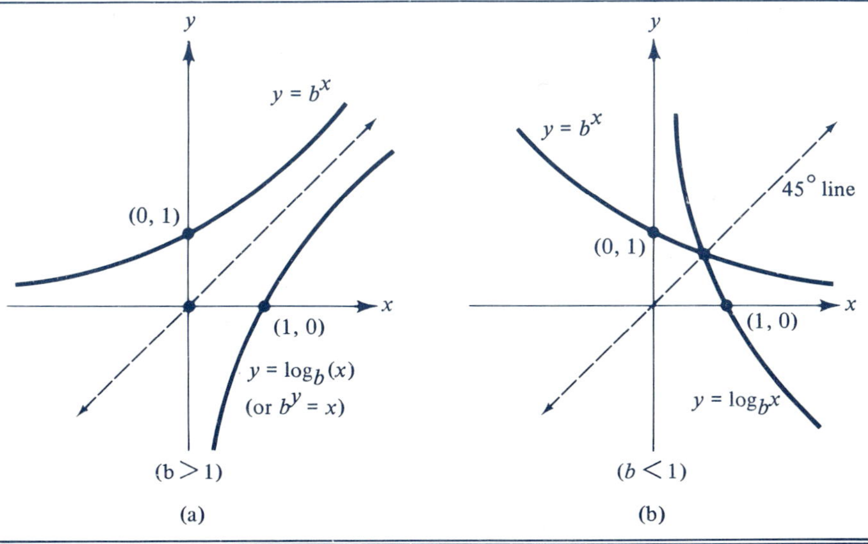

Example 2.20 Determine the value of x in each of the following:

a. $\log_8 x = 2$. b. $\log_{81} 9 = x$.
c. $\log_3 x = -3$. d. $\log_x 12 = \frac{1}{2}$.

Solution: Again, since $y = \log_b x$ is equivalent to $b^y = x$, we obtain

a. $8^2 = x$ or $x = 64$.
b. $81^x = 9$ or $(3^4)^x = 3^2$ or $3^{4x} = 3^2$. Thus $4x = 2$ and $x = \frac{1}{2}$.
c. $3^{-3} = x$ or $x = \dfrac{1}{27}$.
d. $x^{1/2} = 12$ or $(x^{1/2})^2 = 12^2$. Hence, $x = 144$.

Example 2.21 At the end of how many years will the amount of \$1000 invested at 6% compound interest per year become \$6500?

Solution: Since the amount T obtained at the end of x years whenever an amount A is invested at i percent compound interest per year is

$$T = A(1 + i)^x$$

we have $6500 = 1000(1 + 0.06)^x$ or $6.5 = (1.06)^x$. Thus to obtain the required number of years, x, we must solve the equation

$$6.5 = (1.06)^x$$

or equivalently

$$x = \log_{1.06} (6.5)$$

However, to find x requires the evaluation of the logarithm of 6.5 to the base 1.06. Since this base results in an awkward approach to the solution of the problem, we shall postpone it until the properties of the logarithmic functions and the change of base method are established.

Example 2.22 Let us assume that the rate of change in the size of the world population is proportional to the population size at any time t. Let us also suppose that the world population is growing at an annual rate of about 2%; that is, the coefficient of increase is 0.02. Then it is known that the population at any subsequent time t is given by

$$P = P_0(2.718)^{0.02t}$$

where P_0 is the population at the beginning $(t = 0)$. If the world population is estimated at 3 billion people, find the year at which this population will be doubled.

Solution: Since $P = P_0(2.718)^{0.02t}$ gives the population at any time t, and since we want t for which $P = 2P_0$, we must solve the equation

$$2\not{P}_0 = \not{P}_0(2.718)^{0.02t} \qquad \text{or} \qquad 2 = (2.718)^{0.02t}$$

Thus the definition of the logarithmic function yields

$$0.02t = \log_{2.718}(2) \qquad \text{or} \qquad t = \frac{1}{0.02}\log_{2.718}(2)$$

Here also the solution requires the evaluation of a logarithm whose base is indeed awkward. Consequently, we defer the evaluation of t for consideration immediately after the development of the logarithmic properties.

The last two examples have clearly demonstrated the need to evaluate the logarithm of numbers relative to different bases. However, since any positive number other than 1 can be used as the base of the system of logarithms, this evaluation is obviously a complicated task. Fortunately, of the infinitely many possible systems of logarithms only two are of any importance. The first is the so-called system of common logarithms or logarithms to the base 10. The other is the system known as natural logarithms or logarithms with base $e = 2.718. \ldots$ Common logarithms are almost always denoted by $\log_{10} x$ or simply $\log x$ and are particularly applicable whenever numerical calculations must be carried out. On the other hand, natural logarithms are denoted by $\log_e x$ or simply $\ln x$ and are especially useful in theoretical work. The changing of the logarithm of a certain number from one base to another is provided for in Theorem 2.4. Since $y = \log_b x$ is equivalent to $b^y = x$, the properties of the logarithmic function, also stated in this theorem, can be derived from the properties of exponents.

Theorem 2.4 For any two positive numbers A and B and any number $b > 0$ and $b \neq 1$ we have

a. $\log_b 1 = 0$. b. $\log_b b = 1$.
c. $\log_b (AB) = \log_b A + \log_b B$, product rule.
d. $\log_b (A/B) = \log_b A - \log_b B$, quotient rule.
e. $\log_b (A)^n = n \log_b A$, power rule.
f. $b^{\log_b x} = x$.
g. $\log_a x \log_b a = \log_b x$ or $\log_a x = \log_b x/\log_b a$, change of base rule.

Proof: Although the proof of this theorem is quite simple, we shall only present the proof of part (c) and leave the other parts as an exercise for the reader. Thus let

$$x = \log_b A \qquad \text{and} \qquad y = \log_b B$$

Then

$$b^x = A \qquad \text{and} \qquad b^y = B$$

or multiplying corresponding sides of the last two equations we obtain

$$b^{x+y} = AB$$

This, of course, implies that

$$\log_b (AB) = x + y \qquad \text{or} \qquad \log_b (AB) = \log_b A + \log_b B$$

Incidentally, tables of logarithms to the bases 10 and $e = 2.718\ldots$ are available. Therefore in subsequent discussions we shall refer to the tables, rather than evaluating on our own logarithms of numbers to different bases. The use of such tables is usually described in the introduction to each table.

Example 2.23 Solve for x each of the following equations:

a. $\log_2 (x - 1) + \log_2 (x + 1) = 3.$
b. $\log_5 (2x + 4) - \log_5 (x - 1) = 1.$

Solution: Part (c) of Theorem 2.1 implies that

$$\log_2 (x + 1) + \log_2 (x + 1) = \log_2 (x - 1)(x + 1)$$

Thus the equation becomes

$$\log_2 (x - 1)(x + 1) = 3 \qquad \text{or} \qquad (x - 1)(x + 1) = 2^3 = 8$$

Hence, $x^2 - 1 = 8$ or $x^2 = 9$ or $x = \pm 3$ or $x = 3$ (Why?)

b. Similarly, part (d) of the same theorem yields

$$\log_5 (2x + 4) - \log_5 (x - 1) = \log_5 \frac{2x + 4}{x - 1}$$

Consequently, the given equation becomes

$$\log_5 \left(\frac{2x + 4}{x - 1}\right) = 1 \qquad \text{or} \qquad \frac{2x + 4}{x - 1} = 5^1 = 5$$

Thus

$$2x + 4 = 5x - 5 \qquad \text{or} \qquad 9 = 3x \qquad \text{or} \qquad x = 3.$$

Example 2.24

Complete the solution of Examples 2.21 and 2.22.

Solution

a. Example 2.21. Since

$$\log_a x = \frac{\log_b x}{\log_b a}$$

(Theorem 2.4, part g) we have

$$x = \log_{1.06} (6.5) = \frac{\log_{10} (6.5)}{\log_{10} (1.06)}$$

Thus using logarithmic tables with base 10, we can evaluate $\log_{10} (6.5)$ and $\log_{10} (1.06)$ to obtain

$$x = \frac{\log_{10} (6.5)}{\log_{10} (1.06)} = \frac{0.8129}{0.0253} \qquad \text{or} \qquad x = 32.13$$

Therefore the required number of years for \$1000 to become \$6500 is about 32 years.

b. Example 2.22. Similarly,

$$t = \frac{1}{0.02} \log_{2.718} (2)$$

becomes

$$t = \frac{1}{0.02} \frac{\log_{10} (2)}{\log_{10} (2.718)}$$

which, by the use of tables, results in

$$t = \frac{1}{0.02} \frac{0.3010}{0.4343} \qquad \text{or} \qquad t = 34.62$$

Thus, the number of years required to double the existing world population is about $t = 35$, and this event will occur in the year 2008.

Exercises

1. Solve each of the following equations:

a. $\log_2 (x - 1) + \log_2 (x + 2) = 2$.

b. $\log_5 (2x - 4) - \log_5 (x + 1) = 1$.

c. $3^{4x-3} = 27$.

d. $2^{x^2+4x} = \frac{1}{8}$

e. $2^{2x} + 2^{x+2} - 5 = 0$.

f. $\log_{64} x = \frac{1}{2}$.

g. $\log_{1/2} (64) = x$.

h. $\log_x (125) = 3$.

Ans. (a) 2; (e) 0; (f) 8; (g) -6.

2. Graph the following functions and their inverses.

   a. $y = 3^x$ b. $y = \log_{10} x$.

   c. $y = (1/5)^x$. d. $y = \log_6 x$.

   e. $y = \ln (x - 1)$. f. $y = \log_{10} (2x - 4)$.

3. Work Example 2.21 if the $1000 were invested at 6% semi-annually.
4. Work Example 2.22 if the population increases by 50%.
5. Graph the habit strength function

   $$H(n) = 100(1 - 2.718^{-an}) \qquad \text{for} \qquad a = 1, 2, \text{and } 3$$

6. If the original bacterial population is 50, find the population at $t = 2$ and $t = 6$. When would the bacterial population double and triple? (Hint: use $k = 0.01$ and Example 2.22).
7. Explain why in defining the exponential function $y = b^x$ we require $b > 0$ and $b \neq 1$.
8. Prove the remaining parts of Theorem 2.4.
9. Use logarithmic tables to calculate the following:

   a. $3\ 6.789\ (5.67)^{20}$. b. $\dfrac{5\ 67.3}{4\ 1.26}$.

10. Solve the following equations:

    a. $21^{x+1} = 6$. b. $3^{x2} = 51$

References for Supplementary Readings

1. J. W. Bishir, and D. W. Drews, *Mathematics in the Behavioral and Social Sciences* (New York: Harcourt Brace Jovanovich, 1970).
2. J. E. Howell and D. Teichroew, *Mathematical Analysis for Business Decisions*, 2nd rev. ed. (Homewood, Ill.: Irvin, 1971).
3. K. W. Anderson and D. W. Hall, *Sets, Sequences, and Mappings* (New York: Wiley, 1963).
4. R. A. Good, *Introduction to Mathematics* (New York: Harcourt Brace Jovanovich, 1966).

CHAPTER 3

Elementary Combinatorial Analysis

3.1 Sampling Theory and Counting

In probability theory,[1] in statistical analysis, in econometrics, and in mathematical modeling it is often necessary to determine the number of elements in a set and calculate the number of ways in which the elements of the set can be arranged into subsets; that is, one of the first steps in the analysis of a problem is the determination of all logical possibilities and the number of the different possible outcomes. For example, if there are 17 alternate shipping routes from Los Angeles to Chicago and 12 routes from Chicago to Dallas, then there are $17 \times 12 = 204$ routes from Los Angeles to Dallas that pass through Chicago. Similarly, if in a squadron of 40 members the commander chooses the pilot whom he would most like to have fly support for him and also the pilot whom he would least like to have, then the commander has $40 \times 39 = 1560$ possible choices. Likewise, many other problems require that we develop principles of counting elements of sets. We state two of these principles in the following theorems where the symbol $n(A)$ denotes the number of elements in the set A and is commonly known as the "cardinality[2] of the set." Two more counting principles occur quite frequently, and in subsequent sections are described as "permutations" and "combinations."

Theorem 3.1 For any two subsets A and B of a universal set U, the following are true:

a. $n(A \cup B) = n(A) + n(B)$ if and only if A and B are disjoint sets $(A \cap B = \emptyset)$.
b. The product $n(A)n(B)$ represents the number of ordered pairs (a, b) in the Cartesian product $A \times B$; that is, $n(A \times B) = n(A)n(B)$.

The application of part (b) of Theorem 3.1 to the above problem establishes the number of routes from Los Angeles to Dallas through Chicago as 204. Furthermore, the above theorem, together with "mathematical induction," results in the proof of the following important theorem.

Theorem 3.2

a. If $A_1, A_2, \ldots, A_n$ are a finite number of mutually disjoint sets, then

$$n(A_1 \cup A_2 \cup \ldots \cup A_n) = n(A_1) + n(A_2) + \cdots + n(A_n)$$

b. If there are n_1 ways to perform a task T_1, if (no matter how task T_1 is performed) there are then n_2 ways of performing task T_2, if (no matter how tasks T_1 and T_2 are performed) there are n_3 ways of performing task T_3, and so forth, then there are $n_1 \times n_2 \times \cdots \times n_k$ ways of performing tasks $T_1, T_2, \ldots, T_k$ in the given order.

3.2 Permutations

If we consider the set $A = \{a, b, c\}$, we can form the following six different symbols

abc, acb, bac, bca, cab, cba

These symbols we call "permutations" of the letters a, b, and c. Clearly, the order of the letters in the symbols is significant; that is, *abc* and *cba* are two different symbols although both contain the same letters.

The paths of the tree shown in Figure 3.1 exhibit all possible permutations of these three objects. The concept of permutations of any set is formally defined in Definition 3.1.

DEFINITION 3.1 A set of n elements arranged in a definite order, $(a, b) \neq (b, a)$, is called a permutation of the set. The symbol

$P(n, k)$ or simply (n, k) denotes the number of permutations of n different elements taken k at a time.

An application of Theorem 3.2 establishes a formula for evaluating the number $P(n, k)$ of permutations of n objects taken k at a time as follows. In arranging k of the n objects we may choose any one of the n objects for the first position, $n - 1$ or $(n - 2) + 1$ for the second position, $n - 2$ or $(n - 3) + 1$ for the third position, and so on, until we reach the kth position for which we may choose any one of the remaining $(n - k) + 1$ objects. Hence the performance of the k tasks, namely, choosing an object for each position, can be done in

$$P(n, k) = n(n - 1)(n - 2) \ldots (n - k + 1)$$

ways. Since

$$n! = [n(n - 1)(n - 2) \ldots (n - k + 1)](n - k)!$$

division by $(n - k)!$ yields

$$\frac{n!}{(n - k)!} = n(n - 1)(n - 2) \ldots (n - k + 1)$$

or

$$P(n, k) = \frac{n!}{(n - k)!}$$

The last result constitutes the essence of Theorem 3.3.

Theorem 3.3 The number of permutations of n different

Figure 3-1

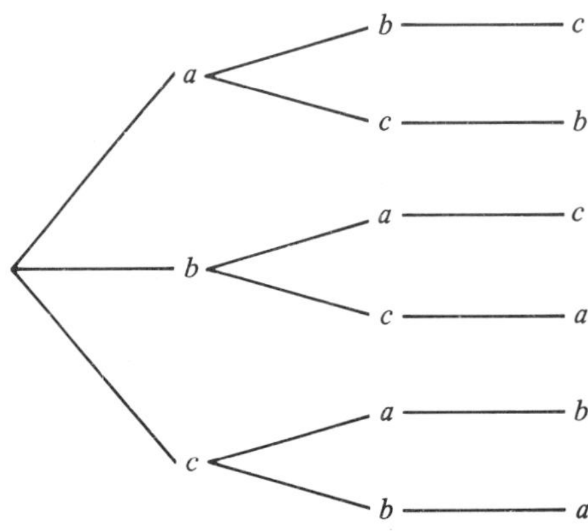

things taken k at a time is given by the formula

$$P(n, k) = \frac{n!}{(n-k)!}$$

Example 3.1 A class of 12 members is to be seated on seven chairs and a bench that accommodates five persons. In how many ways can the bench be occupied?

Solution To answer the question, we must find the number of permutations of 12 different persons taken five at a time; that is, we must calculate $P(12, 5)$. Since

$$P(12, 5) = \frac{12!}{(12-5)!} \quad \text{or} \quad P(12, 5) = \frac{12!}{7!}$$

we have

$$P(12, 5) = 12 \cdot 11 \cdot 10 \cdot 9 \cdot 8 = 95{,}040$$

Thus the bench can be occupied in 95,040 different ways.

Another way of approaching this question would be to consider the number of ways each of five tasks (the bench accommodates five persons) can be performed and apply Theorem 3.2. Since the bench accommodates five persons, we may choose any one of the 12 persons for the first place on the bench, 11 for the second position, 10 for third, 9 for the fourth, and 8 for the fifth (see table below). Thus the number of ways to perform the tasks of seating the people is

$$12 \cdot 11 \cdot 10 \cdot 9 \cdot 8 = \frac{12!}{7!} = \frac{12!}{(12-5)!} = 95{,}040$$

Spaces	*A*	*B*	*C*	*D*	*E*
Persons	12	11	10	9	8

|——→ Bench ←——|

Example 3.2 In how many ways can the letters in the word "love" be arranged?

Solution The word "love" contains four different letters. Hence, we seek the number of permutations of four different objects taken four at a time. This number must be

$$P(4, 4) = \frac{4!}{(4-4)!} = \frac{4!}{0!} = \frac{4\cdot 3\cdot 2\cdot 1}{1} = 24$$

Example 3.3 In how many ways can three people who enter a waiting room containing five chairs be seated?

Solution The first person may choose any one of the five chairs, and after he is seated, the second has a choice of any one of the remaining four chairs, and then the third has a choice of three chairs. Hence the number of ways the three people can be seated is $5\cdot 4\cdot 3 = 60 = P(5, 3)$. In fact $P(n, k)$ is also the number of ways k objects can be arranged in a definite order in n positions if $n > k$.

Example 3.4 A certain car manufacturer makes four different models A, B, C, and D. If models A and B come in any of four body styles, C and D come only as sedans or hardtops; and if each car can come in one of nine colors, then how many different types of cars are produced by this manufacturer?

Solution Since the models A and B come in any of four body styles and in any one of nine colors, we must have $4 \times 9 = 36$ distinguishable types in each of the models A and B. Similarly, models C and D each must have $2 \times 9 = 18$ distinguishable types. Thus the manufacturer produces

$$2 \times 36 + 2 \times 18 = 108$$

different car types in all.

We often deal with permutations of elements that are not all distinct; that is, we may be permuting n elements, k of which are alike. In such cases the symbol $P(n, k)$ is not applicable. For example, the three different letters in the word "one" taken two at a time can be arranged (or permuted) in six ways since $P(3, 2) = 6$. However, the letters in the word "too" can be permuted in only three ways. (Why?) Analogously, the number of permutations of n elements, k of which are alike is less than $n!$. This number can be obtained if we first arrange the $n - k$ different elements in all possible ways in the n positions, and then fill in the empty spaces with the k objects that are alike. Thus we have

$$P(n, n-k) = \frac{n!}{(n-(n-k))!} = \frac{n!}{k!}$$

for the number of permutations of n elements, k of which are alike.

This formula and its generalization are formally stated in the following important theorem.

Theorem 3.4

a. The number of permutations of n elements k of which are alike is given by

$$P(n, n - k) = \frac{n!}{k!}$$

b. If in a set of n elements there are k subsets the first of which contains m_1 elements that are alike, the second contains m_2 elements that are alike, and so forth, then the number of permutations of the n element taken n at a time is

$$\frac{n!}{m_1!\, m_2! \ldots m_k!}$$

Example 3.5 In how many ways can the letters of the word "Tennessee" be arranged?

Solution Since we have four e's, two n's, and two s's in the nine letters of the word "Tennessee," the number of ways the letters can be arranged is given by part (b) of Theorem 3.4; that is,

$$\frac{9!}{4!\, 2!\, 2!} = 3780$$

is the number of ways the letters of this word can be arranged.

Permutations are frequently cyclical or circular; that is, the arrangement of n elements (or objects) may be such that the nth element is adjacent to the first. This, of course, is the case if the n objects are placed on a circle. Then any element selected as the first lies between the second and the nth elements. To obtain all different arrangements in a cyclical or circular permutation of n objects, we leave a particular object in an invariant or fixed position and permute the remaining $n - 1$ of them. Thus the number $(n - 1)!$ represents all circular permutations of n elements.

Exercises

1. In how many ways can six people be seated on a bench?

   Ans. 720

2. With six signal flags of different colors, how many different signals can be made by displaying three flags one above the other?

3. If no two books are alike, in how many ways can two red, three green, and four blue books be arranged on a shelf so that all the books of the same color are together?

   Ans. 1728

4. How many two-digit numbers can be formed with the digits 0, 3, 4, 7 if no repetition in any of the numbers is allowed?

   Ans. 9

5. How many three-digit numbers can be formed from the digits 1, 2, 3, 4, 5 if no digit is repeated in any number?

   Ans. 60

6. How many three-digit numbers can be formed from the digits 3, 4, 5, 6, 7 if digits are allowed to be repeated?

   Ans. 125

7. How many even numbers of four different digits each can be formed from the digits 3, 5, 6, 7, 9?

8. How many integers are there between 100 and 1000 in which no digit is repeated?

   Ans. 648

9. How many numbers between 100 and 1000 can be written with the digits 0, 1, 2, 3, 4 if no digit is repeated in any number?

10. How many of the arrangements of the letters of the word "logarithm" begin with a vowel and end with a consonant?

    Ans. 90,720

11. In how many ways can three girls and three boys be seated in a row, if no two girls and no two boys are to occupy adjacent seats?

    Ans. 72

12. A salesman is going to call on five customers. In how many different sequences can he do this if he

    a. Calls on all five in one day?
    b. Calls on three in one day and two the next day?

    Ans. (a) 120; (b) 120

13. A psychologist wishes to select a sample so that he will have at least 20 people in each possible subdivision obtained by classifying according to

    a. Sex.
    b. Residence (urban or rural).
    c. Race (white or nonwhite).
    d. Marital status (married, divorced, or single).

What is the minimum sample size needed to meet his requirements.

Ans. 480

14. An experimenter has a diagnostic questionnaire consisting of 10 yes or no questions. How many diagnostic categories can be identified with this instrument?

Ans. 1024

15. A company buys a certain electronic component from three venders. In how many ways can it place six orders: two with vender X, two with vender Y, and two with vender Z?

Ans. 90

16. A company has five officers and five directors. Two of the directors are officers. List the possible memberships of a committee of four men who are either officers or directors in terms of the number of members who are

    a. Just officers.
    b. Just directors.
    c. Both officers and directors.

    How many ways are there of obtaining a committee of four consisting of

    a. Two who are just officers, one who is just a director, and one who is an officer and a director?
    b. Two who are just officers and two who are officers and directors.

3.3 Sampling and Combinations

When one chooses a collection of elements from a given set, we say he is sampling the set. If k elements of a set consisting of n distinct elements are selected, then we say that a k sample of the set was obtained. To visualize this process of sampling we may think of a box containing n objects distinguished either by numbers or colors or shapes. Then to k sample means to fill k positions with objects picked from the box. The number of distinct samples obtained depends on what is meant by "distinct"; that is, it depends on the methods or criteria used to judge the samples as different. For example, is the order in which objects are drawn important? Does it matter whether objects are replaced after being drawn? Specifically, let A be the set consisting of the letters a, b, and c and obtain all 2-samples of A. To do this, we must consider the following possibilities of sampling:

a. Sampling in which order is important, that is, $(a, b) \neq (b, a)$, and replacement of the first object is allowed. In other words symbols such as (a, a) are permitted. These considerations yield the following nine 2-samples of the set A:

(a, a), (a, b), (a, c), (b, a), (b, b), (b, c), (c, a), (c, b), and (c, c)

b. Sampling in which order is important, but the first element is not replaced; that is, symbols such as (a, a) are not permitted. Under these restrictions we obtain the following six different 2-samples:

(a, b), (a, c), (b, a), (b, c), (c, a), and (c, b)

c. Sampling in which order is not important, $(a, b) = (b, a)$, but the first object is replaced or (a, a) is permitted. These assumptions result again in the following six 2-samples of the set A:

(a, a), (a, b), (a, c), (b, b), (b, c), and (c, c)

d. Sampling in which order is not important, $(a, b) = (b, a)$, and the first object is not replaced or (a, a) is not permitted. These requirements of our sampling restrict the number of 2-samples of A to the following three only:

(a, b), (a, c), and (b, c)

In generalizing the previous processes of classification in sampling and in counting the different samples, we obtain the useful Theorem 3.5.

Theorem 3.5 In a k sampling of an n set (set of n elements), we obtain the following:

a. If the order in which objects are drawn is important, that is, if $(a, b) \neq (b, a)$, then

   i. If the objects are replaced [that is, we allow (a, a)], we obtain n^k different k samples.
   ii. If the objects are not replaced, we have $n(n - 1)(n - 2) \ldots (n - k + 1)$ different k samples. Clearly, for $n = k$ this number becomes $P(n, n) = n!$ which represents the number of permutations of n objects taken n at a time.

b. If the order in which objects are drawn is not important, that is, if $(a, b) = (b, a)$ and if the objects are not replaced, that is, if

(a, a) is not allowed, then the number of k samples is given by

$$\frac{n!}{(n-k)!\,k!}$$

The example employed in the classification of sampling at the beginning of Section 3.3 provides a good illustration of Theorem 3.5. Specifically, we found the number of 2-samples in the 3-set $A = \{a, b, c\}$ to be the following:

a. $3^2 = 9$, when order was important and replacement was allowed.
b. $3(3-1) = 6$, when order was important but replacement was not allowed.
c. $3!/[(3-2)!\,2!] = 3$, when order was not important and replacement was not allowed.

Example 3.6 Find the number of all possible 3-samples of the 5-set $A = \{a, b, c, d, e\}$.

Solution In applying the different parts of Theorem 3.5, we obtain the following:

a. $5^3 = 125$ is the number of 3-samples of A obtained under the assumptions that order was important and replacement was allowed.
b. $5(5-1)(5-3+1) = 5\cdot4\cdot3 = 60$ is the number of 3-samples of A resulting from the hypothesis that order is important and replacement is not allowed.
c. $5!/[(5-3)!\,3!] = (5\cdot4)/2 = 10$ is the number of 3-samples of A if order is not important and replacement is not allowed.

DEFINITION 3.2 The k samples of any n set A obtained if order is not important and replacement is not allowed are called combinations of A or simply combinations of n things (or objects) taken k at a time.

Clearly, the combinations of n elements taken k at a time are all the unordered k samples without replacement, and their number, $n!/[(n-k)!\,k!]$ is given by part (b) of Theorem 3.5. This number is usually denoted by

$$C\binom{n}{k} = \frac{n!}{(n-k)!\,k!} \qquad \text{or} \qquad \binom{n}{k} = \frac{n!}{(n-k)!\,k!}$$

Theorem 3.6 The number of combinations of n things taken

k at a time is given by

$$\binom{n}{k} = \frac{n!}{(n-k)!\,k!}$$

Example 3.7 The manager of a certain business with 24 employees decides to form four groups of 6 employees each. In how many ways can the first group be chosen?

Solution Since there is no ordering of the employees in the 6-groups formed, and since once an employee is selected to be in a certain 6-group he cannot be selected again (no replacement), we are concerned with the combinations of 24 employees taken 6 at a time; that is, we need to calculate the number

$$\binom{24}{6} = \frac{24!}{(24-6)!\,6!}$$

Thus the number of ways the first group can be chosen is

$$\binom{24}{6} = \frac{24!}{18!\,6!} \qquad \text{or} \qquad \binom{24}{6} = 134{,}596$$

Example 3.8 In a public opinion poll a subject is asked to express agreement or disagreement with each of six items. What is the number of possible response patterns in which the subject agrees with at most three items?

Solution If A_k is the subset of response patterns in which the subject agrees with exactly $k = 0, 1, 2, 3, 4, 5, 6$ items, then A_k contains

$$\binom{6}{k} \text{ elements} \qquad \text{or} \qquad n(A_k) = \binom{6}{k}$$

Thus the number of possible response patterns in which the subject agrees with at most three items is the number of elements in the subset $A_0 \cup A_1 \cup A_2 \cup A_3$. Hence.

$$\begin{aligned} n(A_0 \cup A_1 \cup A_2 \cup A_3) &= n(A_0) + n(A_1) + n(A_2) + n(A_3) \\ &= \binom{6}{0} + \binom{6}{1} + \binom{6}{2} + \binom{6}{3} \\ &= 1 + 6 + 15 + 20 = 42 \end{aligned}$$

is the required number.

Exercises

1. If $5P\binom{n}{3} = 24C\binom{n}{4}$, find n.

Ans. 8

2. How many straight lines are determined by

   a. 6 points?
   b. n points, no three of which lie in the same straight line?

Ans. (a) 15; (b) $\dfrac{n(n-1)}{2}$

3. A student is allowed to choose five questions out of nine. In how many ways can he choose them?

Ans. 126

4. How many different sums of money can be formed by taking two of the following: a cent, a nickel, a dime, a quarter, a half-dollar?

5. How many different committees of two men and one woman can be formed from

   a. Seven men and four women?
   b. Five men and three women?

Ans. (a) 84; (b) 30

6. From five physicists, four chemists, and three mathematicians a committee of six is to be chosen so as to include three physicists, two chemists, and one mathematician. In how many ways can this selection be made?

Ans. 180

7. How many words of two vowels and three consonants may be formed (considering any set a word) from the letters of the word

   a. "Stenographic"? b. "Facetious"?

8. In how many ways can eight women form a committee if at least three women are to be on the committee?

Ans. 219

9. How many baseball nines can be chosen from 13 candidates if A, B, C, D are the only candidates for two positions and can play no other position?

Ans. 216

10. How many different committees including three Democrats and two Republicans can be chosen from eight Republicans and ten Democrats?

11. At a meeting, after everyone had shaken hands once with everyone else, it was found that 45 handshakes were exchanged. How many were at the meeting?

Ans. 10

12. In a public opinion poll a person is asked to express agreement or disagreement with each of 10 items. Find the number of elements in the subset A_k of response patterns in which he agrees with exactly k items for each k. Moreover, find the number of possible response patterns in which he agrees with at most three items.

Ans. $\binom{10}{k}$, 176

13. A police station sends teams of 3 men on emergency calls. If the station has 15 men, how many different emergency calls can it answer?

14. Five candidates for branch manager of a certain bank are ranked according to their scores on certain tests considered relevant to job success.

    a. How many ways can this be done if no two candidates receive the same rank?
    b. How many ways can the candidates be ranked if no two receive the same rank, and candidate A is always ranked above B?

Ans. (b) 60

15. A sample of 6 radios is selected from a lot of 25. The sampling agreement provides that the entire lot will be rejected if 2 or more defective radios are found in the sample.

    a. How many different unordered samples could be drawn?
    b. If 8 of the 25 radios are defective, what percent of the samples would lead to rejection?

3.4 Sums and Products

In this section we shall develop abbreviated ways called sigma notations and pi notations, to write finite or infinite sums and products of a recursive nature. Subsequently, we shall consider the study of arithmetic, geometric, and other special types of useful series. The capital Greek letters "$\sum$" and "$\prod$" shall designate sums and products, respectively.

The sum $a_1 + a_2 + \cdots + a_n$ can be written in the abbreviated form

$$\sum_{k=1}^{n} a_k = a_1 + a_2 + \cdots + a_k$$

Similarly, the product $a_1a_2 \ldots a_n$ can be written as

$$\prod_{k=1}^{n} a_k = a_1a_2 \ldots a_n$$

Analogously, for infinite sums and products we have

$$\sum_{k=1}^{\infty} a_k = a_1 + a_2 + \cdots + a_n + a_{n+1} + \cdots$$

and

$$\prod_{k=1}^{\infty} a_k = a_1a_2 \ldots a_n \ldots$$

In the above notation, a_k represents some function in which k, called the "index" or "subscript", represents the domain variable. The numbers $k = 1, 2, \ldots, n$ constitute the range of summation or product, respectively. For example, if k is the index, then

$$\sum_{k=0}^{5} \binom{5}{k} = \binom{5}{0} + \binom{5}{1} + \binom{5}{2} + \binom{5}{3} + \binom{5}{4} + \binom{5}{5}$$

or

$$\sum_{k=0}^{5} \binom{5}{k} = \frac{5!}{(5-0)!\,0!} + \frac{5!}{(5-1)!\,1!} + \frac{5!}{(5-2)!\,2!} + \frac{5!}{(5-3)!\,3!} + \frac{5!}{(5-4)!\,4!} + \frac{5!}{(5-5)!\,5!}$$

Thus,

$$\sum_{k=0}^{5} \binom{5}{k} = 1 + 5 + 10 + 10 + 5 + 1 \qquad \text{or} \qquad \sum_{k=0}^{5} \binom{5}{k} = 32$$

and the range of the summation is from $k = 0$ to $k = 5$.

Similarly, if i is the index, then for fixed n we have

$$\sum_{i=0}^{n} \binom{n}{i} x^{n-i}y^{i} = \binom{n}{0} x^{n} + \binom{n}{1} x^{n-1}y + \binom{n}{2} x^{n-2}y^{2} + \cdots + \binom{n}{n} y^{n}$$

To demonstrate further the meaning of the summation sign $\sum$ and that of the product sign $\prod$ we consider the following illustrative examples.

$$\sum_{i=1}^{n} (2i - 1) = (2.1 - 1) + (2.2 - 1) + \cdots + (2n - 1)$$

$$= 1 + 3 + 5 + \cdots + (2n - 1) \qquad \textbf{(3.1)}$$

Here the function is $a_i = 2i - 1$, the domain variable or index is i, and the range is from $i = 1$ to $i = n$.

$$\sum_{k=0}^{n} 2^{k} = 2^{0} + 2^{1} + 2^{2} + \cdots + 2^{n} = 1 + 2 + 4 + \cdots + 2^{n} \qquad \textbf{(3.2)}$$

where the function is $a_k = 2^{k}$, the index is k, and the range is from $k = 0$ to $k = n$.

$$\sum_{m=3}^{6} \frac{m-1}{m+1} = \frac{3-1}{3+1} + \frac{4-1}{4+1} + \frac{5-1}{5+1} + \frac{6-1}{6+1}$$

$$= \frac{2}{4} + \frac{3}{5} + \frac{4}{6} + \frac{5}{7} = \frac{521}{210} \qquad \textbf{(3.3)}$$

where $a_m = (m - 1)/(m + 1)$ is the function, m the index, and from $m = 3$ to $m = 6$ is the range.

$$\sum_{k=0}^{\infty} (\tfrac{1}{2})^{k} = (\tfrac{1}{2})^{0} + (\tfrac{1}{2})^{1} + (\tfrac{1}{2})^{2} + (\tfrac{1}{2})^{3} + \cdots + (\tfrac{1}{2})^{n} + \cdots, \qquad \textbf{(3.4)}$$

where again $a_k = (\frac{1}{2})^{k}$ is the function, k the index, and the range is from 0 to ∞.

$$\prod_{i=1}^{4} (i)^{2} = (1)^{2} \cdot (2)^{2} \cdot (3)^{2} \cdot (4)^{2} = 576 \qquad \textbf{(3.5)}$$

where $a_i = (i)^2$ is the function, i is the index, and from 1 to 4 is the range.

$$\prod_{k=0}^{n}\left(\frac{k+1}{k+2}\right) = \frac{0+1}{0+2}\cdot\frac{1+1}{1+2}\cdot\frac{2+1}{2+2}\cdot\frac{3+1}{3+2}\cdots\frac{n}{n+1}\cdot\frac{n+1}{n+2}$$

or

$$\prod_{k=0}^{n}\left(\frac{k+1}{k+2}\right) = \frac{1}{\not 2}\cdot\frac{\not 2}{\not 3}\cdot\frac{\not 3}{\not 4}\cdot\frac{\not 4}{\not 5}\cdots\frac{\not n}{\not n+\not 1}\cdot\frac{\not n+\not 1}{n+2} = \frac{1}{n+2} \tag{3.6}$$

The proof of the following often useful formulas is offered as an exercise in the subsequent section on mathematical induction.

$$\sum_{k=1}^{n} (2k-1) = n^2 \tag{3.7}$$

$$\sum_{k=1}^{n} k = \frac{n(n+1)}{2} \tag{3.8}$$

$$\sum_{k=1}^{n} k^2 = \frac{n(n+1)(2n+1)}{6} \tag{3.9}$$

$$\sum_{k=1}^{n} k^3 = \frac{n^2(n+1)^2}{4} \tag{3.10}$$

Example 3.9 An investment yielding fixed periodic payments is called an "annuity." The present value of an annuity is the amount that must be invested initially in order to provide the payments. If one wishes to receive A dollars annually for the next n years from an annuity on which the remaining principal is compounded annually at $i\%$, then one must find the present value of the annuity.

Solution Since P_k dollars invested at $i\%$ compound interest will amount to $P_k(1+i)^k$ dollars in k years, the amount that must be invested now in order to obtain A dollars in k years is given by

$$A = P_k(1+i)^k \qquad \text{or} \qquad P_k = \frac{A}{(1+i)^k}$$

Hence the present value of this annuity is

$$P = P_1 + P_2 + \cdots + P_n \qquad \text{or} \qquad P = \sum_{k=1}^{n} P_k = \sum_{k=1}^{n} \frac{A}{(1+i)^k}$$

The present value of the annuity is, as we shall see subsequently, represented by the sum of a special "series" called "geometric progression."

Example 3.10 In economics it is usually assumed that if \$1 is spent in goods and services, total income increases by \$1. If a proportion $k(0 < k < 1)$ of increase in income is in turn spent for goods and services, total income will be further increased by k dollars, and so on indefinitely. Now, assuming that the same proportion k of income is spent, find the total additional income produced by the expenditure of \$1.

Solution Since the original spending of \$1 leads to the induced expenditure of $(k + k^2 + k^3 + \cdots)$ dollars and therefore increase of income by the same amount of dollars, we have

$$1 + k + k^2 + k^3 + \cdots, = \sum_{i=0}^{\infty} k^i$$

for the total additional income generated by the spending of \$1. As always, the symbol ∞ indicates that the sum has infinitely many terms.

Again we recognize that the total income can be obtained, as will be pointed out, by the sum of a special "geometric progression."

Example 3.11 In advertising or in propaganda, air-dropped leaflets have been used to reach a large number of people. If a constant ratio R of leaflets actually survives a given time period—when the rest are lost, destroyed, or not obtained and read by people—if each survived leaflet until the nth time period is reached by k persons, and if N leaflets are initially dropped, how many persons will be reached by the leaflets?

Solution Initially or during the time period $n = 0$, N leaflets reached Nk persons; in the next time period $(n = 1)$ NR leaflets will survive and will be reached by NR^1k persons. Similarly, for $N = 2$, NR^2 leaflets survive and are read by NR^2k persons; and so on indefinitely. Thus the total number of persons reached is

$$Nk + NkR + NkR^2 + \cdots = \sum_{n=0}^{\infty} NkR^n = Nk \sum_{n=0}^{\infty} R^n$$

where again the sum

$$\sum_{n=0}^{\infty} R^n$$

is that of a special "series" called "geometric progression."

The basic rules of operation for sums parallel the associative and distributive properties of numbers and are easily obtained from the basic definition of the summation. Thus

For each constant c,

$$\sum_{k=1}^{n} c = c + c + \cdots + c = nc$$

that is,

$$\sum_{k=1}^{6} 4 = 6 \cdot 4 = 24 \tag{3.11}$$

For a constant c and a function x_k, we have

$$\sum_{k=1}^{n} cx_k = c \sum_{k=1}^{n} x_k \tag{3.12}$$

For example,

$$\sum_{k=1}^{3} 4k = 4 \sum_{k=1}^{3} k = 4(1 + 2 + 3) = 24$$

For any two functions x_k and y_k, we have

$$\sum_{k=1}^{n} (x_k + y_k) = \sum_{k=1}^{n} x_k + \sum_{k=1}^{n} y_k \tag{3.13}$$

For instance,

$$\begin{aligned}\sum_{k=1}^{5} (2k^2 + 3k + 1) &= \sum_{k=1}^{5} 2k^2 + \sum_{k=1}^{5} 3k + \sum_{k=1}^{5} 1 \\ &= 2 \sum_{k=1}^{5} k^2 + 3 \sum_{k=1}^{5} k + 5 \cdot 1 \\ &= 2(1^2 + 2^2 + 3^2 + 4^2 + 5^2) \\ &\quad + 3(1 + 2 + 3 + 4 + 5) + 5 \\ &= 2(55) + 3(15) + 5 = 160\end{aligned}$$

Similarly

$$\begin{aligned}\sum_{k=1}^{3} (k^2 - 3k) &= \sum_{k=1}^{3} k^2 - \sum_{k=1}^{3} 3k \\ &= (1^2 + 2^2 + 3^2) - 3(1 + 2 + 3) = 14 - 18 \\ &= -4\end{aligned}$$

For any function x_k, we have

$$\sum_{k=1}^{m} x_k + \sum_{k=m+1}^{n} x_k = \sum_{k=1}^{n} x_k \tag{3.14}$$

That is,

$$\sum_{k=1}^{10} x_k + \sum_{k=11}^{20} x_k = \sum_{k=1}^{20} x_k$$

Exercises

1. If $x_1 = 1$, $x_2 = 3$, $x_3 = -2$, $x_4 = 4$, and $x_5 = 6$, find

   a. $\sum_{i=1}^{4} x_i^2$.
   b. $\sum_{i=1}^{5} (2x_i + 3)$.
   c. $\sum_{i=1}^{4} (x_i^2 - 2x_{i+1})$.
   d. $\sum_{i=1}^{4} (x_i + 1)(x_i - 2)$.

   Ans. (b) 39; (d) 16

2. Rewrite the following expressions in $\sum$ notation:

   a. $1 + \dfrac{x}{2} + \dfrac{x^2}{4} + \dfrac{x^3}{8} + \dfrac{x^4}{16} + \dfrac{x^5}{32}$.
   b. $1 \cdot 2 + 2 \cdot 3 + 3 \cdot 4 + 4 \cdot 5 + 5 \cdot 6$.
   c. $1 + \frac{1}{2} + \frac{1}{4} + \frac{1}{8} + \frac{1}{16} + \cdots$.
   d. $(x - \bar{x})^2 + (x - \bar{x})^2 + (x - \bar{x})^2 + (x - \bar{x})^2$.
   e. $\log 2 + \log 3 + \log 4 + \log 5 + \log 6$.
   f. $1 + 2 + 3 + 4 + \cdots + n$.
   g. $1^2 + 2^2 + 3^2 + 4^2 + \cdots + n^2$.
   h. $1^3 + 2^3 + 3^3 + 4^3 + \cdots + n^3$.
   i. Write the formula that gives the sums expressed in (f), (g), and (h).

   Ans. (a) $\sum_{k=0}^{5} \left(\dfrac{x}{2}\right)^k$ (c) $\sum_{k=0}^{\infty} \left(\dfrac{1}{2}\right)^k$ (e) $\sum_{k=2}^{6} \log k$

3. Mr. X has \$20,000 with which to purchase a 6% annuity compounded annually. What is the amount he will receive annually if he elects to receive income payments for

   a. 2 years?
   b. 5 years?
   c. 10 years?
   d. n years?

4. Compute the following sums:

a. $\sum_{k=2}^{6} (k - 3)$. b. $\sum_{k=1}^{6} (k^2 - 2k)$. c. $\sum_{k=2}^{5} (k + 1)(k - 1)$

Ans. (c) 50

5. What amount must be invested at 5% interest in order to guarantee 20 yearly payments of \$10,000?

Ans. 124,622.73

3.5 Progressions and Series

If $a_1, a_2, a_3, \ldots, a_n, \ldots$ is an infinite sequence, then the sum

$$S = \sum_{k=1}^{\infty} a_k$$

is called an "infinite series,"[3] and

$$\begin{aligned} S_1 &= a_1 \\ S_2 &= a_1 + a_2 \\ &\vdots \\ S^n &= a_1 + a_2 + \cdots + a_n \end{aligned}$$

are called "partial sums" of the series. Typical examples of series are provided by Examples 3.10 and 3.11. By the sum of an infinite series

$$\sum_{k=1}^{\infty} a_k$$

we shall mean the "limit" (if it exists) of the nth partial sum,

$$S = \sum_{k=1}^{n} a_k,$$

as "n approaches infinity."

Since the understanding and application of the limit concept requires more mathematical sophistication than we have assumed about the reader, we shall restrict ourselves for the time being to finite sums and those infinite ones that are relatively easy to calculate.

ARITHMETIC PROGRESSIONS

The sequences of numbers 1, 3, 5, 7, 9, 11, . . . and 10, 7, 4, 1, −2, −5, −8, . . . have a common characteristic, namely, that the value of the difference of any term minus its predecessor remains the same; that is, any term of the sequence is obtained by adding the same number to its predecessor. In the first sequence we have

$$3 - 1 = 5 - 3 = 7 - 5 = \cdots = 2$$

and in the second we have

$$7 - 10 = 4 - 7 = 1 - 4 = -2 - 1 = \cdots = -3$$

This characteristic property of certain sequences generally classifies them as specified in Definition 3.3.

DEFINITION 3.3 A sequence of numbers of the form

$$a, a + d, a + 2d, \ldots, a + (n - 1)d$$

is said to constitute an arithmetic progression or simply an A.P. The numbers a and $l_n = a + (n - 1)d$ are called the first and last or nth term of the A.P., respectively. The number d is called the common difference of the A.P.

Arithmetic and "geometric" progressions existed in ancient times. In fact, the Rhind Papyrus exhibits problems involving arithmetic progressions and a curious problem, number 79 of the same papyrus, refers to a geometric progression in its statement: "Sum the geometric progression of five terms, if its first term is 7 and the common ratio or multiplier is also 7."

Theorem 3.7 If a is the first term of an A.P. and if d is its common difference, then

a. The nth term of the A.P. is $l_n = a + (n - 1)d$.
b. The sum of its first n terms is $S_n = [(a + l_n)/2]n$.

Proof The definition of an A.P. and a simple step of "mathematical induction" prove that the nth term of the A.P. is given by

$$l_n = a + (n - 1)d$$

that is, the nth term is given by the sum of the first term and the common difference d multiplied by the number of terms reduced by one. On the other hand the sum of the first n terms of the arithmetic progression can be written either as

$$S_n = a + (a + d) + (a + 2d) + (a + 3d) + \cdots + [a + (n - 1)d] \tag{3.15}$$

or

$$S_n = [a + (n - 1)d] + [a + (n - 2)d] + \cdots + (a + 2d) + (a + d) + a \qquad \textbf{(3.16)}$$

Thus adding corresponding sides of Equations (3.15) and (3.16), we obtain

$$2S_n = (a + [a + (n - 1)d]) + (a + [a + (n - 1)d]) + \cdots + (a + [a + (n - 1)d])$$

or

$$2s_n = n[a + (a + (n - 1)d)].$$

Therefore the sum of the first n terms of any A.P. is given by

$$S_n = \frac{[a + a + (n - 1)d]}{2} \cdot n \qquad \text{or} \qquad S = \left(\frac{a + l_n}{2}\right) n$$

where $l_n = a + (n - 1)d$ is the nth term of the A.P.

Example 3.12 Determine the number x such that the numbers a, x, b are consecutive terms in an arithmetic progression.

Solution The numbers a, x, and b will be consecutive terms in an A.P. if and only if

$$x - a = b - x \qquad \text{or} \qquad x = \frac{a + b}{2}$$

The number $(a + b)/2$ is called the arithmetic mean of a and b. More generally, the number

$$\frac{a_1 + a_2 + a_3 + \cdots + a_n}{n} = \frac{1}{n} \sum_{k=1}^{n} a_k$$

is called the arithmetic mean of the numbers

$$a_1, a_2, a_3, \ldots, a_n$$

Example 3.13 If a man earns a salary of \$3000 the first year and if he receives an increase of \$100 per year for the next 19 years, find the total amount earned.

Solution Since the increase of salary remains the same, the man's yearly salaries constitute a sequence of numbers that is an A.P. with first term \$3000, common difference, \$100, and 20 terms. Thus the total salary earned is the sum of the first 20 terms of this A.P., namely,

$$S_{20} = \frac{3000 + [3000 + (19 \times 100)]}{2}.20 \qquad \text{or} \qquad S_{20} = \$79{,}000$$

Example 3.14 Insert six arithmetic means between 6 and 34.

Solution The problem requires that we find six numbers a_1, a_2, a_3, a_4, a_5, and a_6 such that the sequence of the numbers 6, a_1, a_2, a_3, a_4, a_5, a_6, and 34 forms an arithmetic progression. Since the first and last terms are given as 6 and 34, respectively, and since

$$l_n = a + (n - 1)d$$

where $n = 8$, we have

$$34 = 6 + 7d \qquad \text{or} \qquad d = 4$$

Therefore the six arithmetic means are 10, 14, 18, 22, 26, and 30.

GEOMETRIC PROGRESSIONS

In the sequences of numbers 2, −4, 8, −16, 32, −64, . . . and 1, $\frac{1}{2}$, $\frac{1}{4}$, $\frac{1}{8}$, $\frac{1}{16}$, $\frac{1}{32}$, . . . each number is the product of the preceding one multiplied by the same number; that is, in the first sequence the multiplier is −2 and that of the second is $\frac{1}{2}$. Clearly the constant multiplier in either sequence is the quotient of any term in the sequence divided by the immediately preceding one. Thus

$$-\frac{4}{2} = \frac{8}{-4} = \frac{-16}{8} = \cdots = -2,$$

and

$$\frac{1/2}{1} = \frac{1/4}{1/2} = \frac{1/8}{1/4} = \cdots = \frac{1}{2}$$

are the multipliers of the first and second sequence, respectively. Sequences associated with such multipliers are classified as the following definition specifies.

DEFINITION 3.4 A sequence of numbers such as a, ar, ar^2, ar^3, . . .,ar^{n-1} is said to form a geometric progression or simply a G.P. The numbers a and $l_n = ar^{n-1}$ are the first and last or nth term of the G.P., respectively. The number r is called the common ratio of the G.P.

For additional and relevant examples of geometric progressions we refer the reader to Examples 3.9, 3.10, and 3.11.

Theorem 3.8 If a is the first term of a geometric progression of numbers and if r is its common ratio, then

a. The nth term of the G.P. is $l_n = ar^{n-1}$.
b. The sum of its first n terms is given by $S_n = (a - ar^n)/(1 - r)$

Proof (a) That the nth term of the geometric progression is $l_n = ar^{n-1}$ follows from its definition and a simple application of "mathematical induction." (b) The sum

$$S_n = a + ar + ar^2 + \cdots + ar^{n-1}$$

can be written as

$$rS_n = ar + ar^2 + ar^3 + \cdots + ar^{n-1} + ar^n$$

Thus subtracting rS_n from S_n we obtain

$$S_n - rS_n = (a + ar + \cdots + ar^{n-1}) - (ar + ar^2 + \cdots + ar^{n-1} + ar^n)$$

or

$$(1 - r)S_n = a - ar^n$$

that is

$$S_n = \frac{a - ar^n}{1 - r}$$

The geometric series

$$\sum_{k=1}^{\infty} ar^{k-1} = a + ar + ar^2 + \cdots$$

is an infinite (infinitely many terms) geometric progression with first and nth terms a and ar^{n-1}, respectively, and common ratio r. The sum of infinite geometric progressions whose common ratio is, in absolute value, larger than or equal to one does not exist. However, for those with ratio whose absolute value is less than one, we have Theorem 3.9.

Theorem 3.9 If

$$\sum_{k=1}^{\infty} ar^{k-1}$$

is an infinite geometric progression whose ratio satisfies the inequality $|r| < 1$, then its sum exists, and is given by

$$S_\infty = \frac{a}{1 - r}$$

The proof of the above theorem depends on the following two facts. First,

$$S_n = \frac{a - ar^n}{1 - r}$$

can be written as

$$S_n = \frac{a}{1 - r} - \frac{a}{1 - r}(r^n)$$

and, second, the "limit" of r^n "as n approaches infinity" is equal to zero if and only if $|r| < 1$.

Example 3.15 Find the sum $\sum_{k=1}^{\infty}(\frac{1}{3})^k$

Solution Since

$$\sum_{k=0}^{\infty}\left(\frac{1}{3}\right)^k k = 1 + \frac{1}{3} + \frac{1}{9} + \frac{1}{27} + \cdots$$

is an infinite geometric progression with ratio $r = \frac{1}{3}$ and $a = 1$, Theorem 3.9 implies that the sum is

$$S = \sum_{k=0}^{\infty}\left(\frac{1}{3}\right)^k = \frac{1}{1 - \frac{1}{3}} = \frac{1}{\frac{2}{3}} = \frac{3}{2}$$

Example 3.16 Find the rational number represented by 1.333 . . ..

Solution Since $1.333\cdots = 1 + 3/10 + 3/100 + 3/1000 + \cdots$ and since the numbers 3/10, 3/100, 3/1000, . . . constitute an infinite geometric progression with ratio $r = 1/10$ and $a = 3/10$, Theorem 3.9 implies that the sum

$$1.333\cdots = 1 + \frac{3}{10} + \frac{3}{100} + \frac{3}{1000} + \cdots = 1 + \frac{3/10}{1 - 1/10}$$

$$= 1 + \frac{3/10}{9/10} = 1 + \frac{1}{3} = \frac{4}{3}$$

Thus, 1.333 . . . is the rational number 4/3.

Example 3.17 Determine the number x so that the numbers a, x, b are consecutive terms in a G.P.

Solution Since a, x, and b will be consecutive terms of a G.P. if and only if $x/a = b/x$, we find that

$$x^2 = ab \qquad \text{or} \qquad x = \pm \sqrt{ab}$$

DEFINITION 3.5 By the geometric mean of two numbers a and b we mean the number $x = \sqrt{ab}$. More generally, by the geometric mean of any n numbers $a_1, a_2, a_3, \ldots,$ and a_n we shall mean the number

$$x = \sqrt[n]{a_1a_2a_3 \ldots a_n}$$

Example 3.18 Insert three numbers (geometric means) a_1, a_2, and a_3 between 1 and 1/16 so that 1, a_1, a_2, a_3, and 1/16 are consecutive terms of a G.P.

Solution Since the first term is $a = 1$, and since the fifth term is $l_5 = 1/16$, we have

$$l_5 = ar^{5-1} \qquad \text{or} \qquad 1/16 = 1r^4$$

that is, $r = \pm\frac{1}{2}$, and the three terms are either

$$a_1 = \tfrac{1}{2},\ a_2 = \tfrac{1}{4},\ a_3 = \tfrac{1}{8}$$

or

$$a_1 = -\tfrac{1}{2},\ a_2 = \tfrac{1}{4},\ a_3 = -\tfrac{1}{8}$$

Exercises

1. In each of the following series, find the nth term and the sum of the first n terms for the indicated value of n.

   a. 1, 7, 13, 19, 25, . . . , and $n = 100$.
   b. 2, 3, 9/2, 27/4, . . . , and $n = 10$.
   c. 6, −12, 24, −48, 96, . . . , and $n = 9$.
   d. −26, −24, −22, −20, . . . , and $n = 40$.
   e. 3, 9/2, 6, 15/2, 9, . . . , and $n = 37$.
   f. 1, $\frac{1}{2}$, $\frac{1}{4}$, $\frac{1}{8}$, . . . , and $n = 10$.

   Ans. (a) 595; 29,800; (c) 1536; 1026; (e) 57; 1110

2. An arithmetic progression has first term 4 and last term 34. If the sum of its terms is 247, find the number of terms and the common difference.

3. An arithmetic progression consisting of 49 terms has last term 28. If the common difference is $\frac{1}{2}$, find the first term and the sum of the terms.

   Ans. 4, 784

4. A geometric progression has first term 3 and last term 48. If each term is twice the previous term, find the number of terms and the sum of its terms.

5. In a G.P. consisting of four terms, if the ratio is positive, the sum of the first two terms is 10, and if the sum of the last two terms is $22\frac{1}{2}$, find the terms of the G.P.

6. Find three numbers in an A.P. whose sum is 48 and the sum of their squares is 800.

Ans. $\{12, 16, 20\}$ or $\{20, 16, 12\}$

7. If \$10 is saved the first month of a certain year, \$20 the second month, \$30 the third month, and so on, find the sum that will accumulate at the end of

   a. The first year. b. Second year.

Ans. (a) 780; (b) 3000

8. A tank contains a saltwater solution in which 972 pounds of salt is dissolved. One third of the solution is drawn off and the tank is filled with pure water. After being stirred until the solution is uniform, one third of the mixture is again drawn off and the tank is again filled with water. If this process is performed as many times as indicated below, what weight of salt remains in the tank?

   a. Four times. b. Ten times.
   c. Infinitely many times.

9. Insert four arithmetic means between 9 and 24.

Ans. 12, 15, 18, 21

10. Insert four geometric means between 2 and 8.

11. Insert between 5 and 26 a number of arithmetic means such that the sum of the A.P. will be 124.

12. The geometric mean of two numbers is 8. If one of the numbers is 6, find the other number.

Ans. 32/3

13. If the third term of a G.P. is 144 and the sixth is 486, find the first term and the ratio of the G.P.

Ans. $a = 64$, $r = 3/2$

14. Find the sum of the following infinite series.

   a. $3 + 1 + \frac{1}{3} + \cdots$. b. $4 + 2 + 1 + \cdots$.
   c. $1 + 1/4 + 1/16 + \cdots$. d. $6 - 2 + \frac{2}{3} - \cdots$.

e. $4 - 8/3 + 16/9 - \cdots$. f. $1 + 1/10 + 1/100 + \cdots$.

Ans. (b) 8; (f) 10.9

15. The sum of the first two terms of a decreasing G.P. is 4, and the sum to infinity is 9/2. Write the first three terms of the G.P.

16. The successive distances traveled by a swinging pendulum bob are 36, 24, 16, . . . inches, respectively. Find the distance that the bob will travel before coming to rest.

Ans. 108

17. Find the rational numbers represented by

a. 0.121212 . . . . b. 0.136136136 . . . .
c. 2.090909 . . . . d. 2.424424424 . . . .

Ans. (b) 136/999; (c) 23/11

18. Work the following examples:

Example 3.9, for

$i = 5$ and $A = \$10{,}000$; and $i = 6$

and $A = \$15{,}000$.

Example 3.10, for

$k = \frac{2}{3}$; and $k = \frac{3}{4}$.

Example 3.11, for

$R = \frac{2}{3}, N = 10{,}000,$ and $K = 100$

and $R = \frac{3}{4}, N = 100{,}000,$ and $K = 1000$.

3.6 Mathematical Induction and the Binomial Theorem

The construction of mathematical models frequently requires a sequence of experiments in which the investigator is led to believe there exists a general principle that accounts for the repeated occurrences of certain results. For example, if we add the first two odd numbers we obtain 2^2, the first three 3^2, and the first four 4^2. Thus if we should repeat this experiment, we might conclude that the sum of the first n odd numbers is n^2 for all n. Similarly, repeated multiplications of $a + b$ by itself result in the following:

$(a + b)^0 = 1$	1
$(a + b)^1 = a + b$	1 1
$(a + b)^2 = a^2 + 2ab + b^2$	1 2 1
$(a + b)^3 = a^3 + 3a^2b + 3ab^2 + b^3$	1 3 3 1
$(a + b)^4 = a^4 + 4a^3b + 6a^2b^2 + 4ab^3 + b^4$	1 4 6 4 1
$(a + b)^5 = a^5 + 5a^4b + 10a^3b^2 + 10a^2b^3 + 5ab^5 + b^5$	1 5 10 10 5 1
....	....

In attempting to generalize the preceding experiment, we observe the following patterns in the expansion of $(a + b)^n$ when n is a natural number.

i. There are $n + 1$ terms; the first term is always a^n; the last term is always b^n.
ii. The powers of a are decreasing by one and those of b are increasing by one for each term.
iii. The sum of the powers of a and b in each term of the expansion is always n.
iv. The coefficient of the $k + 1$ term is always equal to the combinations of n things taken k at a time; that is, it is always equal to

$$\binom{n}{k} = \frac{n!}{(n - k)!k!}$$

The coefficients of the terms of the binomial expansion form an interesting pattern, known as "Pascal's triangle." Each of the coefficients different from one may be obtained by adding the number on its left to that on its right in the row immediately above it. Thus, $5 = 1 + 4$, $10 = 6 + 4, \ldots$ (see the beginning of this section). The French mathematician Blaise Pascal (1623–1662) discovered many properties of this triangular array in his work *Traité du Triangle Arithmétique* written in 1653. Yet Pascal did not discover the "arithmetical triangle." In fact it appears in one of the works of the Chinese algebraist Chu Shi-kie and in some European publications more than 100 years before Pascal's time.

Apparently, the expansion of $(a + b)^n$ is given by

$$(a + b)^n = \sum_{k=0}^{n} \binom{n}{k} a^{n-k}b^k \tag{3.17}$$

The method employed in proving the relation in Equation (3.17) is called the "principle of mathematical induction," and the expansion, Equation (3.17), itself, is known as the "binomial theorem."

These two very important mathematical concepts (or tools) were found in Pascal's works on the "arithmetical triangle." However, the generalized form of the binomial theorem for rational values of the exponent was first established by the great English mathematician Sir Isaac Newton (1642–1727), whose work appears in letters he wrote in 1676. The formal statements of the mathematical induction and the binomial theorem are as follows.

Theorem 3.10 *Mathematical Induction.*[4] If a statement concerning the natural numbers n can be proved to have the properties the statement (a) is true for $n = 1$, and (b) is true for $n = k + 1$ whenever it is true for $n = k$; then the statement is true for all natural numbers n.

Theorem 3.11 *The Binomial Theorem* If a and b are any two numbers and if n is a natural number, then

$$(a + b)^n = \sum_{k=0}^{n} \binom{n}{k} a^{n-k} b^k$$

or

$$(a + b)^n = \binom{n}{0} a^n + \binom{n}{1} a^{n-1} b + \cdots + \binom{n}{k} a^{n-k} b^k + \cdots \binom{n}{n} b^n$$

In fact, the binomial expansion of Theorem 3.11 under certain conditions is valid for all values of n, and it is an infinite expansion as the following corollary specifies.

Corollary 3.1 If a and b are any two numbers such that $|a| > |b|$, and if n is any real number, then

$$(a + b)^n = a^n + \frac{n}{1!} a^{n-1} b + \frac{n(n-1)}{2!} a^{n-2} b^2 + \frac{n(n-1)(n-2)}{3!} a^{n-3} b^3 + \cdots$$

Clearly, the expansion of $(a + b)^n$ specified in the above corollary is nothing but that of Theorem 3.11 for all the natural numbers n. This of course is the case since

$$\binom{n}{0} = \frac{n!}{(n-0)!\,0!} = \frac{n!}{n!\,0!} = 1, \quad \binom{n}{1} = \frac{n!}{(n-1)!\,1!} = \frac{n}{1!},$$

$$\binom{n}{2} = \frac{n!}{(n-2)!\,2!} = \frac{n(n-1)}{2!}, \quad \binom{n}{3} = \frac{n!}{(n-3)!\,3!}$$

$$= \frac{n(n-1)(n-2)}{3!}, \quad \text{and so on}$$

Thus whenever n is a natural number, we may apply either Theorem 3.11 or Corollary 3.1 to obtain the binomial expansion of $(a + b)^n$. For all other values of n, however, the binomial expansion "converges" if $|a| > |b|$, and it can be obtained by application of Corollary 3.1 only.

Example 3.19 Prove that $3 + 6 + 9 + \cdots + 3n = [3n(n+1)]/2$.

Solution We must first of all recognize that we have a statement concerning the natural number n, and that the principle of mathematical induction may be applied. In applying this principle we must always employ the following four steps:

i. Prove the statement for $n = 1$ (or 2, or 3).
ii. Inductive hypothesis: Assume the statement to be true for $n = k$.
iii. Problem: Prove the statement for $n = k + 1$.
iv. Conclusion: If steps (i) through (iii) are verified, then we conclude that the statement is true for all natural numbers.

In the case at hand we have

i. For $n = 1$:

$$3 = \frac{3.2}{2}$$

ii. Inductive hypothesis:

$$3 + 6 + 9 + \cdots + 3k = \frac{3k(k+1)}{2}$$

iii. Problem: Prove that

$$3 + 6 + 9 + \cdots + 3k + 3(k+1) = \frac{3(k+1)(k+2)}{2}$$

Proof From the inductive hypothesis we have that

$$3 + 6 + 9 + \cdots + 3k = \frac{3k(k + 1)}{2}$$

Thus substituting it on the left-hand side of the problem in (iii) we obtain

$$\frac{3k(k + 1) + 6(k + 1)}{2}$$

which becomes

$$\frac{3(k + 1)(k + 2)}{2}$$

when added. Hence,

$$3 + 6 + 9 + \cdots + 3k + 3(k + 1) = \frac{3(k + 1)(k + 2)}{2}$$

iv. Thus

$$3 + 6 + 9 + \cdots + 3n = \frac{3n(n + 1)}{2}$$

is true for all n.

Example 3.20 Apply the binomial theorem to expand the following:

a. $(x + y)^5$. b. $(x - 3)^4$. c. $(1 + x)^{-1}$.

Solution Since n is a natural number, in both (a) and (b) a direct application of the binomial theorem and the formula

$$\binom{n}{k} = \frac{n!}{(n - k)!\,k!}$$

yield

a.

$$(x + y)^5 = \binom{5}{0}x^5 + \binom{5}{1}x^4y + \binom{5}{2}x^3y^2 + \binom{5}{3}x^2y^3 + \binom{5}{4}xy^4 + \binom{5}{5}y^5$$

or

$$(x + y)^5 = x^5 + 5x^4y + 10x^3y^2 + 10x^2y^3 + 5xy^4 + y^5$$

b.

$$(x - 3)^4 = \binom{4}{0}x^4 + \binom{4}{1}x^3(-3) + \binom{4}{2}x^2(-3)^2 + \binom{4}{3}x(-3)^3 + \binom{4}{4}(-3)^4$$

Here $a = x$, $b = -3$, and $n = 4$. Thus

$$(x - 3)^4 = x^4 - 12x^3 + 54x^2 - 108x + 81$$

c. Since $a = 1$, $b = x$, and $n = -1$, the alternate form of the binomial theorem presented in Corollary 3.1 yields a "convergent" infinite expansion if $1 > |x|$, that is, an expansion that adds up to a finite number.

Thus under this assumption we obtain

$$(1 + x)^{-1} = 1^{-1} + \frac{-1}{1!}1^{-1-1}x + \frac{(-1)(-1-1)}{2!}1^{-1-2}x^2 + \frac{(-1)(-1-1)(-1-2)}{3!}1^{-1-3}x^3 + \cdots$$

or

$$(1 + x)^{-1} = 1 - x + x^2 - x^3 + \cdots$$

Example 3.21 Find the term that contains x^{-1} in the expansion of $(1/x - 2)^6$.

Solution Since $a = 1/x$, $b = -2$, and $n = 6$, the general term of this binomial expansion appears as

$$\binom{n}{k}a^{n-k}b^k = \binom{6}{k}\left(\frac{1}{x}\right)^{6-k}(-2)^k = \binom{6}{k}\frac{(-2)^k}{x^{6-k}}$$

$$= (-2)^k\binom{6}{k}x^{k-6}$$

Consequently, the term containing x^{-1} must be that for which the power $k - 6$ of x is equal to -1. Thus the required term is obtained when $k - 6 = -1$ or $k = 5$, that is,

$$(-2)^5 \binom{6}{5} x^{-1} = 192x^{-1}$$

is the required term.

Exercises

Using mathematical induction, prove the following:

1. $1 + 3 + 5 + \cdots + (2n - 1) = n^2.$

2. $1 + 3 + 3^2 + \cdots + 3^{n-1} = \dfrac{3^n - 1}{2}.$

3. $1^3 + 2^3 + 3^3 + \cdots + n^3 = \dfrac{n^2(n + 1)^2}{4}.$

4. $a + ar + ar^2 + \cdots + ar^{n-1} = \dfrac{a(r^n - 1)}{r - 1}, \qquad r \neq 1.$

5. $\dfrac{1}{1.2} + \dfrac{1}{2.3} + \dfrac{1}{3.4} + \cdots + \dfrac{1}{n(n + 1)} = \dfrac{n}{n + 1}.$

6. $1 + 2 + 3 + \cdots + n = \dfrac{n(n + 1)}{2}.$

7. $1^2 + 2^2 + 3^2 + \cdots + n^2 = \dfrac{n(n + 1)(2n + 1)}{6}.$

8. $1^3 + 2^3 + 3^3 + \cdots + n^3 = \dfrac{n^2(n + 1)^2}{4}.$

9. $1.3 + 2.3^2 + 3.3^3 + \cdots + n.3^n = \dfrac{(2n - 1)3^{n+1} + 3}{4}.$

10. $\dfrac{1}{2 \cdot 5} + \dfrac{1}{5 \cdot 8} + \dfrac{1}{8 \cdot 11} + \cdots + \dfrac{1}{(3n - 1)(3n + 2)} = \dfrac{n}{6n + 4}.$

11. $\dfrac{1}{1 \cdot 2 \cdot 3} + \dfrac{1}{2 \cdot 3 \cdot 4} + \dfrac{1}{3 \cdot 4 \cdot 5} + \cdots + \dfrac{1}{n(n + 1)(n + 2)}$

$$= \frac{n(n + 3)}{4(n + 1)(n + 2)}.$$

12. Obtain the following binomial expansions:

    a. $(x + \frac{1}{2})^6$. b. $(x - 2)^5$. c. $(y + 3)^4$.
    d. $(x + 1/x)^5$. e. $(x^2 - y^3)^4$. f. $(a - 2b)^6$.

13. Find and simplify the indicated term in each of the following expansions:

    a. The fifth term of $(a - b)^7$.
    b. The seventh term of $(x^2 - 1/x)^9$.
    c. The middle term of $(x - 1/x)^8$.
    d. The eighth term of $(x/2 - 2y)^{16}$.

    Ans. (a) $35a^3b^4$; (c) 70

14. Obtain the term containing x^3 in the binomial expansion of $(x^2 + 1/x)^2$.

    Ans. No term

15. Write the term independent of x in the expansion of $(\sqrt{x} + 1/3x^2)^{10}$.

    Ans. 5

16. Write and simplify the first four terms of the infinite expansion of each of the following:

    a. $(1 + 2x)^{-1}$. b. $(1 + x)^{1/3}$. c. $(4 + x)^{1/2}$.
    d. $(1 - x)^{-3}$. e. $(1 - 3x)^{-1/3}$. f. $(1 + 1/a)^{-2/3}$.

    Ans. (a) $1 - 2x + 4x^2 - 8x^3 + \cdots$;
    (c) $2 + x/4 - x^2/64 + x^3/512 - \cdots$;
    (d) $1 + 3x + 6x^2 + 10x^3 + \cdots$

References for Supplementary Readings

1. S. Goldberg, *Probability, an Introduction* (Englewood Cliffs, N.J.: Prentice-Hall, 1960).
2. S. Lipschutz, *Theory and Problems of Set Theory*, Schaum's Outline Series (New York: Schaum, 1964, Chapter 9).
3. G. B. Thomas, *Calculus and Analytic Geometry*, 4th ed., (Reading, Mass.: Addison-Wesley, 1968, Chapter 18).
4. C. R. Wylie Jr., *Foundation of Geometry*, (New York: McGraw-Hill, 1962, Introduction).

CHAPTER 4

Probability Theory

4.1 Introduction

In the area of testing and reliability, which constitutes the main concern of statistics, experimentation centers around the process of "randomly" picking and testing items from a shipment to determine whether or not these items are defective. The decision to accept or reject the shipment depends largely on the ratio of defective items found to the total of those tested. Experimentation of some sort comprises the major aspect of many studies. This is particularly true of studies in the behavioral, social, and management sciences. For instance, when the economist analyzes the national economy, the sociologist studies group behavior, and the psychologist conducts a learning experiment, all are experimenting and observing behavior whose outcome is seemingly unpredictable.

In these and other areas of experimentation, certain real numbers, called "probabilities," are associated with the outcomes of each experiment. Thus we are speaking of the probability that a defective item was picked, the probability that the cost of living will rise, the probability that the experimental group will exhibit a certain type of behavior, or the probability that the subject will learn a task. Furthermore, we are speaking of the probability that a 3 comes on top of an "unloaded" rolled die, the probability that a one-eyed jack is drawn from a "well-shuffled" deck of cards, or the probability that in the crossing of two strains the offspring will possess a certain characteristic. The classical games of poker and bridge and the birthday paradox provide further insight into the probability concept.

The theory of probability originates with Chevalier de Méré, a gambler, who in 1654 asked his friend Blaise Pascal to solve a problem known as the problem of points, that is, how to "split the pot" in an unfinished dice game. Pascal's correspondence with Pierre Fermat on this problem became the basis of a new mathematical subject known as "probability theory."

The French mathematician Pierre Simon de Laplace (1749–1827) also contributed a great deal to the advancement of probability theory primarily through his work *Théorie analytique des probabilié's* published in 1912. He referred to probability as a science that began with games and became one of the most important concerns and applications of human knowledge.

4.2 Sample Spaces and Events

Suppose an experiment consists of observing two-person communication links that are established among three individuals. Once established, a communication is assumed to persist for the duration of the experiment. Moreover, suppose that A, B, and C designate the three individuals and a, b, c the respective links AB, AC, and BC. Then the following set

$$S = \{\text{none}, a, b, c, ab, ac, bc, abc\}$$

contains for its elements all possible outcomes of this experiment; that is, S contains all possible two-person communication links established among A, B, and C. Sets such as S are formally defined in Definition 4.1.

DEFINITION 4.1 By a sample space S of an experiment we shall mean the set of all possible outcomes in that experiment. The elements in S shall be called sample points or samples or just points. Any subset of S shall be referred to as an event of the experiment. Moreover, an event shall be called elementary if and only if it contains only one point. The empty set and the sample space S shall be called the impossible and the certain (or sure) event, respectively.

Example 4.1 In rolling a die one and only one of the numbers 1, 2, 3, 4, 5, or 6 comes out on top of the die. Thus the sample space of this experiment is the set

$$S = \{1, 2, 3, 4, 5, 6\}$$

The event that an even number occurs (comes out on top of the die) is the subset

$$E = \{2, 4, 6\} \quad \text{of} \quad S$$

Similarly, the events of an odd number or of a prime number occurring are the following subsets of S.

$$O = \{1, 3, 5\} \quad \text{and} \quad P = \{2, 3, 5\}, \quad \text{respectively}$$

Example 4.2 In the experiment of tossing a coin and rolling a die simultaneously, all possible outcomes constitute the set

$$S = \{H1, H2, H3, H4, H5, H6, T1, T2, T3, T4, T5, T6\}$$

where an element such as $H3$ or $T2$ means heads on the coin and 3 on the die, or tails on the coin and 2 on the die, respectively. Thus S is a sample space of this experiment.

The following subsets of S

$$\begin{aligned} E_H &= \{H2, H4, H6\}, & E_T &= \{T2, T4, T6\} \\ O_H &= \{H1, H3, H5\}, & O_T &= \{T1, T2, T3\} \\ P_H &= \{H2, H3, H5\}, & P_T &= \{T2, T3, T5\} \end{aligned}$$

are events of the experiment. These events occur, respectively, as follows:

Heads and an even number,	Tails and an even number.
Heads and an odd number,	Tails and an odd number.
Heads and a prime number,	Tails and a prime number.

Example 4.3 Let R and F denote that the stock market rises and fails to rise, respectively, and let us observe the behavior of the stock market, say, for three consecutive days. Then, since the market either rises or fails to rise on any given day, the set

$$S = \{FFF, FFR, FRF, FRR, RFF, RFR, RRF, RRR\}$$

is a sample space of this experiment. Two events of this experiment are the following:

a. The stock market rose exactly twice in the three days. This of course constitutes the subset

$$A = \{RRF, RFR, FRR\} \quad \text{of } S$$

b. The stock market rose at least twice in the three days. This constitutes the subset

$$B = \{RRR, RRF, RFR, FRR\} \qquad \text{of } S$$

4.3 The Algebra of Events

Since the sample space S of an experiment and its events are nothing other than a set and all of its subsets, we may think of the sample space S as the universal set and the events that may occur in a performance of the experiment as subsets of S. Thus events constitute an algebra of sets with the usual set operations of union, intersection, and complementation. Therefore we may speak of the union, intersection, or complementation of events as specified in Definition 4.2.

DEFINITION 4.2 If S is a sample space of an experiment and if A, B, and C are events that may occur in a performance of the experiment, then

a. $A \cup B$ is the event that occurs if and only if either A or B occurs (or both).
b. $A \cap B$ is the event that occurs if and only if both A and B occur.
c. A' is the event that occurs if and only if A does not occur.

DEFINITION 4.3 Two events A and B are said to be mutually exclusive if and only if $A \cap B = \emptyset$, that is, if and only if B does not occur whenever A occurs.

Since events may be considered as sets that satisfy an algebra of sets under union, intersection, and complementation, they must also satisfy all theorems about sets. Hence, Theorem 4.1 (among others) must be true.[1]

Theorem 4.1 If A, B, and C are events, then

$$A \cap (B \cup C) = (A \cap B) \cup (A \cap C)$$

and

$$A \cup (B \cap C) = (A \cup B) \cap (A \cup C) \tag{4.1}$$

$$(A \cup B)' = A' \cap B' \qquad \text{and} \qquad (A \cap B)' = (A' \cup B') \tag{4.2}$$

$$A \cup (B \cup C) = (A \cup B) \cup C$$

and

$$A \cap (B \cap C) = (A \cap B) \cap C \tag{4.3}$$

$$A \cup B = B \cup A \qquad \text{and} \qquad A \cap B = B \cap A \tag{4.4}$$

Example 4.4 As an illustration of the above definitions consider the sets S, E, O, P of Example 4.1.

Illustration O is the event that an odd number occurs on the die, and E is the event that an even number occurs on the die. Thus

$$O \cup E = \{1, 3, 5, 2, 4, 6\}$$

is the event that either an odd or an even number occurs on the die.

Similarly, the event

$$O \cap E = \emptyset$$

is the event that both an even number and an odd number will occur. This of course is impossible, and therefore the events O and E are mutually exclusive or

$$O \cap E = \emptyset$$

The event that both an even number and a prime number will occur is the event

$$E \cap P = \{2\}$$

Finally, the event that an even number does not occur on the die is

$$E' = \{1, 3, 5\} \qquad \text{(the complement of } E\text{)}$$

Exercises

1. If A, B, and C are events, then draw Venn diagrams for the events $A \cap B$, $A \cup (B \cap C)$, and $(A \cup B)'$. Translate these events into ordinary language.

2. If A, B, and C are events, then find an expression and draw the Venn diagram for the following events:

   a. A but not B occurs.
   b. Either A or B, but not both occur.
   c. A and B but not C occurs.
   d. Only A occurs.

3. If we toss a coin three times and observe the sequence of heads

H and tails T that appears, then find the sample space and the following events.

a. Two or more heads appear consecutively.
b. All tosses are the same.
c. Only one tail appears.

Ans. 8; (a) 3; (b) 2; (c) 3

4. From a group of three social workers and three welfare workers, three people are assigned to a welfare study committee.

   a. List the elements of the sample space S where each element corresponds to a possible committee.
   b. Find the event (subset of S) containing those elements that correspond to committees in which there are three social workers only.
   c. Find the event (subset of S) containing those elements representing committees that contain a majority of welfare workers.

   Ans. (a) 9; (b) 1; (c) 4

5. A company wishes to buy three lots of material. If there are four domestic and three foreign suppliers from which to choose, write a sample space for this experiment, and then determine how many elements denote purchases from one domestic and two foreign suppliers.

   Ans. 35, 12

6. Company X maintains an inventory subject to certain conditions of supply and demand. The company can supply only 0, 150, or 300 units in a given time period. In the same time period, demand may be only 75, 150, or 300 units. Using ordered pairs (x, y) where x and y represent supply and demand, respectively, write a sample space for this experiment. Moreover, write the events described as follows:

   a. Demand exceeds supply.
   b. Supply exceeds demand.
   c. Demand equals supply.
   d. Demand exceeds supply, and demand differs from supply by 100 units.

7. How many distinct events are contained in a sample space having

   a. Two points?
   b. Three points?
   c. Four points?
   d. n points?

   Ans. (d) 2^n

4.4 Finite Probability Spaces

We say that a sample space is finite if and only if it contains finitely many sample points. Otherwise we say the sample space is infinite. Infinite sample spaces are encountered frequently in real-life situations. However, the study of such spaces requires much greater mathematical sophistication than we have assumed throughout this textbook. Thus in the subsequent sections we shall confine ourselves exclusively to the study of finite sample spaces and their applications.

DEFINITION 4.4 We say that a finite sample space

$$S = \{e_1, e_2, \ldots, e_n\}$$

is a finite probability space if and only if a real number p_i is assigned to each sample point e_i in S. The number p_i is called the probability of e_i for each

$$i = 1, 2, \ldots, n$$

and it satisfies the properties.

$$p_i \geq 0 \qquad \text{for each } i = 1, 2, \ldots, n \tag{4.5}$$

$$p_1 + p_2 + \cdots + p_n = 1 \tag{4.6}$$

Moreover, S is said to be equiprobable whenever every element of S is assigned the same probability.

The probability of an event E, denoted by $P(E)$, is the sum of the probabilities of all the sample points in E. For example, let

$$S = \{HH, HT, TH, TT\}$$

be a sample space for the experiment of tossing two coins and let us assign the real number $\frac{1}{4}$ as the probability of each outcome. Then the sample space S becomes a finite probability space, since

$$\tfrac{1}{4} > 0, \qquad P(HH) = P(HT) = P(TH) = P(TT) = \tfrac{1}{4},$$

and

$$\tfrac{1}{4} + \tfrac{1}{4} + \tfrac{1}{4} + \tfrac{1}{4} = 1$$

The probability of the event E that exactly one head appears when the two coins are tossed is obtained by adding the probabilities of the sample points HT and TH since each shows exactly one head. Thus

$$P(E) = P(HT) + P(TH) = \tfrac{1}{4} + \tfrac{1}{4} = \tfrac{1}{2}$$

Similarly, if we assign the probabilities $\frac{1}{2}$ to HH, $\frac{1}{8}$ to HT, $\frac{1}{8}$ to TH, and $\frac{1}{4}$ to TT, then S becomes a finite probability space once again since each probability is a positive number, and

$$\tfrac{1}{2} + \tfrac{1}{8} + \tfrac{1}{8} + \tfrac{1}{4} = 1$$

However, the probability of E becomes

$$P(E) = P(HT) + P(TH) = \tfrac{1}{8} + \tfrac{1}{8} = \tfrac{1}{4}$$

which is different from the previous probability of E. Now assigning the probabilities $\frac{1}{2}$ to HH, $\frac{1}{4}$ to HT, $\frac{1}{4}$ to TH, and $\frac{1}{2}$ to TT does not make S a probability space since

$$\tfrac{1}{2} + \tfrac{1}{4} + \tfrac{1}{4} + \tfrac{1}{2} = \tfrac{6}{4}$$

which is greater than 1 and therefore violates the definition of a probability space.

At this point it might seem reasonable to ask which one of the probabilities of the event E is the correct one. Since the correctness depends on which of the assigned probabilities best approximates the true physical nature of the coins, either answer may be valid. The coins may or may not be "balanced." Thus the above example clearly indicates that the definition of a probability space does not specify how to assign probabilities to events; however, the definition certainly restricts the possible assignments that can be made. Estimates based on experience or assumptions that seem to fit best the physical requirements are used to assign probabilities to events. For example, the mortality tables used by insurance companies to estimate the probability that an insured person will survive x years are based on estimates suggested by experience. Thus records are kept on a large number of people to estimate the probability that a person n years old will live to be $n + x$, and this probability is estimated to be the ratio of the number of people living at age $n + x$ to the number of people living at age n.

The application of probability theory to mortality tables began around 1662 in England with the work of John Graunt and Edmund Halley. It was not until 1699, however, that the first life insurance company was established in London.

Another interpretation of the probability concept is involved in betting situations. Suppose that a certain statement has been assigned probability p. Furthermore suppose a person wishes to make a bet that the statement will in fact turn out to be true. He then offers to pay x dollars if the statement turns out to be false, provided he receives y dollars if, in fact, it happens to be true. Clearly, he would like to know the values of x and y that would make the game "fair." Thus let us suppose that in a large number of such bets our bettor wins y a fraction p of the time and loses x a fraction $1 - p$ of the time, then his average winning per bet is $yp - x(1 - p)$. Since a game is said to be fair if and only if the average winning is zero, this betting game would be fair when $yp - x(1 - p) = 0$ or $x/y = p/(1 - p)$. The ratio x/y (or $x{:}y$) is said to give odds in favor of the statement. For example, if in a certain horse race a horse has been given 4/5 probability of winning the race, then the odds for a fair bet would be x to y where

$$\frac{x}{y} = \frac{p}{1 - p} = \frac{4/5}{1 - 4/5} = \frac{4/5}{1/5} = \frac{4}{1}$$

that is, the odds are 4/5 to 1/5 or 4 to 1. Therefore a fair bet would be to pay $4 if the horse loses and to receive $1 if it wins.

Theorem 4.2 If S is an equiprobable space of n points, and if E is an event of m points, then

a. The probability of any point in S is $1/n$.
b. The probability of E is $P(E) = m/n$.

To express the idea that a space is equiprobable we shall often be using expressions such as "at random," "randomly," "equally likely," "same likelihood," or other synonyms.

Theorem 4.3 The probability $P(A)$ of any event A in a finite probability space satisfies the following:

a. For every event A, $0 \leq P(A) \leq 1$.
b. If S is the sample space, then $P(S) = 1$.
c. If A and B are mutually exclusive events, then $P(A \cup B) = P(A) + P(B)$.

Part (c) of this theorem can be generalized, via mathematical induction, into Corollary 4.1.

Corollary 4.1 If $E_1, E_2, \ldots, E_n$ are n pairwise mutually exclusive ($E_i \cap E_j = \emptyset$ for $i \neq j$) events, then

$$P(E_1 \cup E_2 \cup \cdots \cup E_n) = P(E_1) + P(E_2) + \cdots + P(E_n)$$

Now, using the properties of Theorem 4.3, we can easily prove the following useful theorem.

Theorem 4.4 If A and B are any two events of an experiment whose sample space is S, then

$$P(\emptyset) = 0 \quad \text{and} \quad P(S) = 1 \tag{4.7}$$

$$P(A') = 1 - P(A) \tag{4.8}$$

$$P(A \cap B') = P(A) - P(A \cap B) \tag{4.9}$$

if $A \leq B$, then

$$P(A) \leq P(B) \tag{4.10}$$

$$P(A \cup B) = P(A) + P(B) - P(A \cap B) \tag{4.11}$$

To illustrate these theorems, we will consider the following examples.

Example 4.5 Let a card be randomly drawn from an ordinary deck of 52 cards, and let A, and B be the events

a. The card is a heart.
b. The card is either an ace, king, or queen, respectively.
   Find the probabilities $P(A)$, $P(B)$, and $P(A \cap B)$.

Solution Since "randomly drawn" implies the space is equiprobable; since an ordinary deck contains 13 hearts and 12 cards of either an ace, a king, or a queen; and since $A \cap B$ has only three elements, namely, the ace, king, and queen of hearts; then by Theorem 4.2 we obtain

$$P(A) = \frac{\text{number of hearts in the deck}}{\text{number of cards in the deck}} = \frac{13}{52} = \frac{1}{4}$$

$$P(B) = \frac{\text{number of aces, kings, and queens}}{\text{number of cards in the deck}} = \frac{12}{52} = \frac{3}{13}$$

and

$$P(A \cap B) = \frac{\text{number of aces, kings, and queens of hearts}}{\text{number of cards in the deck}} = \frac{3}{52}$$

Example 4.6 Let two items be randomly picked from a lot containing 10 items of which 3 are defective, and let A, B, and C be the events.

a. Both items are defective.
b. Both items are nondefective.
c. At least one item is defective, respectively.

   Compute the probabilities $P(A)$, $P(B)$, and $P(C)$.

Solution The outcomes of this experiment are the

$$\binom{10}{2} = 45$$

ways in which two items can be picked from a lot of 10. Moreover, the following are true:

a. The event A can occur in

$$\binom{3}{2} = 3$$

ways since there are three ways in which two defective items can be picked from a lot of three defective ones.

b. The event B can occur in

$$\binom{7}{2} = 21$$

ways since there are seven nondefective items and we pick two at a time.

c. The event C is the complement of B. (Why?)

Thus Theorems 4.2, 4.3, and 4.4 yield that

$$P(A) = \frac{3}{45} = \frac{1}{15}, \qquad P(B) = \frac{21}{45} = \frac{7}{15}$$

and

$$P(C) = P(B') = 1 - P(B) = 1 - \frac{7}{15} = \frac{8}{15}$$

Example 4.7 Of 100 welfare recipients, 30 are high school dropouts, 20 are females, and 10 are both females and high school dropouts. If a recipient is selected at random, find the probability that it is a female or a high school dropout.

Solution Let A and B be the sets of welfare recipients who are high school dropouts and females, respectively. Then $A \cap B$ is the set of welfare recipients who are both high school dropouts and females, and $A \cup B$ is the set of recipients who are either high school dropouts or females (or both). Since the space is equiprobable, the above theorems imply that

$$P(A) = \frac{30}{100} = \frac{3}{10}$$

$$P(B) = \frac{20}{100} = \frac{1}{5}$$

$$P(A \cap B) = \frac{10}{100} = \frac{1}{10}$$

and

$$P(A \cup B) = P(A) + P(B) - P(A \cap B) = \frac{3}{10} + \frac{1}{5} - \frac{1}{10} = \frac{2}{5}$$

Thus the required probability is

$$P(A \cup B) = \frac{2}{5}$$

Exercises

1. Three cars A, B, and C are in a race. If A is twice as likely to win as B, and B is twice as likely to win as C, then find their respective probabilities of winning.

   Ans. $P(A) = 4/7$, $P(B) = 2/7$, $P(C) = 1/7$

2. If three coins are tossed and the number of heads are observed, then determine the following.

   a. A sample space.
   b. A probability space.
   c. The probability of the event that at least one head appears.
   d. The probability of the event that all heads or all tails appear.

3. Four items are randomly picked from a lot of 10 items of which 3 are defective. Find the probability that

   a. None is defective.
   b. Exactly one is defective.
   c. At least one is defective.

   Ans. (a) 1/6; (b) 1/2; (c) 5/6

4. Three light bulbs are randomly picked from 14 bulbs of which 4 are defective. Find the probability that

   a. None is defective.
   b. Exactly one is defective.
   c. At least one is defective.

5. From each lot of 100 items produced by a certain factory, a sample of 10 items is randomly drawn. The lot shall be accepted if the sample contains no more than one defective item. Find the probability that a lot is accepted if it actually contains 5 defective items.

   Ans. .923

6. An insurance company has 10 district sales managers each supervising 10 salesmen. What is the probability that a group

of 4 randomly picked salesmen contains at least 2 who have the same manager?

Ans. .464

7. In a credit union conference each of six different credit unions were represented by its president and secretary-treasurer (not president). Find the probability that two randomly selected persons will be

   a. Of the same credit union.
   b. A secretary-treasurer and a president.

Ans. (a) 1/11; (b) 6/11

8. Of 30 persons, 10 are men and 20 are women. If half of the men and half of the women have blue eyes, find the probability that a randomly selected person is a man or has blue eyes.

Ans. $\frac{2}{3}$

9. If a card is drawn from a well-shuffled ordinary deck of 52 cards, find the probability that the card is one of the following:

   a. A spade. b. A six. c. A six of spades.
   d. Not an ace. e. Not a king or queen.

Ans. (a) 1/4; (b) 1/13; (c) 1/52; (d) 12/13; (e) 11/13

10. Nine black balls, numbered from 1 to 9, are placed in an urn. Nine red balls similarly numbered are placed in a second urn. Two balls are drawn, one from each urn. Find the probability that the sum of the numbers on the two balls drawn is 10.

Ans. 1/9

11. Two dice are thrown. Find the probability that the sum is nine or at least one of the two faces is even.

Ans. $\frac{3}{4}$

12. A, B, C, and D are candidates for a senate seat in an election. A poll gives C and D equal chances. Furthermore, A is only one half as likely to win as C, and B is three times as likely to win as A. Find the respective winning probabilities. What is the probability that either B or C will win?

Ans. $\frac{1}{8}$, $\frac{3}{8}$, $\frac{1}{4}$, $\frac{1}{4}$

13. Find the probability of drawing the following hands in a poker game

   a. Royal flush. b. Straight flush.
   c. Four of a kind. d. Full house.
   e. Straight.

14. In a bridge game we have the North, East, South, and West players, and each player receives 13 cards. Find the probability in each of the following events.

    a. East receives all four aces.
    b. West receives no aces.
    c. North receives five spades, five clubs, two hearts, and one diamond.
    d. East and West receive all the hearts.
    e. North and South receive all the kings and queens.
    f. Each person receives an ace.

15. Find the probability that n persons have distinct birthdays if a person's birthday can fall on any day with the same probability.

16. Let two items be randomly picked from a lot containing 12 items of which four are defective. What are the odds that

    a. At least one item is defective?
    b. Both items are nondefective?
    c. Both items are defective?

17. a. What are the odds that a six will come out on top of a rolled die?
    b. What is an amount of a fair bet?

    Ans. (a) 1 to 5; (b) \$2 to \$10

18. If a probability of $\frac{3}{4}$ has been assigned to a certain horse winning a race, then find the odds for a fair bet as well as an amount of a fair bet.

4.5 Conditional Probability

Frequently the probability of an event changes as we obtain more information about it or as we impose more requirements on its outcome. For example, the probability of an event may be affected by the occurrence of another event. Specifically, suppose that 30 people were interviewed for a certain job, and suppose their qualifications were broken down by educational background and experience as in Table 4.1. Furthermore suppose that A is the event that a person with a B.A. degree was randomly selected, and B is the event that the person selected has experience in this particular job. Then the probability that the person selected will possess a B.A. degree is computed by taking the ratio of the number of persons possessing

Table 4.1

	B.A. Degree	M.A. Degree	
Experienced	5	6	11
Nonexperienced	10	9	19
Totals	15	15	30

a B.A. degree to the total number of applicants. Thus we obtain

$$P(A) = \frac{15}{30}$$

Similarly, we obtain

$$P(B) = \frac{11}{30}$$

To find the probability that the person selected possesses both a B.A. degree and job experience, we observe that there are 5 such persons out of 30. Thus

$$P(A \cap B) = \frac{5}{30} = \frac{1}{6}$$

However, if we wish to determine the probability that the person selected is experienced, given that he holds a B.A. degree, then we observe that the sample space is reduced to one containing only those people with a B.A. degree. This requirement reduces the space into one of only 15 persons, 5 of whom are experienced. Therefore the probability that the person selected is experienced, given his possession of a B.A. degree, is 5/15 or 1/3. This probability is denoted by $P(B/A)$.

DEFINITION 4.5 If A and B are events and if $P(A) \neq 0$, then the conditional probability of B, given A, is defined as

$$P\left(\frac{B}{A}\right) = \frac{P(A \cap B)}{P(A)} \qquad \text{or} \qquad P(A \cap B) = P(A)P\left(\frac{B}{A}\right)$$

More generally,

$$P(A_1 \cap A_2 \cap \cdots \cap A_n) = P(A_1)P\left(\frac{A_2}{A_1}\right)P\left(\frac{A_3}{A_1 \cap A_2}\right) \cdots P(A_n \mid A_1 \cap A_2 \cap \cdots \cap A_{n-1})$$

Definition 4.5 is often used to obtain $P(A \cap B)$ whenever $P(A)$ and $P(B/A)$ are known. For example, two balls are randomly drawn from an urn containing two white and three red balls. If the first ball is not replaced prior to drawing the second ball, find the probablity that both balls drawn will be red. To obtain this probability, let A and B be the events that a red ball was drawn in the first and second draws, respectively. Then, $P(A \cap B)$ is the required probability. Now, since $P(A) = 3/5$, since $P(B/A) = 2/4$ (because there are two red and two white balls left in the urn), and since Definition 4.5 implies that

$$P(A \cap B) = P(A)P(B/A)$$

we obtain

$$P(A \cap B) = \frac{3}{5} \cdot \frac{2}{4} = \frac{3}{10}$$

Example 4.8 If two cards are drawn without replacement from a well-shuffled deck of cards consisting of only 13 hearts, find the probability that the king is not among them.

Solution Let A and B be the events that the king does not appear on the first and second draws, respectively. Then $P(A) = 12/13$ since 12 out of 13 cards are not kings. Similarly, $P(B/A) = 11/12$ since 11 out of 12 cards are not kings once a nonking card was drawn. Since $A \cap B$ is the event that neither card is the king, the required probability is

$$P(A \cap B) = P(A)P\left(\frac{B}{A}\right) = \frac{12}{13} \cdot \frac{11}{12} = \frac{11}{13}$$

Example 4.9 Let a pair of fair dice be rolled. If the sum is seven, find the probability that one of the dice comes up a 3.

Solution Let E be the event that the sum is seven, and let B be the event that a 3 comes up on one of the dice. Then, the required probability is $P(B/E)$ where

$$E = \{(1, 6), (2, 5), (3, 4), (6, 1), (5, 2), (4, 3)\}$$

and

$$B = \{(4, 3), (3, 4)\}$$

Therefore

$$P\left(\frac{B}{E}\right) = \frac{2}{6} = \frac{1}{3}$$

Yet the probability of B without the requirements of the sum being seven is $P(B) = 11/36$.

Example 4.10 Let C be the set of all married couples with exactly two children, and let a couple be randomly selected. Then

a. If the selected couple has a boy, find the probability that the other child will also be a boy.
b. If their first child is a boy, find the probability that the second will also be a boy.

Solution The sample space for this experiment (assumed equiprobable) is the set

$$S = \{(B, B), (B, G), (G, B), (G, G)\}$$

where B and G stand for boy and girl, respectively, and the first letter in each pair indicates the sex of their first child. Now let A and B be the events that the couple has at least one boy and that the other child is also a boy, respectively. Furthermore, let M and N be the events that the first and second child are both boys. Then the required probabilities of B given A, and N given M are

$$P\left(\frac{B}{A}\right) = \frac{P(A \cap B)}{P(A)} = \frac{1/4}{3/4} = \frac{1}{3}. \tag{4.12}$$

$$P\left(\frac{N}{M}\right) = \frac{P(M \cap N)}{P(M)} = \frac{1/4}{2/4} = \frac{1}{2} \tag{4.13}$$

Exercises

1. In the United States Democrats outnumber Republicans two to one. In the last presidential election all Republicans remained loyal, and enough Democrats crossed over to elect a Republican president by a ratio of 3:2. If a voter is selected at random, what is the probability that he is a Democrat? What is the probability of his being a democrat if it is known that he voted Republican?
Ans. $\frac{2}{3}, \frac{4}{9}$

2. In a hand of bridge a player, say the North, has no kings. Find the probability that his partner (South) also has no kings.
Ans. $\binom{35}{13} / \binom{39}{13}$

3. An experimental group of students consists of 50% male and 50% female students. The record shows that about 5% of the male students fail a certain examination, but only .25% of female students fail the same examination. If student X fails, what is the probability that the student is a boy and not a girl?
Ans. 20/21

4. From a well-shuffled deck of cards, find the probability that one draws the following:

   a. A pair. b. A pair if the first card is ace of spades.

   Ans. (a) 1/17; (b) 1/17

5. Among the car salesmen of a used car lot, 30% have no previous experience in sales, 20% have one year of experience, 20% have two years of experience, 14% have three years of experience, 9% have four years of experience, and 7% have five or more years of experience. If a salesman is randomly selected, what is the probability that he has more than two years of experience? What is the probability that he has more than two years of experience, if it is known that he has at least one year of experience?

   Ans. 3/10, 3/7

6. If a pair of fair dice are rolled and if the sum is six, find the probability that one of the dice is a two. What is the probability of obtaining a two without any restriction on the sum?

   Ans. 2/5, 11/36

7. If two numbers are randomly selected from the numbers 1 through 9 and if their sum is even, then find the probability that both numbers are odd.

   Ans. $\frac{5}{8}$

8. A legislative body consists of 60% Democrats and 40% Republicans. If it is known that 35% of the Democrats and 60% of the Republicans favor a pending piece of legislation, what is the probability that a legislator favoring this legislation is a Democrat? A Republican?

   Ans. 7/15, 8/15

9. A shipment of 12 items contains 4 defective ones. Three items are randomly drawn from this shipment, one after the other. What is the probability that all 3 items are nondefective? What is the probability that at least 1 item is defective?

   Ans. 14/55, 41/55

4.6 Bayes' Theorem and Stochastic Processes

Bayes' theorem is nothing more than a restatement of the equation of conditional probability. However, because of its many important applications, we give it special consideration, and by way of examples attempt to demonstrate the manner in which it is usually applied. The proof of this theorem may be found in many books on probability.[2]

Bayes' Theorem 4.5 If $E_1, E_2, \ldots, E_n$ are events of a sample space S such that

$$\bigcup_{i=1}^{n} E_i = S$$

and

$$E_i \cap E_j = \emptyset \qquad \text{for all } i \neq j$$

then for each i and each event A of S we have

$$P\left(\frac{E_i}{A}\right) = \frac{P(E_i)P(A/E_i)}{\sum_{i=1}^{n} P(A/E_i)P(E_i)}, \qquad i = 1, 2, 3, \ldots, n$$

Example 4.11 It is known that drug therapy is most effective in treating schizophrenics, whereas psychotherapy is most effective in treating neurotics. Thus if schizophrenics and neurotics are the only two types of mental patients entering a hospital, it is expedient that a patient be correctly identified. Therefore a test is administered to all entering patients. Testing experience indicates that 25% of the neurotics and 65% of the schizophrenics pass the test. Furthermore, experience shows that in an undiagnosed group about 40% are actually schizophrenic and 60% are neurotic. If from a large sample of incoming patients one is randomly selected and given a test and if the patient passes the test, what is the probability that he is actually neurotic?

Solution Let E_1 be the event that the selected patient is indeed neurotic, and let A be the event that the patient passed the test. Then the required probability is $P(E_1/A)$. Since the events E_1 and $E_2 = E_1'$ (complement of E_1) satisfy the hypothesis of Bayes' theorem, that is,

$$E_1 \cap E_2 = \emptyset \qquad \text{and} \qquad E_1 \cup E_2 = S \qquad \text{(sample space)}$$

application of the theorem yields

$$P\left(\frac{E_1}{A}\right) = \frac{P(E_1)P(A/E_1)}{P(A/E_1)P(E_1) + P(A/E_2)P(E_2)}$$

where

$$P(E_1) = \frac{60}{100}, \quad P(E_2) = \frac{40}{100}, \quad P\left(\frac{A}{E_1}\right) = \frac{25}{100},$$

and

$$P\left(\frac{A}{E_2}\right) = \frac{65}{100}$$

that is,

$$P\left(\frac{E_1}{A}\right) = \frac{(.6)(.25)}{(.25)(.6) + (.65)(.4)} = \frac{.15}{.41} = \frac{15}{41}$$

Hence, the probability that the patient is neurotic is

$$P\left(\frac{E_1}{A}\right) = \frac{15}{41}$$

On the other hand the probability that he is schizophrenic is

$$1 - \frac{15}{41} = \frac{26}{41}$$

Therefore the unconditional probability

$$P(E_1) = \frac{60}{100}$$

has been reduced to

$$P\left(\frac{E_1}{A}\right) = \frac{15}{41}$$

given the information that the patient passed the test; that is, if it is known that a patient passed the test, the odds

$$\frac{P}{1 - P} = \frac{26/41}{1 - 26/41}$$

are 26 to 15 that he is schizophrenic and he should he given drug therapy instead of psychotherapy. However, the odds that he is neurotic are

$$\frac{P}{1 - P} = \frac{15/41}{1 - 15/41}$$

or 15 to 26.

Example 4.12 Let T_1, T_2, and T_3 be the events representing three distinct proposed economic theories. Without knowing which one represents the behavior of our economic system, let

$$P(T_1) = \tfrac{1}{2}, \qquad P(T_2) = \tfrac{1}{3}, \qquad \text{and} \qquad P(T_3) = \tfrac{1}{6}$$

be the assigned probabilities that these theories are correct, respectively. Finally, let Table 4.2 describe probabilistically the behavior

Table 4.2

Economic Theory	Probability of Cost of Living Going Up, Down, No Change		
	A: Increases	B: Decreases	C: No Change
T_1	1/5	1/5	3/5
T_2	1/5	3/5	1/5
T_3	4/5	1/10	1/10

of our economy for the coming year. Now, if a year later it is established that the cost of living has indeed increased, which one of the economic theories is the most likely to be correct?

Solution Since A is the event that the cost of living increases and since T_i $(i = 1, 2, 3)$ is the event representing the behavior of the ith economic theory, we are interested in the conditional probabilities $P(T_1/A)$, $P(T_2/A)$, and $P(T_3/A)$. Thus applying Bayes' theorem we obtain

$$P\left(\frac{T_1}{A}\right) = \frac{P(T_1)P(A/T_1)}{P(A/T_1)P(T_1) + P(A/T_2)P(T_2) + P(A/T_3)P(T_3)}$$

or

$$P\left(\frac{T_1}{A}\right) = \frac{1/2 \cdot 1/5}{1/5 \cdot 1/2 + 1/5 \cdot 1/3 + 4/5 \cdot 1/6} = \frac{1/10}{9/30} = \frac{3}{9}$$

Similarly,

$$P\left(\frac{T_2}{A}\right) = \frac{P(T_2)P(A/T_2)}{P(A/T_1)P(T_1) + P(A/T_2)P(T_2) + P(A/T_3)P(T_3)}$$

$$= \frac{1/3 \cdot 1/5}{9/30} = \frac{2}{9}$$

and

$$P\left(\frac{T_3}{A}\right) = \frac{P(T_3)P(A/T_3)}{P(A/T_1)P(T_1) + P(A/T_2)P(T_2) + P(A/T_3)P(T_3)}$$

$$= \frac{1/6 \cdot 4/5}{9/30} = \frac{4}{9}$$

Thus given that the cost of living has gone up, T_3 is the economic theory most likely to be correct.

DEFINITION 4.6 By a stochastic process we shall mean a sequence of experiments each of which has finitely many outcomes with given probabilities.

In the following sections we shall concern ourselves only with finite stochastic processes, that is, with finite sequences of experiments in which each experiment has a finite number of outcomes with given probabilities. For convenience we shall describe such processes and compute the probability of any event by using tree diagrams as demonstrated in the following examples.

Example 4.13 The 24 employees of a small company are classified as follows:

Class 1 has 10 employees of whom 4 are inefficient.
Class 2 has 6 employees of whom 1 is inefficient.
Class 3 has 8 employees of whom 3 are inefficient.

If we randomly select one class and one employee of this class, what is the probability that the employee is inefficient?

Solution This problem is actually a stochastic process in which the sequence of experiments consists of the following two experiments.

a. Select one of the three classes.
b. Select one employee from the selected class.

The tree diagram in Figure 4.1 describes this process and gives the probabilities of each tree branch as follows. The branches $C1$, $C2$, and $C3$ indicate the class of the employee selected, and the number $\frac{1}{3}$ is the probability that one out of three classes was picked. The branches $(1I, 1E)$, $(2I, 2E)$, and $(3I, 3E)$ designate the selection of the employee from either the first or second or third class and his inefficiency I or efficiency E. The numbers on the branches represent the probabilities of the corresponding events. Finally, the value of P as indicated to the right of the diagram designates the probability of each sequence of events.

Clearly, the probability of any particular path of the tree diagram is obtained by multiplying the probabilities of the two branches that represent the two events. For instance, in the case of the path $C1I$ we have the event $A \cap B$ where A is the event that the employee was selected from class 1 and B is the event that the selected employee was inefficient given that he came from class 1. Thus the required

Figure 4–1

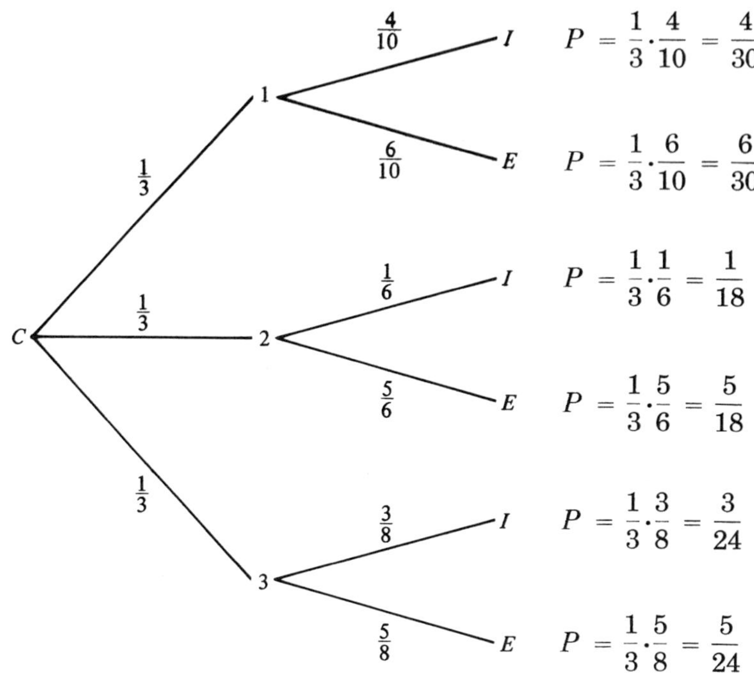

probability is

$$P(A \cap B) = P(A)P\left(\frac{B}{A}\right) = \frac{1}{3} \cdot \frac{4}{10} = \frac{4}{30}$$

Similarly, applying the principles of conditional probability, we obtained the probability of each path of the tree diagram. Now, since there are three mutually exclusive events, each represented by one of the three paths that lead to an inefficient employee, the required probability must be the sum of the probabilities associated with each of the three paths that lead to an inefficient employee. Hence the required probability is

$$P = \frac{4}{30} + \frac{1}{18} + \frac{3}{24} \qquad \text{or} \qquad P = \frac{113}{360}$$

Example 4.14 The three assembly lines A, B, and C produce 50%, 30%, and 20% of the total number of certain auto parts, respectively. The percentages of defective output of these lines are 3%, 4%, and 5%, respectively. If a part is randomly selected, what is the probability that it is defective?

Figure 4-2

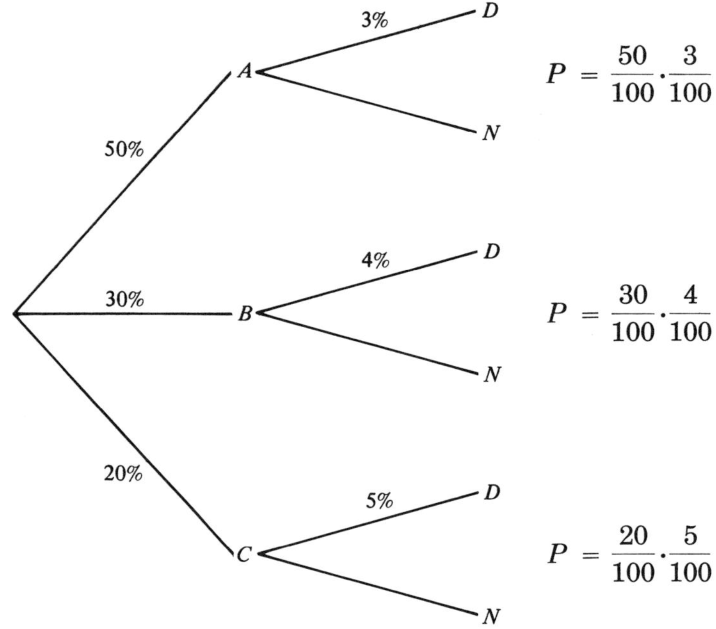

Solution This problem may be considered as a finite stochastic process described in the tree diagram in Figure 4.2. The letters D and N designate defective and nondefective auto parts, respectively.

For further explanation of the tree diagram, we refer the reader to the previous example. Thus the probability that the selected part is defective is given by

$$P = \frac{50}{100} \cdot \frac{3}{100} + \frac{30}{100} \cdot \frac{4}{100} + \frac{20}{100} \cdot \frac{5}{100}$$

or

$$P = \frac{370}{10000} = \frac{37}{1000}$$

that is, 3.7% of the parts are defective.

Exercises

1. Work Example 4.12 by assuming the following:

   a. The cost of living has gone down.
   b. The cost of living remains the same.

2. In the state of California 4% of the men and 1% of the women are taller than 6 feet. Moreover, 60% of the residents of the state are women. If a randomly selected Californian is taller than 6 feet, what is the probability that this person is a woman?

   Ans. 3/11

3. In country X, 40% of the people have brown hair, 25% have brown eyes, and 15% have both brown hair and brown eyes. Now, if a citizen of this country is randomly selected, determine the following.

   a. If he has brown eyes, what is the probability that he does not have brown hair?
   b. If he has brown hair, what is the probability that he also has brown eyes?
   c. What is the probability that he has neither brown hair nor brown eyes?

   Ans. (a) $\frac{2}{5}$; (b) $\frac{3}{8}$; (c) $\frac{1}{2}$

4. In college X, 25% of the boys and 10% of the girls are taking statistics. Furthermore, 60% of the student body are girls. If a randomly selected student is enrolled in statistics, what is the probability that the student is a girl?

   Ans. $\frac{3}{8}$

5. Among the employees of a company, 30% of the men and 20% of the women are college graduates. Furthermore, 60% of the employees are men. If a randomly selected employee is a college graduate, what is the probability that this person is

   (a) A man? (b) A woman?

6. Assembly groups A, B, and C of a certain company produce 25%, 35%, and 40% of a month's production items, respectively. If the defective items turned out by A, B, and C are 1%, 2%, and 3%, respectively, and if a randomly selected item is found to be defective, find the probability that it was produced by

   a. Assembly group A.
   b. Assembly group B.
   c. Assembly group C.

7. An insurance company X classifies automobile drivers as good risks (G), medium risks (M), and bad risks (B). The experience of the company indicates that 20%, 30%, and 50% of its policyholders are classified as G, M, and B, respectively. Furthermore, the probability of having at least one accident per year is .001

for G, .006 for M, and .012 for B. Mr. Smith buys a policy from this company and in six months has an accident. What is the probability that he is

a. Class G driver? b. Class M driver?
c. Class B driver?

Ans. (a) .025; (b) .225; (c) .75

8. Mr. Yam also buys a policy from the insurance company described in Exercise 7, and he has had no accident in two years. What is the probability that he is a G, M, or B class driver? How would he be classified if he had three accidents in one year and if .001, .006, and .012 were the probabilities of one accident per year for G, M, and B drivers, respectively?

9. If in state X the census reveals that 40% of the rural population moves to the city, whereas 20% of the urbanites move to the country each year; and if 60% of the initial population were rural dwellers, what proportion of the total population will be urban dwellers in the long run?

Ans. $\frac{2}{3}$

10. Three members of a four-member discussion group initially favor an issue, while the fourth member is opposed. Suppose that a member is selected at random and given the opportunity to defend his position. Furthermore, suppose that after a given member speaks, one member of the opposition (if any exists) changes his position. After the first member speaks, a member is again randomly selected and given the opportunity to speak. The process continues indefinitely. What is the probability that the members will be in agreement at the end of their third speech? Fourth speech? Fifth speech?

11. A box contains a fair coin, a two-headed coin, and a coin that has probability $\frac{1}{3}$ for heads. If one of the three coins is randomly selected and tossed, what is the probability that heads appears?

Ans. 11/18

12. A man will accept job X with probability $\frac{1}{2}$, job Y with probability $\frac{1}{3}$, and job Z with probability $\frac{1}{6}$. In each case he must decide whether to rent or buy a house. The probabilities of his buying are $\frac{1}{3}$ if he takes job X, $\frac{2}{3}$ if he takes job Y, and 1 if he takes job Z. Given that he buys a house, what are the probabilities of his having accepted each job?

Ans. .3, .4, .3

13. The contents of the urns A and B are as follows:

    a. A contains four red and three blue marbles.
    b. B contains three red and two blue marbles.

    An urn is randomly selected and a marble is drawn and put in the other urn, and then a marble is drawn from the second urn. What is the probability that both marbles drawn are of the same color?

14. A box contains seven radio tubes of which two are defective. The tubes are tested one after the other until the two defective ones are discovered. What is the probability that the process terminated in the

    a. Second test? b. Third test?

    If the process ended on the third test, what is the probability that the first tube is nondefective?

    Ans. (a) 1/42; (b) 1/35

15. Three machines A, B, and C produce 40%, 25%, and 35% of the total numbers of items of a factory. The percentages of defective output of these machines are 2%, 3%, and 4%, respectively. If an item is randomly selected, what is the probability that it is defective?

4.7 Independent Events

In the section on conditional probability we observed that the probability of an event was influenced by the fact that another event had or had not occurred. For instance, in the case of rolling two dice we found the probability of the event "at least one 3 coming up" to be 11/36, whereas the probability of the same event was found to be $\frac{1}{3}$ when we required the sum of the two numbers to be seven. Clearly, the occurrence of the event "the sum of the two numbers is seven" influenced the probability of the event "at least one 3 comes up." However, quite frequently the probability of an event B is not influenced by whether another event A has or has not occurred. When this is the case, we say that the event B is independent of the event A. We formally define independence of events in Definition 4.7.

DEFINITION 4.7 We say that two events A and B are independent if and only if

$$P(A \cap B) = P(A)P(B)$$

that is, the outcome of one event does not influence the outcome of the other. Otherwise, we say that A and B are dependent events.

Example 4.15 A standard die is rolled, and then a card is drawn randomly from a well-shuffled deck of cards. Find the probability of the event that the die shows a 3 or a 6 and the card drawn is a spade.

Solution Let A be the event that either a 3 or a 6 shows on the die, and let B be the event that a spade was drawn from the deck. Then the required probability is $P(A \cap B)$ where obviously A and B are independent events (the occurrence of one does not influence the occurrence of the other). Since 2 out of 6 are the numbers that might occur on the die, and since 13 out of 52 are the spades that might be drawn from the deck of cards, the required probability is

$$P(A \cap B) = P(A)P(B) = \left(\frac{2}{6}\right)\left(\frac{13}{52}\right) = \frac{1}{12}$$

Example 4.16 The probability that person P_1 passes an examination E is $\frac{1}{4}$, and the probability that another person P_2 passes the same examination E is $\frac{2}{5}$. Find the probability that

(a) Both persons will pass the examination.
(b) At least one person will pass the examination.

Solution Let A and B be the events that P_1 and P_2 pass the examination, respectively. Then the required probability is clearly $P(A \cap B)$. Since the passing of the examination E by P_1 influences in no way the performance of P_2, the events A and B must be independent. Thus

$$P(A \cap B) = P(A)P(B) = \frac{1}{4}\cdot\frac{2}{5} = \frac{1}{10}$$

The event that at least one will pass the examination is $A \cup B$, and the required probability is $P(A \cup B)$. Thus

$$P(A \cup B) = P(A) + P(B) - P(A \cap B) = \frac{1}{4} + \frac{2}{5} - \frac{1}{10} = \frac{11}{20}$$

Now, using the previous definition and the principle of mathematical induction, we could prove the following theorem.

Theorem 4.6 The events $A_1, A_2, \ldots, A_n$ are independent if and only if the following are true:

$$P(A_i \cap A_j) = P(A_i)P(A_j) \qquad \text{for all } i \neq j \tag{4.14}$$

$$P(A_1 \cap A_2 \cap \cdots \cap A_n) = P(A_1)P(A_2)\cdots P(A_n) \tag{4.15}$$

Similarly, from the previous definition and the definition of conditional probability the proof of the following theorem may be established.

Theorem 4.7 Two events A and B are said to be independent if and only if

$$P(A) = P\left(\frac{A}{B}\right)$$

Example 4.17 Show that for one roll of a fair die the events

$$A = \{1, 3, 5\} \qquad \text{and} \qquad B = \{2, 3, 6\}$$

are not independent.

Solution Since A contains three of the six possible outcomes,

$$P(A) = \tfrac{3}{6} = \tfrac{1}{2}$$

Similarly, $P(B) = \frac{1}{2}$. However, the event $A \cap B = \{3\}$ contains only one sample point. Thus

$$P(A \cap B) = \tfrac{1}{6} \neq P(A)P(B) = \tfrac{1}{4},$$

and the events A and B are not independent.

The next example indicates that the second part of Theorem 4.6 does not follow from the first part.

Example 4.18

Let

$$S = \{HH, HT, TH, TT\}$$

be the equiprobable space in the experiment of tossing two fair coins. Moreover, let A, B, and C be the following events:

$A = \{HH, HT\}$, that is, heads on the first coin.
$B = \{HH, TH\}$, that is, heads on the second coin.
$C = \{HT, TH\}$, that is, heads on exactly one of the coins.

Then

$$P(A) = P(B) = P(C) = \tfrac{2}{4} = \tfrac{1}{2}$$

and

$$P(A \cap B) = P(A)P(B) = P(A \cap C) = P(A)P(C)$$
$$= P(B \cap C) = P(B)P(C) = \tfrac{1}{2} \cdot \tfrac{1}{2} = \tfrac{1}{4}$$

since each of the events $A \cap B$, $A \cap C$, and $B \cap C$ contains only one out of the four points of S. Therefore the events A, B, and C are pairwise independent, and Equation (4.14) of Theorem 4.6 is satisfied. However,

$$A \cap B \cap C = \emptyset$$

and

$$P(A \cap B \cap C) = P(\emptyset) = 0 \neq P(A)P(B)P(C) = \tfrac{1}{8}$$

that is, Equation (4.15) of Theorem 4.6 does not follow from Equation (4.14).

Exercises

1. If A and B are independent events, show that A' and B' are also independent.

2. A student taking the standardized examinations in mathematics, writing, reading, and english expression estimates that the probability of receiving a passing score X is $1/10$, $3/10$, $5/10$, and $7/10$, respectively. If gradings of these examinations can be regarded as independent events, find the probability that he receives

   (a) No passing score X. (b) Exactly one passing score X.

   Ans. (a) .0945; (b) .8055

3. The decision process of accepting or rejecting a prospective employee is based on his scores on three tests. If S_1, S_2, and S_3 are passing scores with probabilities .6, .3, and .1, respectively, and if S_1, S_2, S_3 are independent events, find the probability that a randomly selected applicant will be rejected.

   Ans. .252

4. An insurance company has five salesmen assigned to separate territories. If the probability that a salesman will make a sale on any given day is .2, what is the probability that at least one of the five salesmen will make a sale on a given day?

   Ans. .673

5. A test has six multiple-choice items each with four choices, only one of which is correct. If a student guesses the answers to each question, and if his guesses are independent, define an appropriate sample space and assign probabilities to each simple event. Furthermore, find the probability that the student guesses correctly.

6. An experimenter needs at least one subject in order to conduct an experiment. If he has a list of volunteers and if each person contacted has probability $\frac{1}{3}$ of showing up, how many persons should he contact to insure that the probability of at least one showing up is at least .99? (Assume that persons contacted form independent events.) What is the probability that the fourth subject contacted is the first to show up for the experiment?

   Ans. $1 - (2/3)^n$, $n = 12$, $8/81$

7. If the probabilities of A and B hitting a target are $\frac{1}{4}$ and $\frac{2}{5}$, respectively, what is the probability that either A or B hits the target whenever both shoot at it? What is the probability that both will hit the target?

   Ans. $11/20$, $1/10$

8. If A is the event that a family has children of both sexes and if B is the event that a family has at most one boy, then show the following:

   a. A and B are independent events if the family has three children.
   b. A and B are dependent if the family has two children.

9. It has been estimated that $\frac{1}{4}$ is the probability that a man will live 10 years beyond retirement age and $\frac{1}{3}$ is the probability that his wife will also live 10 years after his retirement. Find the probability that 10 years after his retirement

   a. Both will be alive.
   b. At least one will be alive.
   c. Neither will be alive.
   d. Only the wife will be alive.

   Ans. (a) $1/12$; (b) $1/2$, (c) $1/2$; (d) $1/4$

10. Box A contains seven items, two of which are defective, and box B contains eight items, three of which are defective. An item is randomly drawn from each box.

    a. What is the probability that both items are nondefective?
    b. What is the probability that one item is defective and one nondefective?
    c. If one item is defective and one is not, what is the probability that the defective item came from box A?

4.8 Bernoulli Trials

We now come to an investigation of probability spaces associated with an experiment repeated finitely many times. For example, whenever three cars, x_1, x_2, and x_3, are in a race, their respective probabilities of winning might be $\frac{1}{2}$, $\frac{1}{3}$, and $\frac{1}{6}$, respectively; that is, the probability space

$$S_1 = \{X_1, X_2, X_3\}$$

where X_i is the event that car x_i wins has probabilities

$$P(X_1) = \tfrac{1}{2}, \qquad P(X_2) = \tfrac{1}{3}, \qquad \text{and} \qquad P(X_3) = \tfrac{1}{6}$$

However, if the cars race twice, then the sample space of the two repeated trials is

$$S = \{(X_1, X_1), (X_1, X_2), (X_1, X_3), (X_2X_1), (X_2, X_2), (X_1X_2), (X_3, X_1), (X_3, X_2), (X_3, X_3)\}$$

where (X_i, X_j) represents the independent events that cars x_i and x_j won the first and second races, respectively, for $i = 1, 2, 3$. Furthermore, the probabilities of the sample points of S are the following:

$$P(X_1, X_1) = P(X_1)P(X_1) = \frac{1}{2}\cdot\frac{1}{2} = \frac{1}{4};$$

$$P(X_1, X_2) = P(X_1)P(X_2) = \frac{1}{2}\cdot\frac{1}{3} = \frac{1}{6}$$

$$P(X_1, X_3) = P(X_1)P(X_3) = \frac{1}{2}\cdot\frac{1}{6} = \frac{1}{12};$$

$$P(X_2, X_1) = \frac{1}{3}\cdot\frac{1}{2} = \frac{1}{6}$$

$$P(X_2, X_2) = \frac{1}{3}\cdot\frac{1}{3} = \frac{1}{9};$$

$$P(X_2, X_3) = \frac{1}{3}\cdot\frac{1}{6} = \frac{1}{18}$$

$$P(X_3, X_1) = \frac{1}{6}\cdot\frac{1}{2} = \frac{1}{12};$$

$$P(X_3, X_2) = \frac{1}{6}\cdot\frac{1}{3} = \frac{1}{18}$$

$$P(X_3, X_3) = \frac{1}{6}\cdot\frac{1}{6} = \frac{1}{36}$$

Therefore the probability of car x_2 winning both the first and second race is

$$P(X_2, X_2) = \tfrac{1}{9}$$

and the probability of x_3 winning the first race and x_1 the second is

$$P(X_3, X_1) = \frac{1}{12}$$

and so on.

Probability spaces associated with a finitely many times repeated experiment are defined in Definition 4.8.

DEFINITION 4.8 By n independent or repeated trials we shall mean the probability space S consisting of ordered n-tuples of elements of some finite probability space S_0 with the probability of an n-tuple defined to be the product of the probabilities of its components.

As an illustration of this definition let

$$S = \{S_1, S_2, \ldots, S_n\}$$

be a finite probability space; let

$$S = \{(S_1, S_2, \ldots, S_n) \mid S_i \in S_0\}$$

and let

$$p(S_1, S_2, \ldots, S_n) = p(S_1)p(S_2) \ldots p(S_n)$$

Then the probability space S constitutes n independent or repeated trials. The previous example of the three racing cars provides a good illustration of independent trials. Yet another way to view a repeated trial is to think of it as a stochastic process whose tree diagram possesses the properties:

i. Every branch point has the same outcomes.
ii. The probability is the same for each branch leading to the same outcome.

Figure 4.3 shows the tree diagram for the three racing cars just given.

DEFINITION 4.9 A sequence of independent, repeated trials is said to be a sequence of Bernoulli trials if and only if the experiment has only two outcomes, *successes* and *failures*, with probability p for a success and probability $q = 1 - p$ for failure.

Repeated trials are certainly common in gambling games such as rolling dice, dealing cards, or tossing coins. However, there are other applications of fundamental interest. For instance, the manufacture of any item may be thought of as repeated trials in which the

Figure 4-3

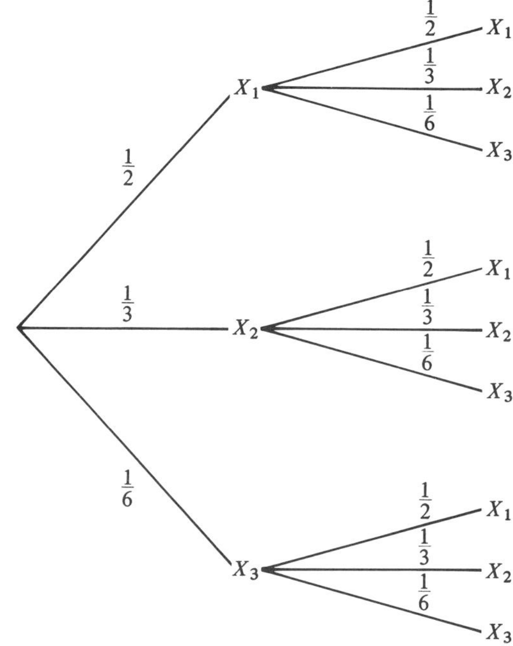

produced item is most of the time nondefective and only occasionally defective. Such applications of probability theory to the quality control of production are both important and extensive. Utilization of Bernoulli trials requires the evaluation of the probability of a certain number of successes specified by the following theorem.[3]

Theorem 4.8 The probability of exactly k successes in a sequence of n Bernoulli trials with probability p is denoted by $B(k, n, p)$, and is given by

$$B(k, n, p) = \binom{n}{k} p^k q^{n-k}$$

Example 4.19 If a fair coin is tossed five times and if heads is called a success, then determine the following probabilities:

a. The probability of exactly two heads occurring.
b. The probability of obtaining at least three heads.
c. The probability of getting no heads.

Solution Clearly, we have a sequence of repeated trials resulting in only two outcomes, heads or tails, that is, successes or failures. Thus the sequence is a sequence of Bernoulli trials with $n = 5$ and $p = q = \frac{1}{2}$. Hence application of Theorem 4.8 yields the following;

a. The probability that exactly two heads $(k = 2)$ occur is given by

$$B\left(2, 5, \frac{1}{2}\right) = \binom{5}{2}\left(\frac{1}{2}\right)^2\left(\frac{1}{2}\right)^3 = 10\left(\frac{1}{2}\right)^5 = \frac{5}{16}$$

b. The probability of at least three heads $(k = 3, 4, 5)$ appearing is obtained by

$$P(k \geq 3) = B(3, 5, \tfrac{1}{2}) + B(4, 5, \tfrac{1}{2}) + B(5, 5, \tfrac{1}{2})$$

or

$$P(k \geq 3) = \binom{5}{3}\left(\frac{1}{2}\right)^3\left(\frac{1}{2}\right)^2 + \binom{5}{4}\left(\frac{1}{2}\right)^4\left(\frac{1}{2}\right) + \binom{5}{4}\left(\frac{1}{2}\right)^5\left(\frac{1}{2}\right)^0$$

or

$$P(k \geq 3) = \frac{10}{32} + \frac{5}{32} + \frac{1}{32} = \frac{1}{2}$$

c. The probability of no heads $(k = 0)$ is

$$B\left(0, 5, \frac{1}{2}\right) = \binom{5}{0}\left(\frac{1}{2}\right)^0\left(\frac{1}{2}\right)^5 \equiv \left(\frac{1}{2}\right)^5 = \frac{1}{32}$$

Consequently, the probability of at least one head is

$$1 - \frac{1}{32} = \frac{31}{32}$$

(Why?)

Example 4.20 A student is taking a true-false examination consisting of 20 questions. If his probability of guessing correctly is $\frac{3}{4}$, and if he is indeed guessing, find the probability of his guessing correctly (a) exactly 10, (b) exactly 20, and (c) at least 15 answers.

Solution Again, this problem constitutes a sequence of Bernoulli trials with the two outcomes "correct" and "incorrect," $n = 20$, $p = \frac{3}{4}$, and $q = \frac{1}{4}$. Therefore, the required probabilities are

$$B\left(10, 20, \frac{3}{4}\right) = \binom{20}{10}\left(\frac{3}{4}\right)^{10}\left(\frac{1}{4}\right)^{10} = \binom{20}{10}\left(\frac{3^{10}}{4^{20}}\right) \tag{4.16}$$

$$B\left(20, 20, \frac{3}{4}\right) = \binom{20}{20}\left(\frac{3}{4}\right)^{20}\left(\frac{1}{4}\right)^{0} = \left(\frac{3}{4}\right)^{20} \tag{4.17}$$

and since "at least 15" implies the number k of successes is greater or equal to 15, the required probability becomes

$$P(k \geq 15) = \sum_{k=15}^{20} \binom{20}{k} \left(\frac{3}{4}\right)^k \left(\frac{1}{4}\right)^{20-k}$$

or

$$P(k \geq 15) = \sum_{k=15}^{20} \binom{20}{k} \frac{3^k}{4^{20}} \tag{4.18}$$

Exercises

1. In a sequence of n Bernoulli trials, prove that the probability of at least one success is $1 - q^n$.

2. If a fair coin is tossed six times and if "heads" is called a success, then determine the following probabilities:

   a. The probability that exactly two heads occur.
   b. The probability that at least four heads come up.
   c. The probability that no heads appear.
   d. The probability that at least one head occurs.

   Ans. (a) 15/64; (b) 11/32; (c) 1/64; (d) 63/64

3. A student is taking a true-false examination consisting of 10 questions. If his probability of guessing correctly is $\frac{2}{3}$ and if he is indeed guessing, determine his probability of guessing correctly

   a. Exactly 8.
   b. Exactly 10.
   c. At least seven answers.

   Ans. (a) .13; (b) .012; (c) .373

4. If 20% of the items produced by a manufacturer are defective and if four items are randomly picked from one of his shipments, determine the following:

   a. The probability that two of the four items are defective.
   b. The probability that three of the four items are defective.
   c. The probability that at least one item is defective.

   Ans. (a) .154; (b) .026; (c) .59

5. If 75% of the population in the country are opposed to a certain war action and 25% are for it, and if only five people are asked

for their opinion, what is the probability that this small poll will show the majority favors the action? What does this problem suggest about opinion polls?

Ans. .106

6. The Kansas City Chiefs football team has a $\frac{2}{3}$ probability of winning whenever it plays. If the team plays four games, find the probability that it wins

   a. Exactly two games.
   b. At least one game.
   c. More than one half of the games.

Ans. (a) 8/27; (b) 80/81; (c) 16/27

7. A company leasing small aircraft has 50 such planes at its disposal. If the company selects its customers so that the probability of a serious accident with any one plane for a five-year period is .001, find the probability that during the next five years at least one plane will be in a serious accident.

8. In a legislative process bills are sent either to a specific committee or to a general committee. Then they are called for action by randomly choosing one of the two committees and asking it to present one bill. If the choosing of the general committee is called a success, find the probability that at least 7 bills remain in the specific committee whenever 10 bills are assigned to each of these two committees.

9. If the probability of a college graduate obtaining a suitable job is $\frac{1}{2}$ on each application, what is the probability that he (she) will receive more than five offers on

   (a) 10 applications?
   (b) 8 applications?
   (c) 4 applications?

Ans. (a) .246; (b) .219; (c) 0

10. If the probability that a college student does not graduate is .3, find the probability that of five randomly selected college students

   (a) One will not graduate.
   (b) Three will not graduate.
   (c) At least one will not graduate.

Ans. (a) .36; (b) .132; (c) .832

4.9 Expected Value, Variance, Deviation

In competitive situations such as gambling (or even marketing a product) the outcomes are usually finitely many mutually exclusive events E_i, $i = 1, 2, 3, \ldots, n$. For each outcome E_i there is an exchange of $|V_i|$ number of dollars that either the player wins or loses. Furthermore, each event E_i occurs with a certain probability P_i. For example, if in the game of rolling a fair die the player wins when a prime number comes up and loses otherwise and if the amount won or lost each time is determined by this number, then the outcomes and the respective probabilities are shown in the following tables.

E_i	2	3	5	1	4	6
P_i	$\frac{1}{6}$	$\frac{1}{6}$	$\frac{1}{6}$	$\frac{1}{6}$	$\frac{1}{6}$	$\frac{1}{6}$

or

V_i	\$2	\$3	\$5	−\$1	−\$4	−\$6
P_i	$\frac{1}{6}$	$\frac{1}{6}$	$\frac{1}{6}$	$\frac{1}{6}$	$\frac{1}{6}$	$\frac{1}{6}$

where the negative numbers indicate the amounts lost due to rolling the nonprime numbers 1, 4, and 6. Similarly, if in the game of tossing a fair coin the player wins \$6 or loses \$2 whenever a tail or head occurs, respectively, then the outcomes and the corresponding probabilities are given by

E_i	Heads	Tails
P_i	$\frac{1}{2}$	$\frac{1}{2}$

or

V_i	−\$2	\$6
P_i	$\frac{1}{2}$	$\frac{1}{2}$

The numbers

$$E_d = 2\cdot\tfrac{1}{6} + 3\cdot\tfrac{1}{6} + 5\cdot\tfrac{1}{6} - 1\cdot\tfrac{1}{6} - 4\cdot\tfrac{1}{6} - 6\cdot\tfrac{1}{6} = -\tfrac{1}{6}$$

and

$$E_c = -2\cdot\tfrac{1}{2} + 6\cdot\tfrac{1}{2} = 2$$

may be interpreted as the average expected gain (loss) of the games "rolling a fair die" and "tossing a fair coin," respectively, in a large number of plays. Such numbers are always associated with chance experiments (games), and in Definition 4.10 we formally define them.

DEFINITION 4.10 If the outcomes of an experiment (or game) are assigned the values $V_1, V_2, V_3, \ldots, V_n$ which occur with probabilities $P_1, P_2, P_3, \ldots, P_n$, respectively, then the number

$$E = P_1V_1 + P_2V_2 + \cdots + P_nV_n$$

is called the expected value[4] of the experiment (or game).

Clearly, if a constant k is added to each outcome V_i, $i = 1, 2, \ldots, n$ of an experiment, then the expected value E becomes $E + k$. (Why?)

The term "expected value" is not to be interpreted as the value that will necessarily occur on a single performance of an experiment. It is, rather, the average expected gain (or loss) per performance of an experiment repeated a large number of times.

DEFINITION 4.11 If E is the expected value of a game, then the game is said to be

a. Fair if and only if $E = 0$.
b. Favorable or unfavorable to the player if and only if E is positive ($E > 0$) or E is negative ($E < 0$).

The terms "favorable" and "unfavorable" when applied to games do not necessarily dictate the decision to play or not to play a given game. In fact, people are willing at times to play unfavorable games. For instance, most people choose to play the game of buying life insurance although such a game may be considered unfavorable.

DEFINITION 4.12 By the variance and standard deviation of an experiment we shall mean the numbers

$$V = (V_1 - E)^2p_1 + (V_2 - E)^2p_2 + \cdots + (V_n - E)^2p_n$$

and

$$D = \sqrt{V},$$

respectively, where E is the expected value, $V_1, V_2, \ldots, V_n$ are the outcomes, and $p_1, p_2, \ldots, p_n$ are the respective probabilities of the experiment.

Example 4.21 In a lottery of 10 prizes 2000 tickets at \$1.50 each were sold. If there is one prize of \$1000, five prizes of \$200 each, and four prizes of \$100 each, find the expected winnings of one ticket, the variance, and standard deviation of the game.

Solution Since the probabilities of winning \$1000, \$200, or \$100 are 1/2000, 5/2000, and 4/2000, respectively, the expected winnings of one ticket are

$$E = \$1000 \frac{1}{2000} + \$200 \frac{5}{2000} + \$100 \frac{4}{2000}$$

or

$$E = \$1.20$$

However, since each ticket costs \$1.50, the expected value of the game is actually \$1.20 − \$1.50 = −\$0.30. Thus the game is unfavorable to the player. Now, using the formulas for variance V and standard deviation D, we find

$$V = (1000 - 0.30)^2 \frac{1}{2000} + (200 - 0.30)^2 \frac{5}{2000} + (100 - 0.30)^2 \frac{4}{2000}$$

or

$$V = 619.28 \qquad \text{and} \qquad D = \sqrt{V} = \sqrt{619.28} = 24.89$$

Example 4.22 If four fair coins are tossed and if N denotes the number of heads occurring in each toss, find the outcomes (values of N) and the expected value of this experiment.

Solution Since four coins are tossed and since N denotes the occurring heads in each toss, we have $N = 0, 1, 2, 3$, or 4 for the outcomes of the experiment. Moreover, since the experiment constitutes a sequence of Bernoulli trials with $N = 4$ and $p = \frac{1}{2}$, the respective probabilities are

$$B\left(0, 4, \frac{1}{2}\right) = \binom{4}{0}\left(\frac{1}{2}\right)^0\left(\frac{1}{2}\right)^{4-0} = \frac{1}{16}$$

$$B\left(1, 4, \frac{1}{2}\right) = \binom{4}{1}\left(\frac{1}{2}\right)^1\left(\frac{1}{2}\right)^{4-1} = \frac{4}{16}$$

$$B\left(2, 4, \frac{1}{2}\right) = \binom{4}{2}\left(\frac{1}{2}\right)^2\left(\frac{1}{2}\right)^{4-2} = \frac{6}{16}$$

$$B\left(3, 4, \frac{1}{2}\right) = \binom{4}{3}\left(\frac{1}{2}\right)^3\left(\frac{1}{2}\right)^{4-3} = \frac{4}{16}$$

and

$$B\left(4, 4, \frac{1}{2}\right) \quad \binom{4}{4}\left(\frac{1}{2}\right)^4\left(\frac{1}{2}\right)^{4-4} = \frac{1}{16}$$

Thus the expected value is given by

$$E = 0\cdot\frac{1}{16} + 1\cdot\frac{4}{16} + 2\cdot\frac{6}{16} + 3\cdot\frac{4}{16} + 4\cdot\frac{1}{16} \qquad \text{or} \qquad E = 2$$

Exercises

1. If E and V are the expected value and the variance of an experiment, then prove the following:

   a. If a constant k is added to each of the outcomes of an experiment, the expected value becomes $E + k$ and the variance remains the same.
   b. If each outcome of an experiment is multiplied by k, then the expected value and the variance become kE and k^2V, respectively.

2. A lottery with 500 tickets gives one prize of \$100, three prizes of \$50 each, and five prizes of \$25 each. Find

   a. The expected winnings of one ticket.
   b. The expected value of the game if the cost of a ticket is \$1.

   Ans. (a) \$0.75; (b) −\$0.25

3. A shipment contains 100 items of which 5 are defective. If a sample of 80 items is randomly selected from the shipment, find the expected number of defective items in this sample.

4. If sex distribution is equally probable, determine the expected number of girls in a family with four children, and the probability that the expected number of girls will occur.

   Ans. 2, $\frac{3}{8}$

5. If in the game of tossing two fair coins a player wins \$6 when two heads occur, \$4 when one head occurs, and \$2 when no head occurs, find

   a. The player's expected winnings.
   b. The amount he must pay to play a fair game.

   Ans. (a) \$4; (b) \$4

6. If in the game of tossing two fair coins a player wins \$5 when two heads occur and \$4 when one head occurs, how much should he lose when no head occurs so that the game is fair?

   Ans. \$13

7. If a shipment contains eight toy guns of which three are defective and if a gun is randomly selected from the shipment and is tested until a nondefective one is chosen, find the expected number of guns to be tested.

   Ans. 81/56

8. A box contains 10 items of which 3 are defective. If 3 items are selected from the box, find the expected number of defective items drawn.

9. If the probability of basketball team A winning any game is $\frac{1}{2}$ and if in a championship series between A and some other team B the first team to win two games in a row or a total of three games wins the series, find the expected number of games in the series.

   Ans. 23/8

10. If on the draw of a card, the dealer will pay \$10 for a two, \$5 for a three or four, and \$3 for a five or six, find the expected value of this game.

    Ans. \$2

References and Supplementary Readings

1. H. Cramer, *The Elements of Probability Theory* (New York: Wiley, 1955).
2. W. Feller, *An Introduction to Probability Theory and Its Applications* (New York: Wiley, 1950).
3. E. Parzen, *Modern Probability Theory and Its Applications* (New York: Wiley, 1960).
4. J. W. Bishir, and D. W. Drews, *Mathematics in the Behavioral and Social Sciences*, Chapter 20 (New York: Harcourt Brace Jovanovich, 1970).

CHAPTER 5

Introduction to Matrices

5.1 Introduction

The subsequent Chapters 6, 7, and 8 will demonstrate that in building mathematical models in the areas of transportation problems, game theory, linear programming, testing theory, and the theory of probability, or in solving systems of linear equations it becomes necessary to work with square or rectangular arrays of numbers commonly known as "matrices." The algebra of matrices was originated by the English mathematician Arthur Cayley (1821–1895) in his work *A Memoir on the Theory of Matrices* published in 1858. However, the term "matrix" was first employed in 1850 by the British mathematician James Sylvester (1814–1897), a friend and collaborator of Cayley in both mathematics and the legal profession.

The following situations demonstrate some typical yet very simple applications of matrices.

a. The game of matching pennies consists of two players R and C who toss a penny independently of each other and, in turn, compare the outcomes. If the outcomes match—that is, if they are both heads or both tails—then player R wins \$5 from player C. If the outcomes do not match—that is, if they are heads and tails or tails and heads—then player C wins \$5 from player R (or R loses \$5). This game can neatly be represented by the square array (or matrix)

$$\begin{array}{c} \\ R_1 \\ R_2 \end{array} \begin{array}{c} \begin{array}{cc} C_1 & C_2 \end{array} \\ \begin{pmatrix} 5 & -5 \\ -5 & 5 \end{pmatrix} \end{array}$$

where player R plays the rows R_1 and R_2 as heads and tails, respectively, while player C plays the columns C_1 and C_2 in a similar fashion.

Another example of a two-person game is provided by two players R and C each holding a 4, a 5, a 6, and a 7 from a deck of cards. In this game each player selects one card and simultaneously the players place the cards face up on a table. If the cards match, then the payoff to each other is zero. If the cards do not match, then the player with the smaller face-value card wins in dollars the sum of the face values of the two cards. Now this game can again be nicely represented by the square array (or matrix)

$$P = \begin{bmatrix} 0 & 9 & 10 & 11 \\ -9 & 0 & 11 & 12 \\ -10 & -11 & 0 & 13 \\ -11 & -12 & -13 & 0 \end{bmatrix}$$

In this game, R plays the rows R_i, $i = 1, 2, 3, 4$ of P, and C plays the columns C_i, $i = 1, 2, 3, 4$ of P as described below.

The selection of the R_i row by R implies the playing of the card $i + 3$, $i = 1, 2, 3, 4$. Similarly, the selection of the C_j column by C implies the playing of the card $j + 3$, $j = 1, 2, 3, 4$. The payoff $(i + 3) + (j + 3)$ is zero if $i = j$, or it is positive for $i < j$ and R wins $(i + 3) + (j + 3)$, or it is negative for $i > j$ in which case R loses to C the amount $(i + 3) + (j + 3)$.

As one might expect, the question raised in this type of game is what row should R select or what kind of move should he make (if there is any) to maximize his winnings no matter what C does? Likewise, what should C do to minimize his losses? These and other more general questions are considered in Chapter 8 on game theory.

b. If a manufacturer produces three products A, B, and C which he sells in two markets, then annual sales volumes are usually indicated by a rectangular array (or matrix) such as

$$Q = \begin{array}{l} \\ \text{Market 1:} \\ \text{Market 2:} \end{array} \begin{array}{c} \begin{array}{ccc} A & B & C \end{array} \\ \begin{pmatrix} 15{,}000 & 5{,}000 & 16{,}000 \\ 6{,}000 & 8{,}000 & 12{,}000 \end{pmatrix} \end{array}$$

where each number in Q denotes the units of a product sold to the corresponding market. Furthermore, if

$$P = \overset{\begin{array}{ccc} A & B & C \end{array}}{(\$2 \quad \$3 \quad \$5)} \qquad \text{and} \qquad C = \overset{\begin{array}{ccc} A & B & C \end{array}}{(\$1.50 \quad \$2.35 \quad \$4.20)}$$

are rectangular arrays representing the unit sales prices and the unit costs, respectively, then as we demonstrate subsequently, appropriate algebraic manipulations of Q, P, and C yield total revenue, total cost, and total profit.

c. In testing theory and factor analysis if each of m students were given a battery of n different tests, the resulting scores could very conveniently be displayed in a rectangular array of m rows (one for each student) and n columns (one for each test). The resulting array or matrix S would, of course, be an $m \times n$, and its general "element" s_{ij} would be the score of the ith student on the jth test.

$$\begin{array}{c} \rightarrow \text{Tests} \leftarrow \\ S = \begin{pmatrix} s_{11} & s_{12} \cdots s_{1n} \\ s_{21} & s_{22} \cdots s_{2n} \\ \cdots & \cdots \\ s_{m1} & s_{m2} \cdots s_{mn} \end{pmatrix}_{m \times n} \end{array} \quad \begin{array}{c} \downarrow \\ \text{Students} \\ \uparrow \end{array}$$

5.2 Properties of Matrices

By an $m \times n$ matrix we shall mean a rectangular array of quantities arranged in m rows and n columns. The numbers m and n are called the "dimensions" of the matrix. Matrices will often be represented by single capital letters. However, in most instances they will be represented by displaying some or all of the constituent quantities in large parentheses or brackets. For example,

$$A = (a_{ij})_{m \times n} = \begin{pmatrix} a_{11} & a_{12} \cdots a_{1n} \\ a_{21} & a_{22} \cdots a_{2n} \\ \cdots & \cdots \\ a_{m1} & a_{m2} \cdots a_{mn} \end{pmatrix} \tag{5.1}$$

is an $m \times n$ matrix; that is, a matrix with m rows—the horizontal lines of elements—and n columns—the vertical lines of elements—of quantities a_{ij} called "elements" of the matrix. In the double subscript notation used in (5.1), the first subscript i, associated with an element a_{ij}, identifies the row, and the second subscript j identifies the column in which the element lies. For example, the element a_{34} lies in the third row and fourth column of the matrix. The line of elements a_{ij} for which $i = j$ (or a_{ii}) constitutes the *principal* or *main diagonal* of an $n \times n$ matrix $A = (a_{ij})$.

The rectangular array

$$A = \begin{pmatrix} 0 & 1 & 3 & 0 \\ 3 & \textcircled{4} & 12 & 6 \\ 7 & 8 & 9 & \textcircled{5} \end{pmatrix}$$

is a 3×4 matrix; that is, A is a matrix with three rows and four columns. The number 4 is the element of A lying in its second row and second column, whereas the element 5 of A lies in the third row and fourth column of the matrix A.

Two matrices $A = (a_{ij})$ and $B = (b_{ij})$ are said to be *equal* if and only if they contain the same number of rows, the same number of columns, and $a_{ij} = b_{ij}$ for all values of i and j. For instance, the matrices

$$A = \begin{pmatrix} 1 & 0 & 2 \\ 1 & 1 & 1 \end{pmatrix} \quad \text{and} \quad B = \begin{pmatrix} 1 & 0 & 2 \\ 1 & 1 & 1 \end{pmatrix}$$

are equal, and

$$A = \begin{pmatrix} x & y \\ z & t \end{pmatrix}$$

would be equal to

$$B = \begin{pmatrix} a & b \\ c & d \end{pmatrix}$$

if and only if

$$x = a, \qquad y = b, \qquad z = c, \qquad \text{and} \qquad t = d$$

A matrix consisting of only one row (or column) is called a "row" (or column) matrix. Both row and column matrices are often called *vectors*. By the *transpose* of an $m \times n$ matrix $A = (a_{ij})$ we shall mean the $n \times m$ matrix obtained from A by interchanging its rows and columns, and we shall denote this matrix by the symbol A^t.

An $m \times n$ matrix A is said to be a *square matrix* if and only if $m = n$, that is, if and only if it has the same number of rows and columns. Thus the matrices

$$P = (\$2 \quad \$3 \quad \$5) \qquad \text{and} \qquad C = (\$1.50 \quad \$2.35 \quad \$4.20)$$

representing the unit sales prices and the unit costs in the previous example, respectively, are, in fact, vectors. The transpose of the square matrix

$$P = \begin{bmatrix} 0 & 9 & 10 & 11 \\ -5 & 0 & 11 & 12 \\ -10 & -11 & 0 & 13 \\ -11 & -12 & -13 & 0 \end{bmatrix}$$

which describes the payoffs in one of the two-person games is

$$P^t = \begin{bmatrix} 0 & -5 & -10 & -11 \\ 9 & 0 & -11 & -12 \\ 10 & 11 & 0 & -13 \\ 11 & 12 & 13 & 0 \end{bmatrix}$$

A square matrix in which every element below the main diagonal is zero is called *upper triangular* and one in which every element above the main diagonal is zero is called *lower triangular.* A square matrix, however, in which each element not on the main diagonal is zero is called a *diagonal matrix,* and is often denoted by

$$\begin{pmatrix} a_{11} & 0 & \cdots 0 \\ 0 & a_{22} & \cdots 0 \\ \cdots & \cdots & \cdots \\ 0 & 0 & \cdots a_{nn} \end{pmatrix}$$

The *identity* matrix I is a diagonal matrix whose elements in the main diagonal are all ones. For example,

$$I = \begin{pmatrix} 1 & 0 & 0 \\ 0 & 1 & 0 \\ 0 & 0 & 1 \end{pmatrix}$$

is a 3×3 identity matrix. The zero matrix 0 is any matrix whose elements are all equal to zero. That is

$$0 = \begin{pmatrix} 0 & 0 & 0 \\ 0 & 0 & 0 \\ 0 & 0 & 0 \end{pmatrix}$$

is a 3×3 zero matrix. Clearly, there exist infinitely many identity (or unit) or zero matrices. A square matrix $A = (a_{ij})$ such that $a_{ij} = a_{ji}$ is said to be symmetric. Thus A is symmetric if and only if $A = A^t$.

To clarify further the definition of certain special matrices, we consider the following examples.

a. The square matrices

$$A = \begin{pmatrix} 2 & 1 & 3 \\ 0 & 1 & 4 \\ 0 & 0 & 6 \end{pmatrix} \quad \text{and} \quad B = \begin{pmatrix} 1 & 0 & 0 \\ 2 & 0 & 0 \\ 3 & 6 & 1 \end{pmatrix}$$

are clearly upper and lower triangular, respectively.

b. Since in the square matrix

$$D = \begin{pmatrix} 1 & 0 & 0 \\ 0 & 3 & 0 \\ 0 & 0 & 4 \end{pmatrix}$$

every element off the diagonal is zero, the matrix D is a diagonal matrix.

c. If

$$A = \begin{pmatrix} 1 & 4 & -3 \\ 4 & 2 & 1 \\ -3 & 1 & 3 \end{pmatrix}$$

then clearly $A = A^t$ or A is a symmetric matrix.

Now if we consider the set M of all matrices, together with the identity and the zero matrices, and if we appropriately define an *addition* and *multiplication* on the elements of M, an algebra of matrices can be developed.[1] This algebra, as we shall see, is quite different from the ordinary algebra of numbers. To motivate matrix addition first and then matrix multiplication, let us consider the following examples.

a. A manufacturing firm produces three different products A, B, and C which are sold to two markets M_1 and M_2. If the annual sales volumes for the years 1971 and 1972 are indicated by the matrices

$$V_1 = \begin{array}{c} \\ M_1: \\ M_2: \end{array} \begin{array}{c} \begin{array}{ccc} A & B & C \end{array} \\ \begin{pmatrix} 5{,}000 & 8{,}000 & 10{,}000 \\ 12{,}000 & 6{,}000 & 15{,}000 \end{pmatrix} \end{array}$$

and

$$V_2 = \begin{array}{c} M_1: \\ M_2: \end{array} \begin{pmatrix} 8{,}000 & 9{,}000 & 12{,}000 \\ 14{,}000 & 7{,}000 & 16{,}000 \end{pmatrix}$$

respectively, then at the end of the two years the total sales volumes will be given by the sum of V_1 and V_2. Clearly, this sum must be the matrix

$$V = V_1 + V_2 = \begin{array}{c} \\ M_1: \\ M_2: \end{array} \begin{array}{c} \begin{array}{ccc} A & B & C \end{array} \\ \begin{pmatrix} 5{,}000 + 8{,}000 & 8{,}000 + 9{,}000 & 10{,}000 + 12{,}000 \\ 12{,}000 + 14{,}000 & 6{,}000 + 7{,}000 & 15{,}000 + 16{,}000 \end{pmatrix} \end{array}$$

or

$$V = \begin{pmatrix} 13{,}000 & 17{,}000 & 22{,}000 \\ 26{,}000 & 13{,}000 & 31{,}000 \end{pmatrix}$$

On the other hand the increase of sales from 1971 to 1972 is the difference $V_2 - V_1$ or

$$V_2 - V_1 = \begin{pmatrix} 8{,}000 - 5{,}000 & 9{,}000 - 8{,}000 & 12{,}000 - 10{,}000 \\ 14{,}000 - 12{,}000 & 7{,}000 - 6{,}000 & 16{,}000 - 15{,}000 \end{pmatrix}$$

or

$$V_2 - V_1 = \begin{pmatrix} 3{,}000 & 1{,}000 & 2{,}000 \\ 2{,}000 & 1{,}000 & 1{,}000 \end{pmatrix}$$

b. If the unit sales prices of the same firm and for each year are given by the matrices

$$P_1 = (\$2.00 \quad \$3.00 \quad \$4.00)$$

and

$$P_2 = (\$1.50 \quad \$2.50 \quad \$5.00)$$

respectively, then the total revenue in each market for each year is clearly obtained from the matrices

$$R_1 = \begin{pmatrix} \overset{A}{5{,}000 \times 2} & \overset{B}{8{,}000 \times 3} & \overset{C}{10{,}000 \times 4} \\ 12{,}000 \times 2 & 6{,}000 \times 3 & 15{,}000 \times 4 \end{pmatrix}$$

$$= \begin{pmatrix} 74{,}000 \\ 102{,}000 \end{pmatrix} = V_1 P_1^t$$

and

$$R_2 = \begin{pmatrix} 8{,}000 \times 1.5 & 9{,}000 \times 2.5 & 12{,}000 \times 5 \\ 14{,}000 \times 1.5 & 7{,}000 \times 2.5 & 16{,}000 \times 5 \end{pmatrix}$$

$$= \begin{pmatrix} 94{,}000 \\ 118{,}500 \end{pmatrix} = V_2 P_2^t$$

that is, in 1971 the revenue was \$74,000 from market M_1 and \$102,000 from M_2. Similarly, there was \$94,500 and \$118,500 revenue from M_1 and M_2 in the year 1972.

Now, with these examples in mind, we shall adopt the following definitions of addition and multiplication of matrices.

SUM AND DIFFERENCE OF MATRICES

Let $A = (a_{ij})$ and $B = (b_{ij})$ be two matrices having the same dimensions m and n; then by the sum or difference $A \pm B$ of the matrices A and B we shall mean the $m \times n$ matrix $C = (a_{ij} \pm b_{ij})$ whose elements are the sums or differences of the respective elements of A and B. For example, if

$$A = \begin{pmatrix} 1 & 0 & 1 \\ 2 & 1 & 3 \end{pmatrix} \quad \text{and} \quad B = \begin{pmatrix} -1 & 2 & 1 \\ -5 & 1 & -6 \end{pmatrix}$$

then

$$A + B = \begin{pmatrix} 0 & 2 & 2 \\ -3 & 2 & -3 \end{pmatrix} \quad \text{and} \quad A - B = \begin{pmatrix} 2 & -2 & 0 \\ 7 & 0 & 9 \end{pmatrix}$$

Given any three matrices A, B, and C, together with a zero matrix 0, the following properties hold true for matrix addition.

i. $A + B = B + A$ commutative law.

ii. $A + (B + C) = (A + B) + C$ associative law.

iii. $A + 0 = 0 + A = A$ additive identity property.

iv. For each A there is $-A$ such that $A + (-A) = 0$ additive inverse property.

DEFINITION 5.1 $-A$ is called the additive inverse or the negative of A, and its elements are the negatives of the elements of A.

Example 5.1 Illustrate the commutative, associative, additive identity, and additive inverse properties using the following matrices:

$$A = \begin{pmatrix} 1 & 2 \\ 3 & 4 \end{pmatrix}, \quad B = \begin{pmatrix} -1 & 0 \\ 2 & -3 \end{pmatrix}, \quad \text{and} \quad C = \begin{pmatrix} 1 & 1 \\ 2 & 2 \end{pmatrix}$$

i. $$A + B = \begin{pmatrix} 0 & 2 \\ 5 & 1 \end{pmatrix}, \quad B + A = \begin{pmatrix} 0 & 2 \\ 5 & 1 \end{pmatrix}$$

or

$$A + B = B + A.$$

ii. $$A + (B + C) = \begin{pmatrix} 1 & 2 \\ 3 & 4 \end{pmatrix} + \begin{pmatrix} 0 & 1 \\ 4 & -1 \end{pmatrix} = \begin{pmatrix} 1 & 3 \\ 7 & 3 \end{pmatrix}$$

and

$$(A + B) + C = \begin{pmatrix} 0 & 2 \\ 5 & 1 \end{pmatrix} + \begin{pmatrix} 1 & 1 \\ 2 & 2 \end{pmatrix} = \begin{pmatrix} 1 & 3 \\ 7 & 3 \end{pmatrix}$$

Thus

$$A + (B + C) = (A + B) + C$$

iii. $$A + 0 = \begin{pmatrix} 1 & 2 \\ 3 & 4 \end{pmatrix} + \begin{pmatrix} 0 & 0 \\ 0 & 0 \end{pmatrix} = \begin{pmatrix} 1 & 2 \\ 3 & 4 \end{pmatrix}$$

or

$$A + 0 = A = 0 + A.$$

iv. If $A = \begin{pmatrix} 1 & 2 \\ 3 & 4 \end{pmatrix}$, then

$$-A = \begin{pmatrix} -1 & -2 \\ -3 & -4 \end{pmatrix} \qquad \text{and} \qquad A + (-A) = \begin{pmatrix} 0 & 0 \\ 0 & 0 \end{pmatrix} = 0$$

(matrix)

In the following, A^n shall mean $A \cdot A \cdot A \cdots A$ n times. Since, as we pointed out, both row and column matrices are called "vectors," we have, in fact, introduced the following definition.

DEFINITION 5.2 By an n-dimensional vector V we shall mean an ordered set of n quantities $x_1, x_2, \ldots, x_n$ which are called the "components" or "coordinates" of the vector. The vector V shall be denoted by $V = (x_1, x_2, \ldots, x_n)$ and referred to as an n-tuple.

DEFINITION 5.3 If

$$X = (x_1, x_2, \ldots, x_n) \qquad \text{and} \qquad Y = (y_1, y_2, \ldots, y_n)$$

are any two n-dimensional vectors, then the product

$$X \cdot Y = \sum_{i=1}^{n} x_i y_i$$

is called *scalar product, dot product,* or *inner product* of the vectors X and Y. The number

$$\sqrt{X \cdot X} = \sqrt{\sum_{i=1}^{n} x_i^2}$$

is called the *length* or *absolute value* of the vector X, and X is said to be a *unit vector* if and only if its length is equal to one.

DEFINITION 5.4 Two vectors X and Y are said to be parallel if and only if their components are proportional. On the other hand we say that they are orthogonal or perpendicular if and only if their inner product $X \cdot Y$ is equal to zero.

For instance, the vectors $A = (1, -1, 0)$ and $B = (0, 0, 2)$ are perpendicular (or orthogonal) since

$$A \cdot B = (1)(0) + (-1)(0) + (0)(2) = 0$$

and the vectors $A = (1, 1, 1)$ and $B = (2, 2, 2)$ are parallel since their components are proportional.

The product of a matrix A and a scalar (or number) is the matrix $kA = Ak$ whose elements are the elements of A each multiplied by k. That is, if

$$A = \begin{pmatrix} 1 & 2 \\ 3 & 4 \end{pmatrix}$$

then

$$5A = \begin{pmatrix} 5 & 10 \\ 15 & 20 \end{pmatrix}$$

Two matrices A and B are said to be *conformable* in the order AB or they can be multiplied as A times B if and only if the number of columns in A is equal to the number of rows in B. Thus if A is an $m \times n$ matrix and B is a $p \times q$ matrix, then A and B are *conformable* in the order AB if and only if $n = p$, and the product is an $m \times q$ matrix obtained as follows.

DEFINITION 5.5 If A and B are $p \times q$ and $q \times r$ matrices, respectively, then the product AB is the $p \times r$ matrix

$$C = (c_{ij})$$

where

$$c_{ij} = \sum_{k=1}^{q} a_{ik}b_{kj}, \qquad i = 1, 2, \ldots, p \qquad j = 1, 2, \ldots, r$$

that is, the element c_{ij} in the ith row and jth column of the product matrix C is the inner product of the ith row vector of A and the jth column vector of B.

Clearly, not any two matrices A and B can be multiplied, or if they can be multiplied in the order AB, it is not always true that they can be multiplied in the order BA. For example, the matrices

$$A = \begin{pmatrix} 2 & 3 \\ 1 & -1 \\ 0 & 4 \end{pmatrix} \quad \text{and} \quad B = \begin{pmatrix} 5 & -2 & 4 & 7 \\ -6 & 1 & -3 & 0 \end{pmatrix}$$

cannot be multiplied in the order BA (why?), whereas in the order

AB we obtain the matrix

$$AB = C = \begin{pmatrix} (2)\ (5) + (3)(-6) & (2)\ (-2) + (3)(1) & (2)\ (4) + (3)(-3) & (2)\ (7) + (3)(0) \\ (1)\ (5) + (-1)(-6) & (1)\ (-2) + (-1)(1) & (1)\ (4) + (-1)(-3) & (1)\ (7) + (-1)(0) \\ (0)\ (5) + (4)(-6) & (0)\ (-2) + (4)(1) & (0)\ (4) + (4)(-3) & (0)\ (7) + (4)(0) \end{pmatrix}$$

or

$$AB = C = \begin{pmatrix} -8 & -1 & -1 & 14 \\ 11 & -3 & 7 & 7 \\ -24 & 4 & -12 & 0 \end{pmatrix}$$

Thus matrix multiplication applied to example (b) of our introduction yields the following:

$$PQ^t = (2, 3, 5) \begin{pmatrix} 15{,}000 & 6{,}000 \\ 5{,}000 & 8{,}000 \\ 16{,}000 & 12{,}000 \end{pmatrix} = (\$125{,}000 \quad \$96{,}000) \tag{5.2}$$

for total revenue in each market:

$$CQ^t = (1.50 \quad 2.35 \quad 4.20)\begin{pmatrix} 15{,}000 & 6{,}000 \\ 5{,}000 & 8{,}000 \\ 16{,}000 & 12{,}000 \end{pmatrix}$$

$$= (\$101{,}450 \quad \$78{,}200) \qquad \textbf{(5.3)}$$

for total costs;

$$PQ^t - CQ^t = (125{,}000 \quad 96{,}000) \quad - (101{,}450 \quad 78{,}200)$$

$$= (\$23{,}550 \quad \$17{,}800) \qquad \textbf{(5.4)}$$

for gross profits

As we pointed out, the set of matrices under addition and multiplication constitutes a new mathematical system. In this system Theorem 5.1 can be proved.[2]

Theorem 5.1 For any suitable conformable matrices the associative law of multiplication holds true. That is,

$$A(BC) = (AB)C$$

For example, if

$$A = \begin{pmatrix} 1 & 2 \\ -1 & 3 \end{pmatrix}, \qquad B = \begin{pmatrix} 1 & 0 & -1 \\ 2 & 1 & 0 \end{pmatrix},$$

and

$$C = \begin{pmatrix} 1 & -1 \\ 3 & 2 \\ 2 & 1 \end{pmatrix}$$

then

$$A(BC) = \begin{pmatrix} 1 & 2 \\ -1 & 3 \end{pmatrix}\left[\begin{pmatrix} 1 & 0 & -1 \\ 2 & 1 & 0 \end{pmatrix}\begin{pmatrix} 1 & -1 \\ 3 & 2 \\ 2 & 1 \end{pmatrix}\right]$$

$$= \begin{pmatrix} 1 & 2 \\ -1 & 3 \end{pmatrix}\begin{pmatrix} -1 & -2 \\ 5 & 0 \end{pmatrix}$$

or

$$A(BC) = \begin{pmatrix} 9 & -2 \\ 16 & 2 \end{pmatrix}.$$

Similarly,

$$(AB)C = \left[\begin{pmatrix} 1 & 2 \\ -1 & 3 \end{pmatrix}\begin{pmatrix} 1 & 0 & -1 \\ 2 & 1 & 0 \end{pmatrix}\right]\begin{pmatrix} 1 & -1 \\ 3 & 2 \\ 2 & 1 \end{pmatrix}$$

$$= \begin{pmatrix} 5 & 2 & -1 \\ 5 & 3 & 1 \end{pmatrix}\begin{pmatrix} 1 & -1 \\ 3 & 2 \\ 2 & 1 \end{pmatrix}$$

or

$$(AB)C = \begin{pmatrix} 9 & -2 \\ 16 & 2 \end{pmatrix}$$

Therefore,

$$A(BC) = (AB)C$$

Theorem 5.2 For any suitably conformable matrices the distributive law holds true, that is,

$$C(A + B) = CA + CB$$

For an illustration of Theorem 5.2, let

$$A = \begin{pmatrix} 1 & 2 \\ 3 & 0 \end{pmatrix}, \quad B = \begin{pmatrix} 2 & -1 \\ 3 & 4 \end{pmatrix}, \quad \text{and} \quad C = \begin{pmatrix} 2 & -2 \\ 1 & 3 \\ 4 & -1 \end{pmatrix}$$

then

$$C(A + B) = \begin{pmatrix} 2 & -2 \\ 1 & 3 \\ 4 & -1 \end{pmatrix}\left[\begin{pmatrix} 1 & 2 \\ 3 & 0 \end{pmatrix} + \begin{pmatrix} 2 & -1 \\ 3 & 4 \end{pmatrix}\right]$$

$$= \begin{pmatrix} 2 & -2 \\ 1 & 3 \\ 4 & -1 \end{pmatrix}\begin{pmatrix} 3 & 1 \\ 6 & 4 \end{pmatrix}$$

or

$$C(A+B) = \begin{pmatrix} -6 & -6 \\ 21 & 13 \\ 6 & 0 \end{pmatrix}$$

Similarly,

$$CA + CB = \begin{pmatrix} 2 & -2 \\ 1 & 3 \\ 4 & -1 \end{pmatrix}\begin{pmatrix} 1 & 2 \\ 3 & 0 \end{pmatrix} + \begin{pmatrix} 2 & -2 \\ 1 & 3 \\ 4 & -1 \end{pmatrix}\begin{pmatrix} 2 & -1 \\ 3 & 4 \end{pmatrix}$$

or

$$CA + CB = \begin{pmatrix} -4 & 4 \\ 10 & 2 \\ 1 & 8 \end{pmatrix} + \begin{pmatrix} -2 & -10 \\ 11 & 11 \\ 5 & -8 \end{pmatrix} = \begin{pmatrix} -6 & -6 \\ 21 & 13 \\ 6 & 0 \end{pmatrix}$$

Thus

$$C(A+B) = CA + CB$$

Theorem 5.3 $AB = 0$ does not always imply that either $A = 0$ or $B = 0$. For instance, if

$$A = \begin{pmatrix} 6 & 4 & 2 \\ 9 & 6 & 3 \\ -3 & -2 & -1 \end{pmatrix} \quad \text{and} \quad B = \begin{pmatrix} 0 & 1 & -2 \\ -1 & 0 & 3 \\ 2 & -3 & 0 \end{pmatrix},$$

then

$$AB = \begin{pmatrix} 0 & 0 & 0 \\ 0 & 0 & 0 \\ 0 & 0 & 0 \end{pmatrix}$$

yet $A \neq 0$ and $B \neq 0$.

Theorem 5.4 The commutative law for multiplication does not hold true; that is, $AB \neq BA$.

For example, if

$$A = \begin{pmatrix} 1 & 2 \\ 3 & 4 \end{pmatrix} \quad \text{and} \quad B = \begin{pmatrix} 1 & 1 \\ 4 & 1 \end{pmatrix}$$

then

$$AB = \begin{pmatrix} 9 & 3 \\ 19 & 7 \end{pmatrix} \quad \text{and} \quad BA = \begin{pmatrix} 4 & 6 \\ 7 & 12 \end{pmatrix}$$

that is,

$$AB \neq BA$$

Since the product AB of two nonzero matrices A and B may be zero, and since the commutative law of multiplication fails to be true ($AB \neq BA$), the last two theorems establish some essential differences between the algebra of matrices and that of real numbers.

Theorem 5.5 For the unit and zero matrices I and 0, respectively, we have $AI = IA = A$ and $A0 = 0A = 0$. (Why?)

Theorem 5.6 The transpose of the product of any two matrices A and B is equal to the product of their transposes in the reverse order; that is,

$$(AB)^t = B^tA^t$$

To illustrate Theorem 5.6 let

$$A = \begin{pmatrix} 1 & 2 \\ 3 & 4 \end{pmatrix} \quad \text{and} \quad B = \begin{pmatrix} 4 & 3 \\ 2 & 1 \end{pmatrix}$$

Then

$$AB = \begin{pmatrix} 1 & 2 \\ 3 & 4 \end{pmatrix}\begin{pmatrix} 4 & 3 \\ 2 & 1 \end{pmatrix} = \begin{pmatrix} 8 & 5 \\ 20 & 13 \end{pmatrix} \quad \text{and} \quad (AB)^t = \begin{pmatrix} 8 & 20 \\ 5 & 13 \end{pmatrix}$$

On the other hand, since

$$B^t = \begin{pmatrix} 4 & 2 \\ 3 & 1 \end{pmatrix} \quad \text{and} \quad A^t = \begin{pmatrix} 1 & 3 \\ 2 & 4 \end{pmatrix}$$

we have

$$B^tA^t = \begin{pmatrix} 4 & 2 \\ 3 & 1 \end{pmatrix}\begin{pmatrix} 1 & 3 \\ 2 & 4 \end{pmatrix} \quad \text{or} \quad B^tA^t = \begin{pmatrix} 8 & 20 \\ 5 & 13 \end{pmatrix}$$

Thus

$$(AB)^t = B^tA^t$$

Example 5.2 If a manufacturer produces three products x, y, and z which he sells in the markets M_1 and M_2 and his annual sales volumes are indicated by the matrix

$$S = \begin{array}{c} \\ \end{array}\!\!\begin{pmatrix} \overset{x}{5{,}000} & \overset{y}{7{,}000} & \overset{z}{12{,}000} \\ 3{,}000 & 13{,}000 & 6{,}000 \end{pmatrix} \begin{array}{l} :M_1 \\ :M_2 \end{array}$$

and if the unit sales prices are given by the vector

$$P = \begin{pmatrix} \overset{x}{\$3.50} & \overset{y}{\$1.50} & \overset{z}{\$2.50} \end{pmatrix}$$

then

$$PS^t = (3.50 \quad 1.50 \quad 2.50) \cdot \begin{pmatrix} 5{,}000 & 3{,}000 \\ 7{,}000 & 13{,}000 \\ 12{,}000 & 6{,}000 \end{pmatrix}$$

$$= \begin{pmatrix} \overset{M_1}{\$58{,}000} & \overset{M_2}{\$45{,}000} \end{pmatrix}$$

is the total revenue from each market. On the other hand, if the vector

$$C = \begin{pmatrix} \overset{x}{\$2.50} & \overset{y}{\$1.00} & \overset{z}{\$1.50} \end{pmatrix}$$

represents unit costs, then the total costs are represented by the matrix product

$$CS^t = (2.50 \quad 1.00 \quad 1.50) \begin{pmatrix} 5{,}000 & 3{,}000 \\ 7{,}000 & 13{,}000 \\ 12{,}000 & 6{,}000 \end{pmatrix}$$

$$= (\$37{,}500 \quad \$29{,}500)$$

Thus the gross profits can be obtained from the matrix difference

$$PS^t - CS^t = (P - C)S^t = (\$10{,}500 \quad \$15{,}500)$$

Example 5.3 If an investment company sells 300, 500, 700, and 900 shares of stock for four companies A, B, C, and D at \$20, \$30, \$25, and \$50, respectively, find the total receipt from the stock sale as a matrix product.

Solution Let $S = (300, 500, 700, 900)$ be the vector indicating the sales volume and $P = (\$20, \$30, \$25, \$50)$ be the vector (or matrix) of unit sales prices. Then the total receipt from the stock sale is given by the matrix product

$$SP^t = (300 \quad 500 \quad 700 \quad 900)\begin{pmatrix} 20 \\ 30 \\ 25 \\ 50 \end{pmatrix} = \$83{,}500$$

Example 5.4 Let S be the system consisting of two players A and B who begin with \$2 each and match coins until one or the other has no more money. If the states of the system are defined by the number of dollars in A's possession—specifically, if the system is in the state S_{i+1} whenever A has i dollars $(i = 0, 1, 2, 3, 4)$—find the matrix of one-step transition probabilities. Then, by raising this matrix to the second, third, and fourth powers, find the matrices containing the two-, three-, and four-step transition probabilities for S. What is the probability that A will be ruined in at most four turns? In exactly four turns?

Solution Player A must either win a dollar or lose a dollar on each turn unless A or B is bankrupt, and the probability of each of these events is $\frac{1}{2}$. Hence, if A has i dollars $(i = 1, 2, 3)$—that is, if the system is in the state S_{i+1} $(i = 1, 2, 3)$—the probability of one-step transition to S_i is $\frac{1}{2}$, the probability of one-step transition to S_{i+2} is $\frac{1}{2}$, and the probability of one-step transition to any other state is zero. (Why?) On the other hand if the system is in the state S_1—that is, if A is bankrupt—the system remains in that state; so the probability of one-step transition from S_1 to S_1 is one, and the probability of any other transition from S_1 is zero. Similarly, if the system is in the state S_5—that is, if B is bankrupt—the system remains in that state; hence the probability of one-step transition from S_5 to S_5 is one, and the probability of any other transition is zero. Thus the required matrix of one-step transition probabilities is

$$P = \begin{bmatrix} 1 & 0 & 0 & 0 & 0 \\ \frac{1}{2} & 0 & \frac{1}{2} & 0 & 0 \\ 0 & \frac{1}{2} & 0 & \frac{1}{2} & 0 \\ 0 & 0 & \frac{1}{2} & 0 & \frac{1}{2} \\ 0 & 0 & 0 & 0 & 1 \end{bmatrix}.$$

Consequently, multiplying P by itself we find the matrix of two-step transition probabilities to be

$$P^2 = \begin{bmatrix} 1 & 0 & 0 & 0 & 0 \\ \frac{1}{2} & \frac{1}{4} & 0 & \frac{1}{4} & 0 \\ \frac{1}{4} & 0 & \frac{1}{2} & 0 & \frac{1}{4} \\ 0 & \frac{1}{4} & 0 & \frac{1}{4} & \frac{1}{2} \\ 0 & 0 & 0 & 0 & 1 \end{bmatrix}$$

Similarly, by computing P^3 and P^4, we find

$$P^3 = \begin{bmatrix} 1 & 0 & 0 & 0 & 0 \\ \frac{5}{8} & 0 & \frac{1}{4} & 0 & \frac{1}{8} \\ \frac{1}{4} & \frac{1}{4} & 0 & \frac{1}{4} & \frac{1}{4} \\ \frac{1}{8} & 0 & \frac{1}{4} & 0 & \frac{5}{8} \\ 0 & 0 & 0 & 0 & 1 \end{bmatrix} \quad \text{and} \quad P^4 = \begin{bmatrix} 1 & 0 & 0 & 0 & 0 \\ \frac{5}{8} & \frac{1}{8} & 0 & \frac{1}{8} & \frac{1}{8} \\ \frac{3}{8} & 0 & \frac{1}{4} & 0 & \frac{3}{8} \\ \frac{1}{8} & \frac{1}{8} & 0 & \frac{1}{8} & \frac{5}{8} \\ 0 & 0 & 0 & 0 & 1 \end{bmatrix}$$

to be the matrices of three-step and four-step transition probabilities, respectively. The probability that A is ruined in at most four turns is simply the probability of a four-step transition from S_3 to S_1, namely, $p_{31}^{(4)} = \frac{3}{8}$, since among such transitions are included those in which the system reaches S_1 in less than four steps and then remains there. On the other hand the probability that A is ruined in four terms and not before is the probability that S reaches S_1 in four steps but does not reach it in three or less, namely,

$$p_{31}^{(4)} - p_{31}^{(3)} = \tfrac{3}{8} - \tfrac{1}{4} = \tfrac{1}{8}$$

This example provides a good illustration of the Markov chains to be discussed in Chapter 6.

Exercises

1. Given the matrices

$$A = \begin{pmatrix} 1 & 2 & -3 \\ 5 & 0 & 2 \\ 1 & -1 & 1 \end{pmatrix}, \qquad B = \begin{pmatrix} 3 & -1 & 2 \\ 4 & 2 & 5 \\ 2 & 0 & 3 \end{pmatrix},$$

and

$$C = \begin{pmatrix} 4 & 1 & 2 \\ 0 & 3 & 2 \\ 1 & -2 & 3 \end{pmatrix}$$

a. Compute $A + B$, and $A - B$.
b. Compute $-2A$, $3C$, and $4A - 5B$.
c. Verify that $A + (B - C) = (A + B) - C$.
d. Find a matrix X such that $A + X = B$.
e. Evaluate A^2 and C^2.
f. If I and 0 are the identity and zero matrices, respectively, then show that $AI = A$ and $A0 = 0$.

2. Given

$$A = \begin{pmatrix} 1 & -1 & 1 \\ -3 & 2 & -1 \\ -2 & 1 & 0 \end{pmatrix} \quad \text{and} \quad B = \begin{pmatrix} 1 & 2 & 3 \\ 2 & 4 & 6 \\ 1 & 2 & 3 \end{pmatrix}$$

demonstrate that $AB = 0$ and

$$BA = \begin{pmatrix} -11 & 6 & -1 \\ -22 & 12 & -2 \\ -11 & 6 & -1 \end{pmatrix}$$

hence, $AB \neq BA$.

3. Given

$$A = \begin{pmatrix} 1 & -3 & 2 \\ 2 & 1 & -3 \\ 4 & -3 & -1 \end{pmatrix}, \qquad B = \begin{pmatrix} 1 & 4 & 1 & 0 \\ 2 & 1 & 1 & 1 \\ 1 & -2 & 1 & 2 \end{pmatrix}$$

and

$$C = \begin{pmatrix} 2 & 1 & -1 & -2 \\ 3 & -2 & -1 & -1 \\ 2 & -5 & -1 & 0 \end{pmatrix}$$

show that $AB = AC$. Thus $AB = AC$ does not necessarily imply $B = C$.

4. Given

$$A = \begin{pmatrix} 1 & 1 & -1 \\ 2 & 0 & 3 \\ 3 & -1 & 2 \end{pmatrix}, \qquad B = \begin{pmatrix} 1 & 3 \\ 0 & 2 \\ -1 & 4 \end{pmatrix},$$

and

$$C = \begin{pmatrix} 1 & 2 & 3 & -4 \\ 2 & 0 & -2 & 1 \end{pmatrix}$$

show that $(AB)C = A(BC)$.

5. Using the matrices of Problem 1, show that $A(B + C) = AB + AC$ and $(A + B)C = AC + BC$.

6. Explain why, in general,

$$(A + B)^2 \neq A^2 \pm 2AB + B^2$$

and

$$A^2 - B^2 \neq (A - B)(A + B).$$

7. Given

$$A = \begin{pmatrix} 2 & -3 & -5 \\ -1 & 4 & 5 \\ 1 & -3 & -4 \end{pmatrix}, \qquad B = \begin{pmatrix} -1 & 3 & 5 \\ 1 & -3 & -5 \\ -1 & 3 & 5 \end{pmatrix},$$

and

$$C = \begin{pmatrix} 2 & -2 & -4 \\ -1 & 3 & 4 \\ 1 & -2 & -3 \end{pmatrix}$$

a. Show that $AB = BA = 0$, $AC = A$, $CA = C$.
b. Use the results of (a) to show that

$$ACB = CBA,\ A^2 - B^2 = (A - B)(A + B)$$

and

$$(A \pm B)^2 = A^2 + B^2$$

8. Five stimulus objects O_1, O_2, O_3, O_4, and O_5 are rated on three factors F_1, F_2, and F_3 as follows:

Object:	O_1	O_2	O_3	O_4	O_5
Factors: F_1	1	2	4	1	0
F_2	2	1	3	0	4
F_3	3	4	2	3	1

Compute the distance matrix $D = (d_{ij})$ where d_{ij} is the distance between objects Oi and Oj when an object say O_1 is considered as the vector $O_1 = (1, 2, 3)$.

(Hint: The distance of the vectors $V = (a_1, a_2, \ldots, a_n)$ and $V = (b_1, b_2, \ldots, b_n)$ is given by

$$d = \sqrt{\sum_{i=1}^{n} (a_i - b_i)^2}$$

9. A manufacturer produces three products a, b, and c which he sells in two markets. Annual sales volumes are indicated by the matrix

$$V = \begin{pmatrix} 10{,}000 & 6{,}000 & 2{,}000 \\ 15{,}000 & 12{,}000 & 16{,}000 \end{pmatrix} \begin{matrix} \text{Market 1} \\ \text{Market 2} \end{matrix}$$

(columns: a, b, c)

If the unit sales prices and unit costs are given by the vectors

$$S = \begin{matrix} a & b & c \\ (\$2.50 & \$1.50 & \$1.25) \end{matrix}$$

and

$$C = \begin{matrix} a & b & c \\ (\$1.50 & \$1.00 & \$0.75) \end{matrix}$$

respectively, determine the total revenue in each market, the total costs, and the gross profits.

Ans. 36,000, 75,000; 22,500, 46,500

10. If in a three-industry open economy the input-coefficient matrix is

$$A = \begin{pmatrix} .2 & .3 & .2 \\ .4 & .1 & .2 \\ .1 & .3 & .4 \end{pmatrix}$$

and if the output vector is $V^t = (10, 6, 5)$, find the net output available for the open sector of the economy, that is, find $V^t - AV$.

Ans. 5.2, 0.4, 0.2

11. A company is considering which of three methods of production it should use in producing three goods A, B, and C. The amount of each goods produced by each method is shown in the matrix

$$R = \begin{matrix} & A & B & C & \\ & 2 & 5 & 1 & \text{Method 1} \\ & 6 & 7 & 3 & \text{Method 2} \\ & 2 & 1 & 4 & \text{Method 3} \end{matrix}$$

If $P = (10, 8, 7)$ is the vector whose components represent the profit per unit for each of the goods, what is RP^t? What does it represent?

5.3 Determinants

Associated with each square matrix $A = (a_{ij})$ of order two is a function of the matrix called its *determinant* denoted either by det (A) or $|A|$ or simply by

$$\begin{vmatrix} a_{11} & a_{12} \\ a_{21} & a_{22} \end{vmatrix}$$

The determinant function assigns to each 2×2 square matrix A, of numbers the number

$$\begin{vmatrix} a_{11} & a_{12} \\ a_{21} & a_{22} \end{vmatrix} = a_{11}a_{22} - a_{21}a_{12}$$

called the value of the determinant.

Clearly, the value of the 2×2 determinant A is the difference of the product of the diagonal elements minus the product of the remaining two elements. For example, the value of the determinant of the matrix

$$A = \begin{pmatrix} 4 & -4 \\ -4 & 4 \end{pmatrix}$$

is

$$\det(A) = |A| = \begin{vmatrix} 4 & -4 \\ -4 & 4 \end{vmatrix}$$

$$= (4)(4) - (-4)(-4) = 16 - 16 = 0$$

while that of

$$A = \begin{pmatrix} 2 & -1 \\ 4 & 3 \end{pmatrix}$$

is

$$\begin{vmatrix} 2 & -1 \\ 4 & 3 \end{vmatrix} = (2)(3) - (4)(-1) = 6 + 4 = 10$$

Similarly, the determinant of a 3×3 matrix $A = (a_{ij})$ is a function of the square matrix A and is denoted by either $\det(A)$ or $|A|$ or just

$$\begin{vmatrix} a_{11} & a_{12} & a_{13} \\ a_{21} & a_{22} & a_{23} \\ a_{31} & a_{32} & a_{33} \end{vmatrix}$$

The value of the determinant of A is defined as

$$\begin{vmatrix} a_{11} & a_{12} & a_{13} \\ a_{21} & a_{22} & a_{23} \\ a_{31} & a_{32} & a_{33} \end{vmatrix} = a_{11}a_{22}a_{33} + a_{12}a_{23}a_{31} + a_{13}a_{21}a_{32} - a_{31}a_{22}a_{13} - a_{32}a_{23}a_{11} - a_{33}a_{21}a_{12}$$

and it can be obtained by the application of a particular scheme similar to that used for 2×2 determinants. This scheme is commonly known as "diagonal multiplication," and it consists of repeating on the right of A the first two columns of the determinant and then adding the signed products of the elements on the various diagonals as indicated in the following array.

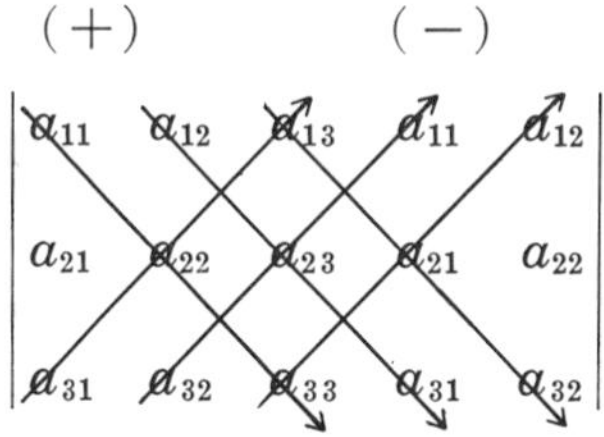

For example, if

$$A = \begin{pmatrix} 0 & 4 & 2 \\ 4 & -2 & -1 \\ 5 & 1 & 3 \end{pmatrix}$$

then the value of the determinant of A is given by the diagonal multiplications in

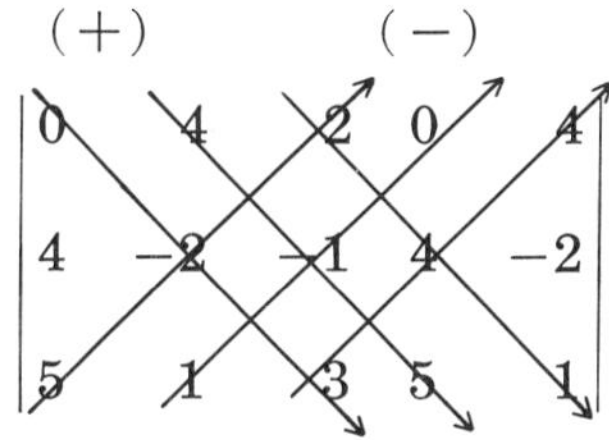

that is,

$$|A| = (0)(-2)(3) + (4)(-1)(5) + (2)(4)(1) - (2)(-2)(5) - (0)(-1)(1) - (4)(4)(3)$$

or

$$|A| = -40$$

It should be clear that the diagonal method employed for the evaluation of 3×3 determinants is not valid for determinants of higher order. Furthermore, no confusion should exist between square matrices and their determinants, for determinants are numbers and matrices are arrays of numbers.

In general, by the determinant of an $n \times n$ matrix $A = (a_{ij})$ or simply by an $n \times n$ determinant or determinant of order n we shall mean a function of n^2 quantities designated by either det (A) or $|A|$ or just

$$\begin{vmatrix} a_{11} & a_{12} & \cdots & a_{1n} \\ a_{21} & a_{22} & \cdots & a_{2n} \\ \cdots & \cdots & \cdots & \cdots \\ a_{n1} & a_{n2} & \cdots & a_{nn} \end{vmatrix} \tag{5.5}$$

As in the case of matrices, the numbers a_{ij} that appear in (5.5) are called "elements" of the determinant. The horizontal lines of elements are called "rows," and the vertical lines of elements are called "columns" of the determinant. The elements $a_{11}, a_{22}, \cdots, a_{nn}$ constitute the principal or main diagonal of the determinant $|A|$.

The determinant formed by the $(n-1)^2$ elements that remain whenever we delete, say, the ith row and the jth column of any determinant of order n is called the *minor* of the element a_{ij} (the element that lies on the ith row and the jth column), and it is denoted by M_{ij}. The *cofactor* A_{ij} of the a_{ij} element of an $n \times n$ determinant is defined as

$$A_{ij} = (-1)^{i+j} M_{ij}$$

For example, in the 3×3 determinant

$$\begin{vmatrix} a_{11} & a_{12} & a_{13} \\ a_{21} & a_{22} & a_{23} \\ a_{31} & a_{32} & a_{33} \end{vmatrix}$$

the minors of a_{22} and a_{23} are the 2×2 determinants

$$\begin{vmatrix} a_{11} & a_{13} \\ a_{31} & a_{33} \end{vmatrix} \quad \text{and} \quad \begin{vmatrix} a_{11} & a_{12} \\ a_{31} & a_{32} \end{vmatrix}$$

respectively, whereas the cofactors of a_{22} and a_{23} are

$$(-1)^{2+2} \cdot \begin{vmatrix} a_{11} & a_{13} \\ a_{31} & a_{33} \end{vmatrix} = \begin{vmatrix} a_{11} & a_{13} \\ a_{31} & a_{33} \end{vmatrix}$$

and

$$(-1)^{2+3} \cdot \begin{vmatrix} a_{11} & a_{12} \\ a_{31} & a_{32} \end{vmatrix} = - \begin{vmatrix} a_{11} & a_{12} \\ a_{31} & a_{32} \end{vmatrix}$$

respectively.

Since it can be shown[3] that the sum is the same for all rows and for all columns whenever the elements of any row or any column of n-order determinant are multiplied by their respective cofactors and then added, we may define the value (or expansion) of an n-order determinant as follows.

DEFINITION 5.6 The value of the $n \times n$ determinant

$$|A| = \begin{vmatrix} a_{11} & a_{12} & \cdots & a_{1n} \\ a_{21} & a_{22} & \cdots & a_{2n} \\ \cdot\cdot & \cdot\cdot & \cdot\ \cdot & \cdot\cdot \\ a_{n1} & a_{n2} & \cdots & a_{nn} \end{vmatrix}$$

is equal to the sum of the products of the elements of any row or column and their respective cofactors. That is,

$$|A| = \sum_{j=1}^{n} a_{ij}A_{ij} = \sum_{i=1}^{n} a_{ij}A_{ij}$$

Clearly, Definition 5.6 makes a determinant of order n depend on n determinants of order $n - 1$, each of which in turn depends on $n - 1$ determinants of order $n - 2$, and so on, until the expansion involves only determinants of order 2 or 3 whose values can be determined by previous definitions. Thus this process leads to the value of any $n \times n$ determinant.

To illustrate Definition 5.6, let us consider the matrix

$$A = \begin{pmatrix} 1 & -2 & 3 \\ -4 & 6 & 0 \\ 2 & 1 & -3 \end{pmatrix}$$

and evaluate its determinant

$$|A| = \begin{vmatrix} 1 & -2 & 3 \\ -4 & 6 & 0 \\ 2 & 1 & -3 \end{vmatrix}$$

using (a) the first row of $|A|$ and (b) the second column of $|A|$.

a. Since the cofactors of the elements 1, -2, and 3 of the first row of $|A|$ are

$$(-1)^{1+1}\begin{vmatrix} 6 & 0 \\ 1 & -3 \end{vmatrix}, \qquad (-1)^{1+2}\begin{vmatrix} -4 & 0 \\ 2 & -3 \end{vmatrix},$$

and

$$(-1)^{1+3}\begin{vmatrix} -4 & 6 \\ 2 & 1 \end{vmatrix}$$

or -18, -12, and -16, respectively, Definition 5.6 yields

$$|A| = (1)(-18) + (-2)(-12) + (3)(-16) = -42$$

for the value of the determinant of the matrix A.

b. Similarly, since the cofactors of the elements -2, 6, and 1 of the second column of A are

$$(-1)^{1+2}\begin{vmatrix} -4 & 0 \\ 2 & -3 \end{vmatrix}, \qquad (-1)^{2+2}\begin{vmatrix} 1 & 3 \\ 2 & -3 \end{vmatrix},$$

and

$$(-1)^{3+2}\begin{vmatrix} 1 & 3 \\ -4 & 0 \end{vmatrix}$$

or -12, -9, and -12, respectively, we obtain

$$|A| = (-2)(-12) + (6)(-9) + (1)(-12) = -42$$

Thus in both cases the value of the determinant is the same as suggested by Definition 5.6.

Since the same value is obtained whether we expand a determinant in terms of the elements of an arbitrary row or an arbitrary column, the following theorems are true.

Theorem 5.7 If all the elements in any row or column of a determinant are zero, the value of the determinant is zero.

Proof If we expand the given determinant, in terms of the row or column of zeros, then each term in the expansion contains a zero factor. Hence, the entire expansion is zero.

Theorem 5.8 If each element in one row or in one column of a determinant is multiplied by k, the determinant is multiplied by k.

Proof If we expand the given determinant in terms of the row or column whose elements were multiplied by k, then each term in the expansion contains k as a factor. If k is then factored from the expansion, the result is simply k times the expansion of the original determinant.

Theorem 5.9 If $|A|$ is any determinant and if $|B|$ is the determinant obtained from $|A|$ by interchanging any two rows or two columns of $|A|$, then $|B| = -|A|$.

Theorem 5.10 If corresponding elements of two rows or of any two columns of a determinant are proportional, then the value of the determinant is zero.

Theorem 5.11 The value of a determinant is not changed if the elements of any row (or column) are modified by adding to them the same multiple of the corresponding elements of any other row (or column).

In the following example we shall illustrate Theorems 5.9, 5.10, and 5.11 only for determinants of the third order, and leave the proof of the general case as an exercise for those students who are inclined to appreciate mathematical rigor and possess the required experience and training.

Example 5.5 By considering the 3×3 determinant

$$|A| = \begin{vmatrix} a_{11} & a_{12} & a_{13} \\ a_{21} & a_{22} & a_{23} \\ a_{31} & a_{32} & a_{33} \end{vmatrix},$$

illustrate (a) Theorem 5.9, (b) Theorem 5.10, and (c) Theorem 5.11.

Illustration

a. Interchanging rows 1 and 3 of $|A|$, we obtain

$$|B| = \begin{vmatrix} a_{31} & a_{32} & a_{33} \\ a_{21} & a_{22} & a_{23} \\ a_{11} & a_{12} & a_{13} \end{vmatrix}.$$

Hence, the application of the diagonal method results in

$$|A| = a_{11}a_{22}a_{33} + a_{12}a_{23}a_{31} + a_{13}a_{21}a_{32} - a_{13}a_{22}a_{31} - a_{11}a_{23}a_{32} - a_{12}a_{21}a_{33}$$

and

$$|B| = \begin{array}{|ccc|cc} \multicolumn{3}{c}{(+)} & \multicolumn{2}{c}{(-)} \\ a_{31} & a_{32} & a_{33} & a_{31} & a_{32} \\ a_{21} & a_{22} & a_{23} & a_{21} & a_{22} \\ a_{11} & a_{12} & a_{13} & a_{11} & a_{12} \end{array} =$$

$$a_{13}a_{22}a_{31} + a_{11}a_{23}a_{32} + a_{12}a_{21}a_{33} - a_{11}a_{22}a_{33} - a_{12}a_{23}a_{31} - a_{13}a_{21}a_{32}$$

Therefore,

$$|B| = -|A|.$$

b. If we assume that columns one and three are proportional, then the diagonal method yields

$$|A| = \begin{array}{|ccc|cc} \multicolumn{3}{c}{(+)} & \multicolumn{2}{c}{(-)} \\ a_{11} & a_{12} & ka_{11} & a_{11} & a_{12} \\ a_{21} & a_{22} & ka_{21} & a_{21} & a_{22} \\ a_{31} & a_{32} & ka_{31} & a_{31} & a_{32} \end{array} = 0 \qquad \text{(Why?)}$$

c. If we modify the second row by adding k times the corresponding elements of the first row, then the application of the diagonal method indicates that the new determinant

$$|B| = \begin{array}{|ccc|cc} a_{11} & a_{12} & a_{13} & a_{11} & a_{12} \\ a_{21} + ka_{11} & a_{22} + ka_{12} & a_{23} + ka_{13} & a_{21} + ka_{11} & a_{22} + ka_{12} \\ a_{31} & a_{32} & a_{33} & a_{31} & a_{32} \end{array}$$

$$\begin{aligned} &= a_{11}a_{22}a_{33} + ka_{11}a_{12}a_{33} + a_{12}a_{23}a_{31} + ka_{12}a_{13}a_{31} + a_{13}a_{21}a_{32} \\ &\quad + ka_{13}a_{11}a_{32} - a_{13}a_{22}a_{31} - ka_{13}a_{12}a_{31} - a_{11}a_{23}a_{32} \\ &\quad - ka_{11}a_{13}a_{32} - a_{12}a_{21}a_{33} - ka_{12}a_{11}a_{33} = |A| \end{aligned}$$

Thus the value of the determinant remains unchanged.

Of course, we could illustrate the above theorems for any other rows or columns. Illustration of the following two theorems shall be left as an exercise for the student.

Theorem 5.12 The sum of the products formed by multiplying the elements of one row (or column) of a determinant by the cofactors of the corresponding elements of another row (or column) is zero.

Theorem 5.13 The determinant of the product of any two square matrices A and B is the product of their determinants. That is,

$$|AB| = |A||B|$$

For example, if

$$A = \begin{pmatrix} 1 & 2 \\ 3 & 4 \end{pmatrix} \quad \text{and} \quad B = \begin{pmatrix} -1 & 2 \\ -3 & 1 \end{pmatrix}$$

then

$$AB = \begin{pmatrix} -7 & 4 \\ -15 & 10 \end{pmatrix} \quad \text{and} \quad |AB| = -70 + 60 = -10$$

Since

$$|A| = -2 \quad \text{and} \quad |B| = 5$$

we have

$$|AB| = -10 = |A||B|$$

Clearly, Theorem 5.11 implies that we may introduce a zero (or zeros) along any row or column of a determinant without altering its value. For example, the determinant

$$|A| = \begin{vmatrix} 1 & 2 & 3 \\ 4 & 5 & 6 \\ 7 & 8 & 9 \end{vmatrix}$$

is equal to the determinant

$$|B| = \begin{vmatrix} 1 & 2 & 3 \\ 0 & -3 & -6 \\ 0 & -6 & -12 \end{vmatrix}$$

which is obtained when we modify the second and third rows of $|A|$ by adding to each of them -4 times and -7 times, respectively, the corresponding elements of its first row. Similarly, the new determinant $|B|$ becomes equal to

$$|C| = \begin{vmatrix} 1 & 0 & 0 \\ 0 & -3 & -6 \\ 0 & -6 & -12 \end{vmatrix}$$

whenever we add to the second and third columns of $|B|$ -2 times and -3 times, respectively, the corresponding elements of its first column. Furthermore, if we modify $|C|$ by adding to its third row -2 times the second row, we reduce it to the equal determinant

$$|D| = \begin{vmatrix} 1 & 0 & 0 \\ 0 & -3 & -6 \\ 0 & 0 & 0 \end{vmatrix}$$

Finally, if we modify $|D|$ by adding to its third column -2 times the second column, we obtain the determinant

$$|E| = \begin{vmatrix} 1 & 0 & 0 \\ 0 & -3 & 0 \\ 0 & 0 & 0 \end{vmatrix} = 0$$

which has the same value with $|A|$. Thus a repeated application of Theorem 5.11 reduces any determinant into one of equal value containing nonzero elements only along the main diagonal. This observation suggests another way of evaluating a determinant, namely, that of reducing the determinant into one that contains nonzero elements only along the diagonal and expanding this rather than the original determinant.

Exercises

1. Show that the determinant

$$\begin{vmatrix} 1 & 2 & 3 & 4 \\ 2 & 4 & 6 & 3 \\ 3 & 8 & 12 & 2 \\ 4 & 16 & 24 & 1 \end{vmatrix}$$

equals zero.

2. Transform the determinant

$$\begin{vmatrix} -2 & 4 & 1 & 3 \\ 1 & -2 & 2 & 4 \\ 3 & 1 & -3 & 2 \\ 4 & 3 & -2 & -1 \end{vmatrix}$$

into an equal determinant having three zeros in the third column, and find its value.

Ans. 364

3. Without changing the value of the determinant

$$\begin{vmatrix} 4 & -2 & 1 & 3 & 1 \\ -2 & 1 & -3 & -2 & -2 \\ 3 & 4 & 2 & 1 & 3 \\ 1 & -3 & 4 & -1 & -1 \\ 2 & -1 & 2 & 4 & 2 \end{vmatrix}$$

obtain four zeros in the fourth column.

4. For the determinant

$$\begin{vmatrix} -1 & 2 & 3 & -2 \\ 4 & -1 & -2 & 2 \\ -3 & 1 & 2 & -1 \\ 2 & 4 & -1 & 3 \end{vmatrix}$$

a. Write the minors and cofactors of the elements in the third row.
b. Express the value of the determinant in terms of minors or cofactors.
c. Find the value of the determinant.

Ans. (c) −38

5. Transform the determinant

$$\begin{vmatrix} -2 & 1 & 2 & 3 \\ 3 & -2 & -3 & 2 \\ 1 & 2 & 1 & 2 \\ 4 & 3 & -1 & -3 \end{vmatrix}$$

into a determinant having three zeros in a row and then evaluate the determinant using its expansion by minors.

Ans. 28

6. Evaluate each of the following determinants:

a. $\begin{vmatrix} 2 & -1 & 3 & 2 \\ -3 & 1 & -2 & 4 \\ 1 & -3 & -1 & 3 \\ -1 & 2 & -2 & -3 \end{vmatrix}$ b. $\begin{vmatrix} 3 & -1 & 2 & 1 \\ 4 & 2 & 0 & -3 \\ -2 & 1 & -3 & 2 \\ 1 & 3 & -1 & 4 \end{vmatrix}$

c. $\begin{vmatrix} 2 & 1 & 3 \\ -1 & 3 & -5 \\ 1 & 4 & -2 \end{vmatrix}$ d. $\begin{vmatrix} 1 & 2 & -1 \\ 3 & 4 & -2 \\ 3 & 1 & -4 \end{vmatrix}$ e. $\begin{vmatrix} 2 & 3 & 6 \\ -3 & 1 & -4 \\ 7 & 5 & 16 \end{vmatrix}$

Ans. (c) 0; (d) 7; (e) 0

7. Expand each of the following determinants:

a. $\begin{vmatrix} a & b & c \\ a^2 & b^2 & c^2 \\ a^3 & b^3 & c^3 \end{vmatrix}$ b. $\begin{vmatrix} 1 & 1 & 1 & 1 \\ 1 & x & y & z \\ 1 & x^2 & y^2 & z^2 \\ 1 & x^3 & y^3 & z^3 \end{vmatrix}$

5.4 Rank and Equivalence of Matrices

Some very important characteristics of matrices are included in the following definitions.

DEFINITION 5.7 By the rank of a matrix A we shall mean the largest k for which there exists a $k \times k$ submatrix of A whose determinant is not zero.

DEFINITION 5.8 Any one of the following simple manipulations on matrices shall be called an *elementary row* or *column operation* (or transformation).

i. The multiplication of each element of a row or column by a non-zero constant.
ii. The interchange of any two rows or of any two columns.

iii. The addition of any multiple of the elements of one row, or one column, to the corresponding elements of another row, or column, respectively.

DEFINITION 5.9 Any two matrices A and B are said to be equivalent if and only if one can be obtained from the other by a series of elementary transformations.

One of the most interesting and useful properties of elementary transformations is that the rank of a matrix remains invariant (does not change) under any sequence of elementary row or column operations. For example, the rank of the matrix

$$A = \begin{pmatrix} 1 & 1 & 1 \\ 2 & 0 & -2 \\ 3 & 1 & -1 \end{pmatrix}$$

is 2. (Why?)

Now, by performing a sequence of row operations we can reduce A into the upper triangular matrix

$$A_1 = \begin{pmatrix} 1 & 1 & 1 \\ 0 & 1 & 2 \\ 0 & 0 & 0 \end{pmatrix} \qquad \text{(how?)}$$

which has rank 2 (why?) and is equivalent to A (why?). Moreover, by a sequence of column operations we can reduce A_1 into the diagonal matrix

$$A_2 = \begin{pmatrix} 1 & 0 & 0 \\ 0 & 1 & 0 \\ 0 & 0 & 0 \end{pmatrix}$$

which again has rank 2 and is equivalent to A.

Question Can we always reduce a matrix A into an equivalent diagonal matrix D?

Theorem 5.14 By means of elementary transformations any matrix A of rank r can be reduced to an equivalent matrix of the form

$$A_n = \begin{pmatrix} 1 & 0 \cdots 0 \cdots 0 \cdots 0 \\ 0 & 1 \quad 0 \cdots 0 \cdots 0 \\ 0 & 0 \quad 1 \quad 0 \cdots 0 \\ & \cdot \\ & \cdot \\ \cdots & \cdots \cdots \cdots \\ & \cdot \\ & \cdot \\ \cdots & \cdots \cdots \cdots \end{pmatrix}$$

where the number of ones in A_n is equal to the rank of A.

The matrix A_n of Theorem 5.14 is called the *normal form* of the matrix A and can obviously be obtained from it by the application of finitely many elementary operations. For instance, if by $aR_i + R_j$ (or $aC_i + C_j$) we mean modify the jth row (or column) by adding to it a times the ith row (or column) and if by R_i/k or C_i/k we mean multiply by $1/k$ the ith row (or column), then the matrix

$$A_n = \begin{pmatrix} 1 & 0 & 0 \\ 0 & 1 & 0 \\ 0 & 0 & 0 \end{pmatrix}$$

constitutes the normal form of the matrix

$$A = \begin{pmatrix} 1 & 1 & 1 \\ 2 & 0 & -2 \\ 3 & 1 & -1 \end{pmatrix}$$

since $-2R_1 + R_2$ and $-3R_1 + R_3$ reduce A into the matrix

$$\begin{pmatrix} 1 & 1 & 1 \\ 0 & -2 & -4 \\ 0 & -2 & -4 \end{pmatrix}$$

and $-R_2 + R_3$ further reduces it into

$$\begin{pmatrix} 1 & 1 & 1 \\ 0 & -2 & -4 \\ 0 & 0 & 0 \end{pmatrix}$$

Now the application of $-C_1 + C_2$ and $-C_1 + C_3$ transforms the last matrix into

$$A_1 = \begin{pmatrix} 1 & 0 & 0 \\ 0 & -2 & -4 \\ 0 & 0 & 0 \end{pmatrix}$$

which in turn becomes the matrix

$$A_n = \begin{pmatrix} 1 & 0 & 0 \\ 0 & 1 & 0 \\ 0 & 0 & 0 \end{pmatrix}$$

as the result of the elementary operations $-2C_2 + C_3$ and $R_2/-2$ on A_1.

DEFINITION 5.10 The matrix that results when an elementary row (or column) operation is applied on the identity matrix is called an "elementary" row (or column) matrix.

Clearly, every elementary matrix is square, is equivalent to an identity matrix, and has rank equal to its dimension. Furthermore, to effect a given elementary row (or column) transformation on a matrix $A_{m\times n}$, we apply the transformation to the identity matrix to form the corresponding elementary row (or column) matrix E and multiply A on the left by E (or on the right by E). For example, the elementary row operation $-4R_1 + R_2$ reduces the matrix

$$A = \begin{pmatrix} 1 & 2 & 3 \\ 4 & 5 & 6 \\ 7 & 8 & 9 \end{pmatrix} \quad \text{into} \quad \begin{pmatrix} 1 & 2 & 3 \\ 0 & -3 & -6 \\ 7 & 8 & 9 \end{pmatrix}$$

However, the same result can be obtained if we perform this row operation, $-4R_1 + R_2$, on the identity matrix to construct an appropriate elementary row matrix such as

$$E = \begin{pmatrix} 1 & 0 & 0 \\ -4 & 1 & 0 \\ 0 & 0 & 1 \end{pmatrix}$$

and then multiply A on the left by E; that is,

$$EA = \begin{pmatrix} 1 & 0 & 0 \\ -4 & 1 & 0 \\ 0 & 0 & 1 \end{pmatrix} \begin{pmatrix} 1 & 2 & 3 \\ 4 & 5 & 6 \\ 7 & 8 & 9 \end{pmatrix} = \begin{pmatrix} 1 & 2 & 3 \\ 0 & -3 & -6 \\ 7 & 8 & 9 \end{pmatrix}$$

Similarly, the elementary column operation $-3C_1 + C_3$ transforms the same matrix A into

$$\begin{pmatrix} 1 & 2 & 0 \\ 4 & 5 & -6 \\ 7 & 8 & -12 \end{pmatrix}$$

and so does multiplication of A on the right by the elementary column matrix

$$E = \begin{pmatrix} 1 & 0 & -3 \\ 0 & 1 & 0 \\ 0 & 0 & 1 \end{pmatrix}$$

The matrix E is obtained when $-3C_1 + C_3$ is applied on the identity matrix. Thus

$$\begin{pmatrix} 1 & 2 & 3 \\ 4 & 5 & 6 \\ 7 & 8 & 9 \end{pmatrix} \begin{pmatrix} 1 & 0 & -3 \\ 0 & 1 & 0 \\ 0 & 0 & 1 \end{pmatrix} = \begin{pmatrix} 1 & 2 & 0 \\ 4 & 5 & -6 \\ 7 & 8 & -12 \end{pmatrix}.$$

Now the application of both elementary operations and elementary matrices may be used to establish the following important theorem.

Theorem 5.15 Two matrices A and B are equivalent if and only if $PAQ = B$ where both P and Q are products of elementary matrices.

In view of Theorem 5.15, we say that two matrices A and B are similar if and only if they are equivalent and the matrices P and Q satisfy the equations $PQ = QP = 1$. Moreover, two matrices are equivalent if and only if they have the same rank.

5.5 Inverse of a Matrix

In intermediate algebra we learned that for any real number a, $a \neq 0$, there is a real number $1/a = a^{-1}$ that is called the "reciprocal" or the multiplicative inverse of a, and that $aa^{-1} = a^{-1}a = 1$. Likewise, we say that the $n \times n$ matrix B is the inverse or reciprocal of the $n \times n$ matrix A, and we denote it by A^{-1} if and only if $BA = AB = I$ (identity matrix). An $n \times n$ matrix A is called *nonsingular* if it has an inverse. Otherwise we say that A is *singular*. It is easy to show[1] that the inverse of a matrix is unique and that the inverse of the inverse of A is A; that is,

$$(A^{-1})^{-1} = A$$

Example 5.6 Find the inverse of the matrix

$$A = \begin{pmatrix} 1 & 2 \\ 3 & 4 \end{pmatrix}$$

To find A^{-1} we are looking for a matrix

$$\begin{pmatrix} x & y \\ z & t \end{pmatrix}$$

such that

$$\begin{pmatrix} 1 & 2 \\ 3 & 4 \end{pmatrix}\begin{pmatrix} x & y \\ z & t \end{pmatrix} = \begin{pmatrix} 1 & 0 \\ 0 & 1 \end{pmatrix}$$

or writing out the matrix product and equating corresponding elements, we are looking for the numbers x, y, z, and t such that

$$\begin{aligned} x + 2z &= 1 \\ y + 2t &= 0 \\ 3x + 4z &= 0 \\ 3y + 4t &= 1 \end{aligned}$$

Now, solving these equations we obtain $x = -2$, $y = 1$, $z = 3/2$, and

$t = -\frac{1}{2}$. Hence the inverse A^{-1} of A is

$$A^{-1} = \begin{pmatrix} -2 & 1 \\ 3/2 & -1/2 \end{pmatrix} \qquad \text{(Why?)}$$

In order to obtain the inverse of a more general matrix we shall define the *adjoint* of an $n \times n$ matrix and use it together with the determinant of the matrix as follows.

DEFINITION 5.11 If A_{ij} is the cofactor of the a_{ij} element in the square matrix $A = (a_{ij})$, then the matrix $(A_{ij})^t$ is called the adjoint of A, and it is denoted by adj $A = (A_{ij})^t$.

Theorem 5.16 The inverse or reciprocal A^{-1} of a nonsingular matrix $A = (a_{ij})$ is given by

$$A^{-1} = \frac{\text{adj } A}{|A|}$$

From the preceeding Theorem 5.16 and the properties of determinants the following important theorems follow.[4]

Theorem 5.17 For any nonsingular matrix A

$$AA^{-1} = A^{-1}A = I \qquad \text{(identity)}$$

Theorem 5.18 For any square matrix A,

$$(\text{adj } A)A = A(\text{adj } A) = |A| I$$

Theorem 5.19 If A is nonsingular, then

$$|A^{-1}| = \frac{1}{|A|}$$

Theorem 5.20 If A is an $n \times n$ nonsingular matrix, if B is an $n \times m$ matrix, and if C is an $m \times n$ matrix, then there exist matrices X_1 and X_2 such that

$$AX_1 = B \qquad \text{and} \qquad X_2A = C$$

Theorem 5.21 For any two nonsingular matrices A and B,

$$(AB)^{-1} = B^{-1}A^{-1}$$

Example 5.7 Illustrate Theorems 5.17 through 5.20 using the matrices

$$A = \begin{pmatrix} 1 & 2 & 3 \\ 1 & 3 & 4 \\ 1 & 4 & 3 \end{pmatrix}, \qquad B = \begin{pmatrix} 1 & 3 & 3 \\ 1 & 4 & 3 \\ 1 & 3 & 4 \end{pmatrix}$$

and

$$C = \begin{pmatrix} 1 \\ 2 \\ 1 \end{pmatrix}$$

Theorem 5.17

a.

$$\text{adj } A = \begin{pmatrix} \begin{vmatrix} 3 & 4 \\ 4 & 3 \end{vmatrix} & -\begin{vmatrix} 1 & 4 \\ 1 & 3 \end{vmatrix} & \begin{vmatrix} 1 & 3 \\ 1 & 4 \end{vmatrix} \\ -\begin{vmatrix} 2 & 3 \\ 4 & 3 \end{vmatrix} & \begin{vmatrix} 1 & 3 \\ 1 & 3 \end{vmatrix} & -\begin{vmatrix} 1 & 2 \\ 1 & 4 \end{vmatrix} \\ \begin{vmatrix} 2 & 3 \\ 3 & 4 \end{vmatrix} & -\begin{vmatrix} 1 & 3 \\ 1 & 4 \end{vmatrix} & \begin{vmatrix} 1 & 2 \\ 1 & 3 \end{vmatrix} \end{pmatrix}^t$$

$$= \begin{pmatrix} -7 & 1 & 1 \\ 6 & 0 & -2 \\ -1 & -1 & 1 \end{pmatrix}^t$$

or

$$\text{adj } A = \begin{pmatrix} -7 & 6 & -1 \\ 1 & 0 & -1 \\ 1 & -2 & 1 \end{pmatrix}$$

On the other hand,

$$|A| = \begin{vmatrix} 1 & 2 & 3 \\ 1 & 3 & 4 \\ 1 & 4 & 3 \end{vmatrix} = -2$$

Thus

$$A^{-1} = \frac{\text{adj } A}{|A|} = \begin{pmatrix} 7/2 & -3 & 1/2 \\ -1/2 & 0 & 1/2 \\ -1/2 & 1 & -1/2 \end{pmatrix}$$

and

$$AA^{-1} = \begin{pmatrix} 1 & 2 & 3 \\ 1 & 3 & 4 \\ 1 & 4 & 3 \end{pmatrix} \begin{pmatrix} 7/2 & -3 & 1/2 \\ -1/2 & 0 & 1/2 \\ -1/2 & 1 & -1/2 \end{pmatrix}$$

$$= \begin{pmatrix} 1 & 0 & 0 \\ 0 & 1 & 0 \\ 0 & 0 & 1 \end{pmatrix} = I = A^{-1}A$$

Theorem 5.18

b.

$$A\ (\text{adj } A) = \begin{pmatrix} 1 & 2 & 3 \\ 1 & 3 & 4 \\ 1 & 4 & 3 \end{pmatrix} \begin{pmatrix} -7 & 6 & -1 \\ 1 & 0 & -1 \\ 1 & -2 & 1 \end{pmatrix}$$

$$= \begin{pmatrix} -2 & 0 & 0 \\ 0 & -2 & 0 \\ 0 & 0 & -2 \end{pmatrix} = (-2) \begin{pmatrix} 1 & 0 & 0 \\ 0 & 1 & 0 \\ 0 & 0 & 1 \end{pmatrix}$$

or

$$A\ (\text{adj } A) = |A|\, I = (\text{adj } A)A$$

Theorem 5.19

c.

$$|A^{-1}| = \begin{vmatrix} 7/2 & -3 & 1/2 \\ -1/2 & 0 & 1/2 \\ -1/2 & 1 & -1/2 \end{vmatrix} = -\frac{1}{2} = \frac{1}{-2} = \frac{1}{|A|}$$

Theorem 5.20

d.

$AX_1 = B$ implies

$$A^{-1}(AX_1) = A^{-1}B \qquad \text{or} \qquad X_1 = A^{-1}B$$

that is,

$$X_1 = \begin{pmatrix} 7/2 & -3 & 1/2 \\ -1/2 & 0 & 1/2 \\ -1/2 & 1 & -1/2 \end{pmatrix} \begin{pmatrix} 1 & 3 & 3 \\ 1 & 4 & 3 \\ 1 & 3 & 4 \end{pmatrix}$$

$$= \begin{pmatrix} 1 & 0 & 7/2 \\ 0 & 0 & 1/2 \\ 0 & 1 & -1/2 \end{pmatrix}$$

Similarly, $X_2A = C$ implies

$$(X_2A)A^{-1} = CA^{-1} \qquad \text{or} \qquad X_2 = CA^{-1}$$

that is,

$$X_2 = \begin{pmatrix} 1 \\ 2 \\ 1 \end{pmatrix} \begin{pmatrix} 7/2 & -3 & 1/2 \\ -1/2 & 0 & 1/2 \\ -1/2 & 1 & -1/2 \end{pmatrix} = (-2, 0, 1)$$

A much simpler method for determining the inverse of the matrix

$$B = \begin{pmatrix} 1 & 3 & 3 \\ 1 & 4 & 3 \\ 1 & 3 & 4 \end{pmatrix}$$

is described in the following table in which the left-hand side is the matrix B and the right-hand side the identity matrix. In this process we use a sequence of elementary row transformations that carry B into the identity. Thus as B is reduced to the identity matrix, the identity is carried into the inverse of B as follows:

$$\begin{array}{c} \qquad B \qquad\qquad I \\ \left[\begin{array}{ccc|ccc} 1 & 3 & 3 & 1 & 0 & 0 \\ 1 & 4 & 3 & 0 & 1 & 0 \\ 1 & 3 & 4 & 0 & 0 & 1 \end{array}\right] \xrightarrow[-R_1+R_3]{-R_1+R_2} \left[\begin{array}{ccc|ccc} 1 & 3 & 3 & 1 & 0 & 0 \\ 0 & 1 & 0 & -1 & 1 & 0 \\ 0 & 0 & 1 & -1 & 0 & 1 \end{array}\right] \xrightarrow{-3R_2+R_1} \end{array}$$

$$\left[\begin{array}{ccc|ccc} 1 & 0 & 3 & 4 & -3 & 0 \\ 0 & 1 & 0 & -1 & 1 & 0 \\ 0 & 0 & 1 & -1 & 0 & 1 \end{array}\right] \xrightarrow{-3R_3+R_1} \left[\begin{array}{ccc|ccc} 1 & 0 & 0 & 7 & -3 & -3 \\ 0 & 1 & 0 & -1 & 1 & 0 \\ 0 & 0 & 1 & -1 & 0 & 1 \end{array}\right].$$

That is,

$$B^{-1} = \begin{pmatrix} 7 & -3 & -3 \\ -1 & 1 & 0 \\ -1 & 0 & 1 \end{pmatrix}$$

That the last matrix is indeed the inverse of B can be demonstrated by multiplying it with B to obtain the identity matrix as follows:

$$\begin{pmatrix} 7 & -3 & -3 \\ -1 & 1 & 0 \\ -1 & 0 & 1 \end{pmatrix} \begin{pmatrix} 1 & 3 & 3 \\ 1 & 4 & 3 \\ 1 & 3 & 4 \end{pmatrix} = \begin{pmatrix} 1 & 0 & 0 \\ 0 & 1 & 0 \\ 0 & 0 & 1 \end{pmatrix}$$

Example 5.8

In two ways, determine the inverse of the matrix

$$A = \begin{pmatrix} 1 & 2 & 4 \\ -1 & 0 & 3 \\ 3 & 1 & -2 \end{pmatrix}$$

Method (a). Let us perform elementary row operations as indicated in the following table.

$$\begin{array}{c} \qquad A \qquad\qquad\qquad I \\ \left[\begin{array}{rrr|rrr} 1 & 2 & 4 & 1 & 0 & 0 \\ -1 & 0 & 3 & 0 & 1 & 0 \\ 3 & 1 & -2 & 0 & 0 & 1 \end{array}\right] \end{array}$$

$$\xrightarrow[-3R_1+R_3]{R_1+R_2} \left[\begin{array}{rrr|rrr} 1 & 2 & 4 & 1 & 0 & 0 \\ 0 & 2 & 7 & 1 & 1 & 0 \\ 0 & -5 & -14 & -3 & 0 & 1 \end{array}\right]$$

$$\xrightarrow{5/2R_2+R_3} \left[\begin{array}{rrr|rrr} 1 & 2 & 4 & 1 & 0 & 0 \\ 0 & 2 & 7 & 1 & 1 & 0 \\ 0 & 0 & 7/2 & -1/2 & 5/2 & 1 \end{array}\right]$$

$$\xrightarrow{2/7R_3} \left[\begin{array}{rrr|rrr} 1 & 2 & 4 & 1 & 0 & 0 \\ 0 & 2 & 7 & 1 & 1 & 0 \\ 0 & 0 & 1 & -1/7 & 5/7 & 2/7 \end{array}\right]$$

$$\xrightarrow{-R_2+R_1} \left[\begin{array}{rrr|rrr} 1 & 0 & -3 & 0 & -1 & 0 \\ 0 & 2 & 7 & 1 & 1 & 0 \\ 0 & 0 & 1 & -1/7 & 5/7 & 2/7 \end{array}\right]$$

$$\xrightarrow[-7R_3+R_2]{3R_3+R_1} \left[\begin{array}{rrr|rrr} 1 & 0 & 0 & -3/7 & 8/7 & 6/7 \\ 0 & 2 & 0 & 2 & -4 & -2 \\ 0 & 0 & 1 & -1/7 & 5/7 & 2/7 \end{array}\right]$$

$$\begin{array}{c} \qquad\quad I \qquad\qquad\qquad A^{-1} \\ \xrightarrow{R_2/2} \left[\begin{array}{rrr|rrr} 1 & 0 & 0 & -3/7 & 8/7 & 6/7 \\ 0 & 1 & 0 & 1 & -2 & -1 \\ 0 & 0 & 1 & -1/7 & 5/7 & 2/7 \end{array}\right] \end{array}$$

Then

$$A^{-1} = \begin{pmatrix} -3/7 & 8/7 & 6/7 \\ 1 & -2 & -1 \\ -1/7 & 5/7 & 2/7 \end{pmatrix}$$

Method (*b*). The determinant of A is

$$|A| = \begin{vmatrix} 1 & 2 & 4 \\ -1 & 0 & 3 \\ 3 & 1 & -2 \end{vmatrix} = 7$$

and

$$\text{adj } A = \begin{pmatrix} \begin{vmatrix} 0 & 3 \\ 1 & -2 \end{vmatrix} & -\begin{vmatrix} -1 & 3 \\ 3 & -2 \end{vmatrix} & \begin{vmatrix} -1 & 0 \\ 3 & 1 \end{vmatrix} \\ -\begin{vmatrix} 2 & 4 \\ 1 & -2 \end{vmatrix} & \begin{vmatrix} 1 & 4 \\ 3 & -2 \end{vmatrix} & -\begin{vmatrix} 1 & 2 \\ 3 & 1 \end{vmatrix} \\ \begin{vmatrix} 2 & 4 \\ 0 & 3 \end{vmatrix} & -\begin{vmatrix} 1 & 4 \\ -1 & 3 \end{vmatrix} & \begin{vmatrix} 1 & 2 \\ -1 & 0 \end{vmatrix} \end{pmatrix}^t \quad \text{or}$$

$$\text{adj } A = \begin{pmatrix} -3 & 7 & -1 \\ 8 & -14 & 5 \\ 6 & -7 & 2 \end{pmatrix}^t$$

or

$$\text{adj } A = \begin{pmatrix} -3 & 8 & 6 \\ 7 & -14 & -7 \\ -1 & 5 & 2 \end{pmatrix}$$

Thus the inverse

$$A^{-1} = \frac{\text{adj } A}{|A|} = \frac{1}{7}\begin{pmatrix} -3 & 8 & 6 \\ 7 & -14 & -7 \\ -1 & 5 & 2 \end{pmatrix}$$

Negative powers for nonsingular matrices are defined as

$$A^{-n} = (A^{-1})^n$$

Consequently

$$A^rA^s = A^{r+s} \qquad \text{and} \qquad (A^r)^s = A^{rs}$$

for all integral values of r and s. However, negative powers of singular matrices are not defined.

Exercises

1. Find the rank of the following matrices:

a. $\begin{pmatrix} 3 & -1 & 0 & 1 \\ 6 & 2 & -1 & -3 \end{pmatrix}$ b. $\begin{pmatrix} 2 & -1 & 3 \\ 1 & -2 & 3 \\ 5 & 0 & 3 \end{pmatrix}$

Ans. (a) 2; (b) 3

2. Determine the rank of the following matrices as a function of k.

a. $\begin{pmatrix} 8(1-k) & -2 & 0 \\ -2 & 3-2k & -1 \\ 0 & -1 & 2(1-k) \end{pmatrix}$

b. $\begin{pmatrix} 1-k & 1 & 1 \\ 1 & 3-k & 3 \\ 2 & 1 & 4-k \end{pmatrix}$

Ans. (a) For $k = 1$ or 2; $r = 2$; otherwise $r = 3$.

3. In two ways evaluate the inverse of the following matrices.

a. $\begin{pmatrix} 1 & 1 & 0 \\ -1 & 2 & 1 \\ 0 & 1 & 3 \end{pmatrix}$ b. $\begin{pmatrix} 2 & 1 \\ -1 & 1 \end{pmatrix}$

c. $\begin{pmatrix} 1 & 2 & 0 \\ 2 & 3 & -1 \\ -1 & -1 & 2 \end{pmatrix}$

Ans. (c) $\begin{pmatrix} -5 & 4 & 2 \\ 3 & -2 & -1 \\ -1 & 1 & 1 \end{pmatrix}$

4. Show that the following pairs of matrices are equivalent.

a. $A = \begin{pmatrix} 0 & 1 & 0 \\ 1 & 2 & 1 \end{pmatrix}$ $B = \begin{pmatrix} 1 & 0 & 0 \\ 0 & 1 & 1 \end{pmatrix}$

b. $A = \begin{pmatrix} 0 & 1 & 0 \\ 1 & 2 & 1 \\ 1 & 1 & 2 \end{pmatrix}$ $B = \begin{pmatrix} 1 & 1 & 0 \\ 0 & 1 & 1 \\ 2 & 1 & 1 \end{pmatrix}$

5. Given the matrices

$$A = \begin{pmatrix} 1 & 2 & 0 \\ 2 & 3 & -1 \\ -1 & -1 & 2 \end{pmatrix}, \quad X = \begin{pmatrix} 2 \\ 1 \\ -1 \end{pmatrix},$$

and

$$Y = \begin{pmatrix} 2 \\ 3 \\ -4 \end{pmatrix}$$

evaluate AX, AY, and X^TAY.

References for Supplementary Readings

1. D. T. Finkbeiner, *Elements of Linear Algebra.* (San Francisco: Freeman, 1972).
2. D. Zelinsky, *A First Course in Linear Algebra* (New York: Academic Press, 1968).
3. C. R. Wylie, Jr., *Advanced Engineering Mathematics*, Chapter 10 (New York: McGraw-Hill, 1966).
4. G. Hadley, *Linear Algebra*, (Reading, Mass.: 1961), Addison-Wesley.

CHAPTER 6

Application of Matrices

6.1 Systems of Linear Equations and Inequalities

The theory of linear equations and inequalities plays a very important and motivating role in the applications of determinants and matrices. In fact, as we shall see in the subsequent section, many problems in the algebra of matrices are equivalent to systems of linear equations or linear inequalities or both.[1] As an illustration let us consider buying the following three kinds of food:

a. Food X containing one unit of vitamin A, three units of vitamin B, and four units of vitamin C.
b. Food Y containing two, three, and five units of vitamins A, B, and C, respectively.
c. Food Z containing no units of vitamin B, and 3 units each of vitamin A and C.

Furthermore, let us assume that we need to have 11 units of vitamin A, 9 units of vitamin B, and 20 units of vitamin C. Naturally, we would like to find out all possible amounts of the three foods that will provide precisely the required units of vitamins. Thus let x, y, and z be the amounts of food X, Y, and Z, respectively, that will provide these vitamin requirements. Then the vitamin content of the foods and the vitamin requirement imply that x, y, and z satisfy the following system of linear equations.

$$\begin{aligned} 1x + 2y + 3z &= 11 \\ 3x + 3y + 0z &= 9 \\ 4x + 5y + 3z &= 20 \end{aligned} \quad \text{or} \quad \begin{pmatrix} 1 & 2 & 3 \\ 3 & 3 & 0 \\ 4 & 5 & 3 \end{pmatrix} \begin{pmatrix} x \\ y \\ z \end{pmatrix} = \begin{pmatrix} 11 \\ 9 \\ 20 \end{pmatrix} \tag{6.1}$$

The solution of the above system will provide the required amounts of foods X, Y, and Z.

Frequently, problems encountered in linear programming and game theory,[2] as well as in linear transformations and characteristic values of matrices, require an understanding of both the theory and solution methods related to linear systems. For instance, suppose a truck rental company has the following two types of trucks:

a. Type X of 20 cubic feet of refrigerated space and 40 cubic feet of nonrefrigerated.
b. Type Y of 30 cubic feet each of refrigerated and nonrefrigerated space.

Moreover, suppose that a food plant wishes to ship 900 cubic feet of refrigerated and 1200 cubic feet of nonrefrigerated produce, respectively. Clearly the plant management would like to know how many trucks of each type should be rented so as to minimize cost if truck type X rents for 30¢ per mile and truck type Y for \$0.40 per mile. Thus let x and y be the number of trucks of type X and Y, respectively, that the plant rents. Then since there are 20 cubic feet and 30 cubic feet of refrigerated space in each type of truck and 900 cubic feet of refrigerated produce for shipment, since there are 40 cubic feet and 30 cubic feet of nonrefrigerated space in each truck and 1200 cubic feet of nonrefrigerated produce must be shipped, and since

$$C = 30x + 40y$$

is the rental cost, the management of the plant must determine the values of x and y which make

$$C = 30x + 40y$$

minimum and satisfy the inequalities

$$20x + 30y \geq 900$$

$$40x + 30y \geq 1200 \qquad \text{(Why?)}$$

$$x \geq 0$$

$$y \geq 0 \qquad \text{(Why?)}$$

Prior to the solution of the previous two problems and the general study of systems of equations and inequalities, we must consider some fundamental terminology and the related concepts of linear dependence and independence of a set of quantities, $Q_1, Q_2, \cdots, Q_n$.

An expression of the form

$$a_1x_1 + a_2x_2 + \cdots + a_nx_n = b$$

shall be called a linear equation over the field R of real numbers if and only if the constants $a_1, a_2, \cdots, a_n$, and b are real numbers, and the variables $x_1, x_2, \cdots, x_n$ are real valued. The numbers a_i, $i = 1, 2$ $\cdots, n$, are called the coefficients of the unknowns x_i, $i = 1, 2, \cdots, n$, respectively, and b is called the constant term or simply the constant of the equation. On the other hand an expression of the form

$$a_1x_1 + a_2x_2 + \cdots + a_nx_n > b \qquad (\text{or } <b)$$

is called a linear inequality. For example, the expressions

$$2x + 3y - 4z = 6 \qquad \text{and} \qquad 2x + 3y - 4z < 6 \qquad (\text{or } >6)$$

are a linear equation and a linear inequality, respectively. The real numbers 2, 3, and -4 are the coefficients of x, y, and z, respectively. Now, a set of values for the unknowns, say

$$x_1 = k_1,\ x_2 = k_2,\ \cdots,\ x_n = k_n,$$

which satisfies the equation or inequality, is known as a solution set or, if written as an ordered n-tuple $(k_1, k_2, \cdots, k_n)$, as a solution vector.

DEFINITION 6.1 A set of quantities $Q_1, Q_2, \ldots, Q_n$ is said to be linearly dependent if and only if the linear combination $c_1Q_1 + c_2Q_2 + \cdots + c_nQ_n$ is equal to zero for at least one nonzero c, say c_i. Otherwise we say that these quantities are linearly independent. Of course, the quantities $Q_1, Q_2, \ldots, Q_n$ represent matrices or vectors or other mathematical entities, whereas $c_1, c_2, \ldots, c_n$ are scalars (or numbers).

Theorem 6.1 If $Q_1, Q_2, \ldots, Q_n$ are linearly dependent, then at least one of the quantities can be expressed as a linear combination of the others.

Proof The linear dependence of $Q_1, Q_2, \ldots, Q_n$ implies that

$$c_1Q_1 + c_2Q_2 + \cdots + c_nQ_n = 0$$

for at least one nonzero c, say c_i. Thus we may divide by c_i and obtain

$$Q_i = -\frac{c_1}{c_i}Q_1 - \frac{c_2}{c_i}Q_2 - \cdots - \frac{c_n}{c_i}Q_n$$

which expresses Q_i as a linear combination of the other Q's.

Example 6.1 Show that the vectors $V = (1, 1)$, $U = (2, -1)$, and $W = (2, 3)$ are linearly dependent, and express V as a linear combination of U and W.

Solution Let us consider the linear combination

$$c_1V + c_2U + c_3W = 0 \qquad \text{(vector)}$$

Then

$$c_1(1, 1) + c_2(2, -1) + c_3(2, 3) = (0, 0)$$

or

$$(c_1 + 2c_2 + 2c_3, c_1 - c_2 + 3c_3) = (0, 0)$$

or

$$\begin{aligned} c_1 + 2c_2 + 2c_3 &= 0 \\ c_1 - c_2 + 3c_3 &= 0 \end{aligned} \tag{6.2}$$

Thus solving for c_1, c_2, and c_3, we find $c_1 = 8$, $c_2 = -1$, and $c_3 = -3$ satisfy the Equations (6.2). Since the linear combination $c_1V + c_2U + c_3W$ is equal to zero for at least one nonzero c, it follows that V, U, and W are linearly dependent, and since $8V - U - 3W = 0$, we can express any one of the three vectors as a linear combination of the other two. Certainly,

$$V = \tfrac{1}{8}U + \tfrac{3}{8}W$$

Example 6.2 Show that the matrices

$$A = \begin{pmatrix} 1 & 0 \\ 0 & 1 \end{pmatrix}, \quad B = \begin{pmatrix} 0 & 2 \\ 0 & 0 \end{pmatrix}, \quad C = \begin{pmatrix} 0 & 0 \\ 3 & 0 \end{pmatrix}, \quad D = \begin{pmatrix} 0 & 0 \\ 0 & 4 \end{pmatrix}$$

are linearly independent.

Solution Let us consider the linear combination $c_1A + c_2B + c_3C + c_4D = 0$ (matrix) and solve for c_1, c_2, c_3, and c_4. This yields

$$c_1\begin{pmatrix}1 & 0\\ 0 & 0\end{pmatrix} + c_2\begin{pmatrix}0 & 2\\ 0 & 0\end{pmatrix} + c_3\begin{pmatrix}0 & 0\\ 3 & 0\end{pmatrix} + c_4\begin{pmatrix}0 & 0\\ 0 & 4\end{pmatrix} = \begin{pmatrix}0 & 0\\ 0 & 0\end{pmatrix}$$

or

$$\begin{pmatrix}c_1 & 2c_2\\ 3c_3 & 4c_4\end{pmatrix} = \begin{pmatrix}0 & 0\\ 0 & 0\end{pmatrix}$$

that is, $c_1 = c_2 = c_3 = c_4 = 0$ is the only solution set for the linear equation

$$c_1A + c_2B + c_3C + c_4D = 0$$

Therefore the matrices A, B, C, and D must be linearly independent.

The preceding two examples clearly indicate that problems of linear dependence and independence are closely related to those of obtaining solutions satisfying simultaneously two or more equations. Such sets of linear equations constitute systems of linear equations whose most general form appears as follows:

$$\begin{aligned} a_{11}x_1 + a_{12}x_2 + \cdots + a_{1n}x_n &= b_1 \\ a_{21}x_1 + a_{22}x_2 + \cdots + a_{1n}x_n &= b_2 \\ a_{m1}x_1 + a_{m2}x_2 + \cdots + a_{mn}x_n &= b_m \end{aligned} \tag{6.3}$$

or simply as

$$AX = B \tag{6.4}$$

where $A = (a_{ij})_{m\times n}$, $X = (x_i)_{n\times 1}$, and $B = (b_i)_{m\times 1}$ are the matrix of coefficients, the vector of unknowns, and the vector of constants, respectively; and m the number of equations, is not necessarily equal to n, the number of unknowns.

Now, we say that the system (6.3) is homogeneous if and only if $b_i = 0$ for all i or $B = 0$. Otherwise, we say the system is nonhomogeneous. The matrix $A = (a_{ij})_{m\times n}$ is called the matrix of coefficients or coefficient matrix of the system, and the matrix $[AB]$ obtained when we adjoin the column of constants $b_1, b_2, \cdots, b_m$ to the coefficient matrix, as an extra column, is called the augmented matrix of the system (6.3). For example, in the nonhomogeneous system

$$\begin{aligned} x + y + z &= 1 \\ 2x - y - 3z &= 4 \end{aligned} \qquad \text{or} \qquad \begin{pmatrix}1 & 1 & 1\\ 2 & -1 & -3\end{pmatrix}\begin{pmatrix}x\\ y\\ z\end{pmatrix} = \begin{pmatrix}1\\ 4\end{pmatrix}$$

the coefficient and augmented matrices are

$$A = \begin{pmatrix} 1 & 1 & 1 \\ 2 & -1 & -3 \end{pmatrix} \quad \text{and} \quad [AB] = \begin{pmatrix} 1 & 1 & 1 & 1 \\ 2 & -1 & -3 & 4 \end{pmatrix}$$

respectively. Similarly, in the homogeneous system

$$\begin{aligned} x + y + z &= 0 \\ x - y - z &= 0 \\ 2x + y + 3z &= 0 \end{aligned} \quad \text{or} \quad \begin{pmatrix} 1 & 1 & 1 \\ 1 & -1 & -1 \\ 2 & 1 & 3 \end{pmatrix} \begin{pmatrix} x \\ y \\ z \end{pmatrix} = \begin{pmatrix} 0 \\ 0 \\ 0 \end{pmatrix}$$

The coefficient and augmented matrices are

$$A = \begin{pmatrix} 1 & 1 & 1 \\ 1 & -1 & -1 \\ 2 & 1 & 3 \end{pmatrix}$$

and

$$[AB] = \begin{pmatrix} 1 & 1 & 1 & 0 \\ 1 & -1 & -1 & 0 \\ 2 & 1 & 3 & 0 \end{pmatrix}$$

respectively.

In the introduction of linear systems, we have emphasized their possible utilizations in solving and modeling many real-life situations. Subsequently, we present two problems whose analysis necessitates knowing whether or not a system has any solutions, on the one hand, and finding the solutions on the other. Moreover, we state several important theorems that relate to solutions of linear systems. As in previous sections, we prove some theorems, whereas others are illustrated only by way of examples.

Problem 6.1 A man has invested \$36,000 in 3%, 4%, and 5% bonds. How much does he have invested in each of these bonds if his investment in the 5% bond is equal to the sum of his investments in the 3% and 4% bonds, and if his yearly income from these investments is \$1600?

Solution Let x, y, and z be his investments in the 3%, 4%, and 5% bonds, respectively. Then, the unknowns x, y, and z satisfy the following system of equations

$$x + y + z = 36{,}000$$

$$z = x + y$$

$$\frac{3x}{100} + \frac{4y}{100} + \frac{5z}{100} = 1600$$

or

$$x + y + z = 36{,}000$$

$$-x - y + z = 0$$

$$3x + 4y + 5z = 160{,}000 \qquad \textbf{(6.5)}$$

To complete the solution we must solve system (6.5) if it has any solutions in x, y, and z. However, the lack of special techniques necessitates its postponement for later consideration.

Problem 6.2 An investor has considered the various securities in which he might invest and has classified them into four types A, B, C, and D according to the extent of risk. To minimize the element of risk, he restricts purchases of types A and B to not more than 30% of his total investment. Because of similar considerations, he wishes to have at least 40% of his total investment in types A and C. Within these restrictions, how can he maximize his dividend income on a \$100,000 investment if the four types of investment have dividend returns as follows:

A: 6%, B: 7%, C: 3%, D: 5%

Solution Let x, y, z, and t be the percentages of his investment in types, A, B, C, and D, respectively. Then he must maximize

$$D = 6x + 7y + 3z + 5t \qquad \text{(total dividends)}$$

subject to the restrictions

$$x + y \leq 30 \qquad \text{(not more than 30\%)}$$

$$x + z \geq 40 \qquad \text{(at least 40\%)}$$

$$x + y + z + t = 100 \qquad \text{(100\% of total investment)}$$

Since $a \leq b$ if and only if there is $x \geq 0$ such that $a + x = b$, the inequalities

$$x + y \leq 30 \qquad \text{and} \qquad x + z \geq 40$$

become the equations

$$x + y + v = 30 \qquad \text{and} \qquad x + z - w = 40$$

for $v, w \geq 0$.
Thus the original problem can be formulated as follows:

Find the solutions of the system

$$\begin{aligned} x + y + v &= 30 \\ x + z - w &= 40 \\ x + y + z + t &= 100 \end{aligned} \tag{6.6}$$

for which the function

$$D = 6x + 7y + 3z + 5t$$

is a maximum. Again, since completion of this solution requires techniques not presently available, we defer it for later consideration.

Clearly, the completion of Problems 6.1 and 6.2 require knowledge of the existence of solutions and the means (methods) of obtaining them. These considerations become the essence of the subsequent theorems and methods of solutions on linear systems of equations.

Theorem 6.2 If X_1 and X_2 are any two solution vectors of the system $AX = 0$, then the linear combination $c_1X_1 + c_2X_2$ is also a solution of $AX = 0$.

Proof If we substitute

$$X = c_1X_1 + c_2X_2$$

in AX, we obtain

$$A(c_1X_1 + c_2X_2) = c_1(AX_1) + c_2(AX_2) = c_10 + c_20 = 0$$

since

$$AX_1 = 0 \qquad \text{and} \qquad AX_2 = 0$$

Thus $c_1X_1 + c_2X_2$ is a solution of the system $AX = 0$.

Theorem 6.3 If $X_1, X_2, \ldots, X_k$ are linearly independent solution vectors of the system $AX = 0$, and if k is the maximum number of such linearly independent vectors, then any solution vector X of $AX = 0$ can be expressed as

$$X = c_1X_1 + c_2X_2 + \cdots + c_kX_k$$

for arbitrary constants $c_1, c_2, \ldots, c_k$.

A linear combination of the maximum number of linearly independent solution vectors of $AX = 0$ is called a *complete solution* of the system $AX = 0$. As a result of Theorem 6.3, any solution of the system $AX = 0$ can be obtained from its complete solution.

Theorem 6.4 If $AX = B$ is a nonhomogeneous system of linear equations, any solution of it can be written in the form $c_1X_1 + c_2X_2 + \cdots + c_kX_k + Y$ where $c_1X_1 + c_2X_2 + \cdots + c_kX_k$ is a complete solution of the related homogeneous system $AX = 0$, and Y is a particular solution of the nonhomogeneous system $AX = B$.

Theorems 6.2, 6.3, and 6.4 assure us that if we have more than one solution for the system $AX = 0$ or $AX = B$, then we can obtain infinitely many other solutions simply by forming arbitrary linear combinations of these solutions. However, the theorems say nothing about methods of solving systems, neither do they say anything about the existence of solutions. Before we turn to these very important questions, let us observe that the homogeneous system $AX = 0$ always has the solution $X = 0$. This solution shall henceforth be referred to as the *trivial solution* of the system, and any other solution of it will be called *nontrivial*. When a system has a solution, we say that the system is *consistent*. Otherwise, we say it is *inconsistent*. With this preliminary background we proceed to state two theorems that assure us of the existence of solution(s), and in the next section we develop general solution methods for linear systems of equations.

Theorem 6.5 The homogeneous system $AX = 0$ of n equations in n unknowns has nontrivial solutions if and only if the determinant $|A|$ of coefficients is zero. The number of linearly independent solutions is equal to $n - r$ where r is the rank of the coefficient matrix.

Theorem 6.6 The system $AX = B$ of m equations in n unknowns has a solution if and only if the coefficient matrix A and the augmented matrix have the same rank, say r. The maximum number of linearly independent solutions of the corresponding homogeneous system $AX = 0$ is $n - r$.

Example 6.3

a. The homogeneous system

$$\begin{aligned} x + y + z &= 0 \\ 2x - 3y + 4z &= 0 \\ 3x - 2y + 5z &= 0 \end{aligned}$$

has a nontrivial solution since the determinant of coefficients is zero; that is,

$$\begin{vmatrix} 1 & 1 & 1 \\ 2 & -3 & 4 \\ 3 & -2 & 5 \end{vmatrix} = 0 \qquad \text{(Why?)}$$

Since the rank of the coefficient matrix

$$\begin{pmatrix} 1 & 1 & 1 \\ 2 & -3 & 4 \\ 3 & -2 & 5 \end{pmatrix}$$

is 2 (why?), the number of linearly independent solution vectors is $3 - 2 = 1$ (why?).

b. The homogeneous system

$$\begin{aligned} x + y + z &= 0 \\ 2x + y + 2z &= 0 \\ 3x + 3y + z &= 0 \end{aligned}$$

has only the trivial solution $x = y = z = 0$ since the determinant of coefficients is not zero; that is,

$$\begin{vmatrix} 1 & 1 & 1 \\ 2 & 1 & 2 \\ 3 & 3 & 1 \end{vmatrix} = 2 \neq 0 \qquad \text{(Why?)}$$

c. The nonhomogeneous system

$$\begin{aligned} x + y + z + t &= 1 \\ 2x - y - z + 3t &= 4 \end{aligned}$$

has solutions since the rank of both the coefficient matrix

$$\begin{pmatrix} 1 & 1 & 1 & 1 \\ 2 & -1 & -1 & 3 \end{pmatrix}$$

and the augmented matrix

$$\begin{pmatrix} 1 & 1 & 1 & 1 & 1 \\ 2 & -1 & -1 & 3 & 4 \end{pmatrix}$$

is 2 (why?). The number of linearly independent solutions of the related homogeneous is $4 - 2 = 2$ (why?).

Exercises

1. Find the coefficient and augmented matrices of the following systems, and determine whether or not each system has a solution.

a. $$\begin{aligned} x - 2y + z &= 0 \\ 2x - y - z &= 0 \\ x + y + z &= 0. \end{aligned}$$

b. $$\begin{aligned} 2x - 3y + 5z &= 0 \\ x + y + z &= 0 \\ 4x - y + 7z &= 0. \end{aligned}$$

c. $$\begin{aligned} 2x - 3y + 5z &= 2 \\ x + y + z &= -1 \\ 4x - y + 7z &= 0. \end{aligned}$$

d. $$\begin{aligned} 5x - 2y - 3z &= 1 \\ -x + 4y - z &= 6 \\ 2x + 10y - 6z &= 4. \end{aligned}$$

e. $$\begin{aligned} x + y + z &= 0 \\ 2x + y - z &= 0 \\ 3x - y + z &= 0. \end{aligned}$$

Ans. (b) yes; (c) no; (e) no

2. Show that the vectors

$$A = (1, 2, 3) \qquad B = (2, -1, 3) \qquad C = (3, 1, 6)$$

are linearly dependent, and find C as a linear combination of A and B.

Ans. $C = A + B$

3. Show that the vectors

$$A = (1, 1, 1) \qquad B = (2, 1, -1) \qquad C = (3, -1, 1)$$

are linearly independent.

6.2 Methods of Solving Systems of Equations

In this section, four of the most popular methods employed in solving linear systems of equations and some demonstrative examples will be presented.

GAUSS REDUCTION

The Gauss reduction method was named in honor of the German mathematician Carl Friedrich Gauss (1777–1855), who is generally acknowledged as one of the greatest mathematicians of all time. Its objective is to obtain zeros below the main diagonal of the coefficient matrix and ones along the main diagonal. Fortunately, this method applies to both homogeneous and nonhomogeneous systems of the form

$$\begin{aligned}
a_{11}x_1 + a_{12}x_2 + \cdots + a_{1n}x_n &= b_1 \\
a_{21}x_1 + a_{22}x_2 + \cdots + a_{2n}x_n &= b_2 \\
\cdots\cdots\cdots\cdots\cdots\cdots\cdots \\
a_{m1}x_1 + a_{m2}x_2 + \cdots + a_{mn}x_n &= b_m
\end{aligned} \tag{6.7}$$

where m and n are not necessarily the same.

Using elementary row operations on the augmented matrix of (6.7), we reduce it to a simpler equivalent system, that is, a system that has the same solutions as (6.7). This of course is the case since the rank of a matrix does not change under elementary row transformations.[1] The reduction process includes the following steps.

Step 1. If necessary, interchange rows in the augmented matrix so that the element in the first row and first column, a_{11}, is not zero, and divide by a_{11} if $a_{11} \neq 1$.

Step 2. For each $i > 1$ apply the elementary row operations

$$-a_{i1}R_1 + R_i \qquad i = 1, 2, \ldots, m$$

that is, modify the ith row by adding to it the corresponding elements of the first row multiplied by $-a_{i1}$ for each $i = 1, 2, \ldots, m$. This sequence of elementary row operations reduces the augmented matrix into one of the form

$$\begin{pmatrix}
1 & b_{12} & b_{13}\cdots b_{1n} & c_1 \\
0 & b_{22} & b_{23} \quad b_{2n} & c_2 \\
\cdot & & & \\
\cdot & & & \\
\cdot & & & \\
0 & b_{m2}\cdots & b_{mn} & c_m
\end{pmatrix} \tag{6.8}$$

and the system (6.7) reduces to the equivalent system

$$\begin{aligned}
x_1 + b_{12}x_2 + \cdots + b_{1n}x_n &= c_1 \\
b_{22}x_2 + \cdots + b_{2n}x_n &= c_2 \\
\cdots \quad \cdot \quad \cdots \\
b_{m2}x_2 + \cdots + b_{mn}x_n &= c_m
\end{aligned} \tag{6.9}$$

Now, we may apply the same steps to the last $m - 1$ rows of the equivalent augmented matrix (6.8) noting that if $b_{22} = 0$, a rearrangement of the last $m - 1$ rows will introduce a nonzero in the place of

b_{22} unless all elements in the remaining rows are zero, which, of course, may be the case at some stage in the process. The result of repeating this process is an equivalent system about which we can determine the existence and character of its solutions. In the following sequence of examples we demonstrate this method.

Example 6.4 Solve the system

$$\begin{aligned} x_1 + 2x_2 \qquad &= 1 \\ 2x_1 + 3x_2 - x_3 &= 2 \\ -x_1 - x_2 + 2x_3 &= 3 \end{aligned}$$

Solution The augmented matrix of the system is

$$A = \left(\begin{array}{ccc|c} 1 & 2 & 0 & 1 \\ 2 & 3 & -1 & 2 \\ -1 & -1 & 2 & 3 \end{array}\right)$$

Now, applying the elementary row operations T_i, $i = 1, 2, 3$, we can reduce A as follows:

T_1: $-2R_1 + R_2$, reduces A into $A_1 = \begin{pmatrix} 1 & 2 & 0 & 1 \\ 0 & -1 & -1 & 0 \\ -1 & -1 & 2 & 3 \end{pmatrix}$

T_2: $1R_1 + R_3$, reduces A_1 into $A_2 = \begin{pmatrix} 1 & 2 & 0 & 1 \\ 0 & -1 & -1 & 0 \\ 0 & 1 & 2 & 4 \end{pmatrix}$

T_3: $1R_2 + R_3$, reduces A_2 into $A_3 = \begin{pmatrix} 1 & 2 & 0 & 1 \\ 0 & -1 & -1 & 0 \\ 0 & 0 & 1 & 4 \end{pmatrix}$

Thus the system becomes

$$\begin{pmatrix} 1 & 2 & 0 \\ 0 & -1 & -1 \\ 0 & 0 & 1 \end{pmatrix} \begin{pmatrix} x_1 \\ x_2 \\ x_3 \end{pmatrix} = \begin{pmatrix} 1 \\ 0 \\ 4 \end{pmatrix} \quad \text{or} \quad \begin{aligned} x_1 + 2x_2 &= 1 \\ -x_2 - x_3 &= 0 \\ x_3 &= 4 \end{aligned}$$

From the third equation it follows that $x_3 = 4$, and from the second that $x_2 = -x_3$ or $x_2 = -4$. Finally, from the first equation we obtain $x_1 = 1 - 2x_2$ or $x_1 = 9$. Hence, the solution vector is $(9, -4, 4)$ or

$$x_1 = 9 \qquad x_2 = -4 \qquad x_3 = 4$$

Example 6.5 Find all the solutions of the system

$$\begin{aligned} x_1 + 2x_2 + x_3 - x_4 + 2x_5 &= 2 \\ x_1 + 4x_2 + 5x_3 - 3x_4 + 8x_5 &= -2 \\ -2x_1 - x_2 + 4x_3 - x_4 + 5x_5 &= -10 \\ 3x_1 + 7x_2 + 5x_3 - 4x_4 + 9x_5 &= 4 \end{aligned}$$

Solution The augmented matrix of this system is

$$A = \left(\begin{array}{rrrrr|r} 1 & 2 & 1 & -1 & 2 & 2 \\ 1 & 4 & 5 & -3 & 8 & -2 \\ -2 & -1 & 4 & -1 & 5 & -10 \\ 3 & 7 & 5 & -4 & 9 & 4 \end{array}\right)$$

Now, applying the elementary row operations T_i, $i = 1, 2, \ldots, 6$, we may reduce A as follows:

T_1: $-1R_1 + R_2$

T_2: $2R_1 + R_2$

T_3: $-3R_1 + R_4$

reduce A into

$$A_1 = \begin{pmatrix} 1 & 2 & 1 & -1 & 2 & 2 \\ 0 & 2 & 4 & -2 & 6 & -4 \\ 0 & 3 & 6 & -3 & 9 & -6 \\ 0 & 1 & 2 & -1 & 3 & -2 \end{pmatrix}$$

Similarly, T_4 (interchange R_2 and R_4) transforms A_1 into

$$A_2 = \begin{pmatrix} 1 & 2 & 1 & -1 & 2 & 2 \\ 0 & 1 & 2 & -1 & 3 & -2 \\ 0 & 3 & 6 & -3 & 9 & -6 \\ 0 & 2 & 4 & -2 & 6 & -4 \end{pmatrix}$$

Finally, T_5: $-3R_2 + R_3$, and T_6: $-2R_2 + R_4$ reduce A into

$$A_3 = \begin{pmatrix} 1 & 2 & 1 & -1 & 2 & 2 \\ 0 & 1 & 2 & -1 & 3 & -2 \\ 0 & 0 & 0 & 0 & 0 & 0 \\ 0 & 0 & 0 & 0 & 0 & 0 \end{pmatrix}$$

Thus the system becomes

$$\begin{pmatrix} 1 & 2 & 1 & -1 & 2 \\ 0 & 1 & 2 & -1 & 3 \\ 0 & 0 & 0 & 0 & 0 \\ 0 & 0 & 0 & 0 & 0 \end{pmatrix} \begin{pmatrix} x_1 \\ x_2 \\ x_3 \\ x_4 \\ x_5 \end{pmatrix} = \begin{pmatrix} 2 \\ -2 \\ 0 \\ 0 \end{pmatrix}$$

or

$$x_1 + 2x_2 + x_3 - x_4 + 2x_5 = 2$$

$$x_2 + 2x_3 - x_4 + 3x_5 = -2$$

From the last system it is evident that we do not have a unique solution. In fact, three of the unknowns in the second equation can be assigned arbitrary values a, b, and c, and the fourth unknown can be determined from them. Hence, if for any numbers a, b, and c,

$$x_3 = a, \qquad x_4 = b, \qquad \text{and} \qquad x_5 = c$$

then

$$x_2 = -2a + b - 3c - 2 \qquad \text{(from second equation)}$$

and

$$x_1 = 3a - b + 4c + 6 \qquad \text{(from first equation)}$$

Therefore a complete solution of the original system is given by

$$(x_1, x_2, x_3, x_4, x_5)^t = a(3, -2, 1, 0, 0)^t + b(-1, 1, 0, 1, 0)^t + c(4, -3, 0, 0, 1)^t + (6, -2, 0, 0, 0)^t$$

Remarks A careful examination of the complete solution of this system reveals, as expected, that

$$a(3, -2, 1, 0, 0)^t + b(-1, 1, 0, 1, 0)^t + c(4, -3, 0, 0, 1)^t$$

is a complete solution of the corresponding homogeneous system, since

$$(3, -2, 1, 0, 0)^t \qquad (-1, 1, 0, 1, 0)^t \qquad (4, -3, 0, 0, 1)^t$$

are three linearly independent particular solutions of this system, and that $(6, -2, 0, 0, 0)^t$ is a particular solution of the nonhomogeneous system under consideration. Clearly these remarks constitute a good illustration of Theorem 6.4.

Example 6.6 Show that the system

$$\begin{aligned} x + 2y - z &= 1 \\ 2x + 4y - 2z &= 4 \\ 3x + 6y - 3z &= 5 \end{aligned}$$

has no solution.

Solution The augmented matrix of this system is

$$A = \begin{pmatrix} 1 & 2 & -1 & 1 \\ 2 & 4 & -2 & 4 \\ 3 & 6 & -3 & 5 \end{pmatrix}$$

Now, the elementary row operations $-2R_1 + R_2$ and $-3R_1 + R_3$ reduce A into the equivalent matrix

$$A_1 = \begin{pmatrix} 1 & 2 & -1 & 1 \\ 0 & 0 & 0 & 2 \\ 0 & 0 & 0 & 2 \end{pmatrix}$$

Thus the system reduces to the following equivalent one

$$\begin{pmatrix} 1 & 2 & -1 \\ 0 & 0 & 0 \\ 0 & 0 & 0 \end{pmatrix} \begin{pmatrix} x \\ y \\ z \end{pmatrix} = \begin{pmatrix} 1 \\ 2 \\ 2 \end{pmatrix}$$

or

$$\begin{aligned} x + 2y - z &= 1 \\ 0x + 0y + 0z &= 2 \end{aligned}$$

which is inconsistent (has no solution) since there are no numbers x, y, and z satisfying the second equation.

Example 6.7 Check whether or not the following systems have a nontrivial solution, and find the solution(s) if it exists.

a. $$\begin{aligned} x + 2y \phantom{{}-z} &= 0 \\ 2x + 3y - z &= 0 \\ -x - y + 2z &= 0 \end{aligned}$$

b. $$\begin{aligned} 2x - y + 3z &= 0 \\ 5x + 2y + 5z &= 0 \\ x + 4y - z &= 0 \end{aligned}$$

Solution

a. The determinant of coefficients in case (a) is

$$|A| = \begin{vmatrix} 1 & 2 & 0 \\ 2 & 3 & -1 \\ -1 & -1 & 2 \end{vmatrix} = \begin{vmatrix} 1 & 0 & 0 \\ 2 & -1 & -1 \\ -1 & 1 & 2 \end{vmatrix}$$

$$= -1 \neq 0 \qquad \text{(Why?)}$$

and that in case (b) is

$$|A| = \begin{vmatrix} 2 & -1 & 3 \\ 5 & 2 & 5 \\ 1 & 4 & -1 \end{vmatrix} = 0 \qquad \text{(Why?)}$$

Therefore the system in (a) has no nontrivial solutions, whereas that in (b) does (Why?).

b. The augmented matrix of the system is

$$A = \begin{pmatrix} 2 & -1 & 3 & 0 \\ 5 & 2 & 5 & 0 \\ 1 & 4 & -1 & 0 \end{pmatrix}$$

and it reduces to an equivalent one as follows:

a. T_1 (interchange rows so that the a_{11} element is one) transforms A into the matrix

$$A_1 = \begin{pmatrix} 1 & 4 & -1 & 0 \\ 2 & -1 & 3 & 0 \\ 5 & 2 & 5 & 0 \end{pmatrix}$$

b. T_2: $-2R_1 + R_2$ and T_3: $-5R_1 + R_3$ reduce A_1 into

$$A_2 = \begin{pmatrix} 1 & 4 & -1 & 0 \\ 0 & -9 & 5 & 0 \\ 0 & -18 & 10 & 0 \end{pmatrix}$$

c. T_4: $-2R_2 + R_3$ transforms A_2 into

$$A_3 = \begin{pmatrix} 1 & 4 & -1 & 0 \\ 0 & -9 & 5 & 0 \\ 0 & 0 & 0 & 0 \end{pmatrix}$$

Hence the original system reduces to the equivalent system

$$\begin{pmatrix} 1 & 4 & -1 \\ 0 & -9 & 5 \\ 0 & 0 & 0 \end{pmatrix} \begin{pmatrix} x \\ y \\ z \end{pmatrix} = \begin{pmatrix} 0 \\ 0 \\ 0 \end{pmatrix}$$

or

$$x + 4y - z = 0$$
$$- 9y + 5z = 0$$

The second equation of this system is satisfied by any value of z. So if we let $z = 9m$, y can be determined from z and the second equation as $y = 5m$ (where m is any number). Now using the first equation we obtain $x = -4y + z$ or $x = -11m$ for any m. Therefore the complete solution set of this system is

$$(x, y, z)^t = m(-11, 5, 9)^t$$

where m is any number, and the vector $(-11, 5, 9)^t$ constitutes a particular nontrivial solution of this system.

CRAMER'S RULE

This method of solving linear systems of equations is known as *Cramer's rule* in honor of the Swiss mathematician Gabriel Cramer (1704–1752). Unfortunately, it applies only to systems with a square nonsingular coefficient matrix as the following theorem specifies.

Theorem 6.7 If the coefficient matrix A of the system $AX = B$ $(B \neq 0)$ of n equations in n unknowns is nonsingular $(|A| \neq 0)$, then the system has the unique solution

$$x_1 = \frac{|D_1|}{|A|}, \qquad x_2 = \frac{|D_2|}{|A|}, \qquad \ldots, \qquad x_n = \frac{|D_n|}{|A|}$$

where D_i is the determinant obtained from A by replacing the ith column of A by the column vector B.

Example 6.8 Using Cramer's method, solve the system

$$\begin{aligned} x + 2y + z \qquad\quad &= 2 \\ 2x \qquad - 2z + t &= 6 \\ 4y + 3z + 2t &= -1 \\ -x + 6y - z - t &= 2 \end{aligned}$$

Solution The values of the different determinants required by

Theorem 6.7 are as follows:

$$|A| = \begin{vmatrix} 1 & 2 & 1 & 0 \\ 2 & 0 & -2 & 1 \\ 0 & 4 & 3 & 2 \\ -1 & 6 & -1 & -1 \end{vmatrix} = -92,$$

$$|D_1| = \begin{vmatrix} \overset{\downarrow}{2} & 2 & 1 & 0 \\ 6 & 0 & -2 & 1 \\ -1 & 3 & 2 & 2 \\ 2 & 6 & -1 & -1 \end{vmatrix} = -184$$

$$|D_2| = \begin{vmatrix} 1 & \overset{\downarrow}{2} & 1 & 0 \\ 2 & 6 & -2 & 1 \\ 0 & -1 & 3 & 2 \\ -1 & 2 & -1 & -1 \end{vmatrix} = -46$$

$$|D_3| = \begin{vmatrix} 1 & 2 & \overset{\downarrow}{2} & 0 \\ 2 & 0 & 6 & 1 \\ 0 & 4 & -1 & 2 \\ -1 & 6 & 2 & -1 \end{vmatrix} = 92$$

and

$$|D_4| = \begin{vmatrix} 1 & 2 & 1 & \overset{\downarrow}{2} \\ 2 & 0 & -2 & 6 \\ 0 & 4 & 3 & -1 \\ -1 & 6 & -1 & 2 \end{vmatrix} = 0$$

Thus

$$x = \frac{|D_1|}{|A|} = \frac{-184}{-92} = 2$$

$$y = \frac{|D_2|}{|A|} = \frac{-46}{-92} = \frac{1}{2}$$

$$z = \frac{|D_3|}{|A|} = \frac{92}{-92} = -1$$

$$t = \frac{|D_4|}{|A|} = \frac{0}{-92} = 0$$

INVERSE METHOD

In addition to Cramer's rule for solving nonhomogeneous linear systems of n equations in n unknowns, there exists the "inverse method" described in Theorem 6.8.

Theorem 6.8 If A is a nonsingular matrix ($|A| \neq 0$), and if B is a nonzero vector, then the system $AX = B$ has a unique solution given by $X = A^{-1}B$.

Example 6.9 Using the inverse method, solve the system,

$$\begin{aligned} 2x + 3y + z &= 9 \\ x + 2y + 3z &= 6 \\ 3x + y + 2z &= 8 \end{aligned}$$

Solution Since the coefficient matrix

$$A = \begin{pmatrix} 2 & 3 & 1 \\ 1 & 2 & 3 \\ 3 & 1 & 2 \end{pmatrix}$$

is nonsingular with inverse

$$A^{-1} = \frac{1}{18}\begin{pmatrix} 1 & -5 & 7 \\ 7 & 1 & -5 \\ -5 & 7 & 1 \end{pmatrix} \qquad \text{(Why?)}$$

the application of Theorem 6.8 yields the solution

$$X = \begin{pmatrix} x \\ y \\ z \end{pmatrix} = A^{-1} \begin{pmatrix} 9 \\ 6 \\ 8 \end{pmatrix}$$

$$\begin{pmatrix} x \\ y \\ z \end{pmatrix} = \frac{1}{18} \begin{pmatrix} 1 & -5 & 7 \\ 7 & 1 & -5 \\ -5 & 7 & 1 \end{pmatrix} \begin{pmatrix} 9 \\ 6 \\ 8 \end{pmatrix}$$

$$(x, y, z)^t = \left(\frac{35}{18}, \frac{29}{18}, \frac{5}{18}\right)^t$$

or

$$x = \frac{35}{18}, \qquad y = \frac{29}{18}, \qquad \text{and} \qquad z = \frac{5}{18}$$

SPECIAL HOMOGENEOUS

A method designed to solve special homogeneous systems is included in Theorem 6.9.

Theorem 6.9 In the homogeneous system $AX = 0$ of n equations in n unknowns if the rank of A is $n - 1$ and if the submatrix obtained from A by omitting the kth row is also of rank $n - 1$, then a complete solution of the system is

$$x_i = cA_{ki}, \qquad i = 1, 2, \ldots, n$$

where c is an arbitrary constant and A_{ki}, $i = 1, 2, \ldots, n$ are the cofactors of the elements of the kth row.

Example 6.10 Solve the following system

$$\begin{aligned} 3x - 2y - z &= 0 \\ x + y + z &= 0 \\ -2x + 3y + 2z &= 0 \end{aligned}$$

Solution Since the matrix of coefficients

$$A = \begin{pmatrix} 3 & -2 & -1 \\ 1 & 1 & 1 \\ -2 & 3 & 2 \end{pmatrix}$$

has rank 2 (why?) as does the matrix

$$B = \begin{pmatrix} 1 & 1 & 1 \\ -2 & 3 & 2 \end{pmatrix}$$

obtained from A by omitting the first row ($k = 1$), the complete solution of the system is given by

$$x_i = cA_{1i}, \qquad i = 1, 2, 3$$

or

$$x_1 = cA_{11}, \qquad x_2 = cA_{12}, \qquad \text{and} \qquad x_3 = cA_{13}$$

or

$$x_1 = c\begin{vmatrix} 1 & 1 \\ 3 & 2 \end{vmatrix}, \qquad x_2 = -c\begin{vmatrix} 1 & 1 \\ -2 & 2 \end{vmatrix}$$

and

$$x_3 = c\begin{vmatrix} 1 & 1 \\ -2 & 3 \end{vmatrix}$$

that is,

$$x_1 = -c, \qquad x_2 = -4c, \qquad \text{and} \qquad x_3 = 5c \qquad \text{for any } c$$

Thus the complete solution of the system is

$$(x_1, x_2, x_3) = c(-1, -4, 5)$$

Exercises

Whenever possible, solve each of the following systems in more than one way. If there is no solution, explain why. Give a geometric interpretation of each problem.

1. $4x + 2y = 5$
   $3x + 4y = 1.$

2. $$\frac{3x-2}{5}+\frac{7y+1}{10}=10$$

$$\frac{x+3}{2}-\frac{2y-5}{3}=3.$$

3. $$\frac{3}{x}-\frac{6}{y}=\frac{1}{6}$$

$$\frac{2}{x}+\frac{3}{y}=\frac{1}{2}$$

Ans. (3) $x = 6, y = 18$

4. $$\begin{aligned} 2x + y - z &= 5 \\ 3x - 2y + 2z &= -3 \\ x - 3y - 3z &= -2. \end{aligned}$$

5. $$\begin{aligned} 3x + y - 2z &= 1 \\ 2x + 3y - z &= 2 \\ x - 2y + 2z &= -10. \end{aligned}$$

6. $$\begin{aligned} 2x + y - z + t &= -4 \\ x + 2y + 2z - 3t &= 6 \\ 3x - y - z + 2t &= 0 \\ 2x + 3y + z + 4t &= -5. \end{aligned}$$

Ans. (6) $(1, -2, 3, -1)$

7. $$\begin{aligned} 2x - 3y + 4z &= 0 \\ x + y - 2z &= 0 \\ 3x + 2y - 3z &= 0. \end{aligned}$$

Ans. (7) Trivial only

8. $$\begin{aligned} 3x + 5y + 3z &= 0 \\ 2x + 3y + 3z &= 0. \end{aligned}$$

9. $$\begin{aligned} 3x + 5y + 3z + t &= 0 \\ x - y - z + 2t &= 0 \\ 5x + 3y + z + 5t &= 0. \end{aligned}$$

10. $$\begin{aligned} x - 3y &= 0 \\ x - 4y &= 1 \\ 2x - 5y &= -1. \end{aligned}$$

Ans. (10) $x = -3, y = -1$

11. $$\begin{aligned} 2x_1 + 2x_2 + 4x_3 - x_4 + 2x_5 &= 3 \\ 3x_1 + 4x_2 + 5x_3 - x_4 - 2x_5 &= 7 \\ x_1 + 3x_2 + 4x_3 + 5x_4 - x_5 &= 4. \end{aligned}$$

12.
$$\begin{aligned} x_1 - 2x_2 + x_3 - 3x_4 &= 0 \\ 2x_1 + x_2 - 3x_3 + x_4 &= 0 \\ 3x_1 + 3x_2 - 2x_3 + x_4 &= 0. \end{aligned}$$

Find the value(s) of k for which the following systems have a nontrivial solution, and solve the systems.

13.
$$\begin{aligned} 5x + ky + z &= 0 \\ x - 4ky + 3z &= 0 \\ 3x + 3ky - z &= 0 \end{aligned}$$

Ans. (13) All values of k

14.
$$\begin{aligned} 3kx + 4y + 3z &= 0 \\ kx + 2y + z &= 0 \\ 2x + 3y + 2z &= 0. \end{aligned}$$

Ans. (14) $k = 1, x = -1, y = 0, z = 1$

If possible, find the solutions of the following problems:

15. The solutions of Problems 6.1 and 6.2.

16. The solutions of the two introductory examples at the beginning of this chapter.

6.3 Inequalities

The general theory of linear inequalities and systems of linear inequalities can very easily be developed parallel to systems of linear equations. This of course is the case, since with the exception of the following three properties, equations and inequalities possess identical characteristics.[3]

Property i. If $x > y$ and $a > 0$, then $ax > ay$ and $x/a > y/a$;
Property ii. If $x > y$ and $a < 0$, then $ax < ay$ and $x/a < y/a$;
Property iii. If $x > y$ and $xy > 0$, then $1/x < 1/y$.

Thus Properties (i) and (ii) specify that when both sides of an inequality are multiplied or divided by a positive number, the order of the inequality remains the same. On the other hand if both sides of an inequality are multiplied or divided by a negative number, its order changes. For example, multiplication and division of both sides of the inequality $10 > 5$ by 3 and -3 yield

a. $(3)(10) > (3)(5)$ or $30 > 15$ (same order).
b. $(-3)(10) < (-3)(5)$ or $-30 < -15$ (order changed).
c. $10/3 > 5/3$ (same order).
d. $10/-3 < 5/-3$ (order changed).

Moreover, since $(5)(3) = 15$ and $(-4)(-10) = 40$ are both positive, Property (iii) implies that $1/5 < 1/3$ and $1/-4 < 1/-10$ whenever $5 > 3$ and $-4 > -10$.

A detailed study of inequalities constitutes in itself a very interesting and extensive branch of mathematics exceeding our particular needs and objectives. However, in order to acquire a few of the fundamental concepts in linear inequalities essential to the subsequent study of linear programming, we shall begin by analyzing some special examples. Prior to considering these examples of inequalities, we undertake the presentation of the essential steps involved in the solutions of algebraic inequalities.

Step 1. Transfer all terms of the inequality to one side.

Step 2. Perform the indicated additions and subtractions, and write the inequality in the form

$$\frac{f(x)}{g(x)} \geq 0 \qquad \text{or} \qquad \frac{f(x)}{g(x)} \leq 0$$

Step 3. Write

$$\frac{f(x)}{g(x)} \geq 0 \qquad \text{or} \qquad \frac{f(x)}{g(x)} \leq 0$$

as

$$\frac{f(x)g(x)}{g^2(x)} \geq 0 \qquad \text{or} \qquad \frac{f(x)g(x)}{g^2(x)} \leq 0$$

Then multiplication of both sides of the inequality by $g^2(x)$ reduces it to

$$f(x)g(x) \geq 0 \qquad \text{or} \qquad f(x)g(x) \leq 0 \qquad \text{(Why?)}$$

Step 4. Factor $f(x)g(x)$ completely.[3] This factorization transforms the inequality into the form

$$(x - r_1)(x - r_2) \ldots (x - r_n) \geq 0 \text{ (or } \leq 0) \tag{6.10}$$

where the factors, $x - r_i$, $i = 1, 2, \ldots, n$ are real or complex, distinct or coincident. Whatever the case may be, this last form enables us to obtain the solution set of the algebraic inequality as the following examples demonstrate.

Example 6.11

Solve the inequality $\dfrac{x+3}{2} \geq \dfrac{2-x}{3}$.

Solution First we transfer all the terms of the inequality to one side, and then perform the indicated operations. Thus we obtain

$$\frac{x+3}{2} - \frac{2-x}{3} \geq 0 \qquad \text{or} \qquad \frac{5x+5}{6} \geq 0$$

Now if we multiply both sides of the inequality by 6, we obtain $5x + 5 \geq 0$. Thus division of both sides by 5 yields the equivalent inequality

$$x + 1 \geq 0 \qquad \text{or} \qquad x \geq -1$$

Therefore the solution set of the given inequality is the set

$$S = \{x \mid x \geq -1\}$$

and its graph is shown in the following.

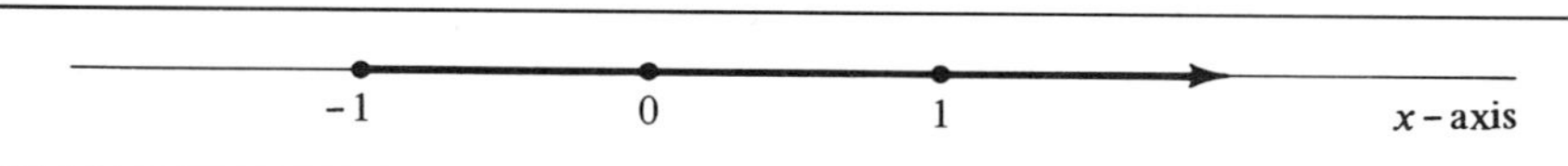

Example 6.12 Find all values of x for which

$$\frac{1}{x-1} < \frac{2}{2x-1}$$

Solution Transferring all terms to one side of the inequality yields

$$\frac{1}{x-1} - \frac{2}{2x-1} < 0$$

Thus if we perform the indicated subtraction, the inequality takes the equivalent form

$$\frac{1}{(x-1)(2x-1)} < 0 \qquad \text{or} \qquad \frac{(x-1)(2x-1)}{(x-1)^2(2x-1)^2} < 0$$

or

$$(x-1)(2x-1) < 0 \qquad \text{(Why?)}$$

To obtain the solution of the last inequality we make use of the self-explanatory Table 6.1. In the table we indicate the fact that the factor, $x - 1$, is zero at $x = 1$, positive for $x > 1$, and negative for $x < 1$. Similarly, we show that $2x - 1$ is zero at $x = \frac{1}{2}$, positive for $x > \frac{1}{2}$, and negative for $x < \frac{1}{2}$. Furthermore we indicate that the product $(x-1)(2x-1)$ is positive for $x > 1$, negative for $\frac{1}{2} < x < 1$, and again positive for $x < \frac{1}{2}$, whereas at $x = 1$ and $x = \frac{1}{2}$ the product, $(x-1)(2x-1)$, is obviously zero.

Table 6.1

x axis	$x-1$	$2x-1$	$(x-1)(2x-1)<0$	
	+	+	+	
1	0			
	−	+	−	$\frac{1}{2}<x<1$
$\frac{1}{2}$		0		
	−	−	+	

Now since we were solving for all the values of x for which $(x-1)(2x-1) < 0$, the solution set must be

$$S = \{x \mid \tfrac{1}{2} < x < 1\}$$

as shown in Table 6.1. Clearly, such tables reveal the sign of every factor in the product defining the inequality, and consequently they become very useful devices in solving any inequality that is transformed into the product form

$$(x - r_1)(x - r_2) \ldots (x - r_k) \geq 0 \qquad \text{or } \leq 0$$

Hence, using the same Table 6.1, we can easily see that the solution set for the opposite inequality

$$(x-1)(2x-1) > 0$$

is the set

$$S_1 = \{x \mid x > 1\} \cup \{x \mid x < \tfrac{1}{2}\} \qquad \text{(see Table 6.1)}$$

Example 6.13 Solve the inequality

$$\frac{2}{x-1} > \frac{2}{2x+1}$$

Solution If we should transfer all the terms of the inequality to one side and perform the indicated operations, we would obtain

$$\frac{2}{x-1} - \frac{2}{2x+1} > 0 \qquad \text{or} \qquad \frac{2x+4}{(x-1)(2x+1)} > 0$$

Thus division of both sides in the last inequality by 2 yields

$$\frac{x+2}{(x-1)(2x+1)} > 0$$

Table 6.2

x axis	$x - 1$	$x + 2$	$2x + 1$	$(x + 2)(x - 1)(2x + 1) > 0$	
	+	+	+	+	$x > 1$
1	0				
	−	+	+	−	
$-\frac{1}{2}$			0		
	−	+	−	+	$-2 < x < -\frac{1}{2}$
−2		0			
	−	−	−	−	

or equivalently

$$\frac{(x + 2)(x - 1)(2x + 1)}{(x - 1)^2(2x + 1)^2} > 0$$

or

$$(x + 2)(x - 1)(2x + 1) > 0 \qquad \text{(Why?)}$$

As in the previous example, Table 6.2 shows that the set

$$S = \{x \mid x > 1\} \cup \{x \mid -2 < x < -\tfrac{1}{2}\}$$

is the solution set of the given inequality.

Example 6.14 Solve the inequality

$$\left|\frac{1}{x - 1}\right| > \left|\frac{2}{x + 1}\right|$$

Solution When an inequality involves the absolute value, we square both sides of the inequality first, and then we follow the same four steps as in the previous cases. Therefore squaring both sides of the inequality and transferring all its terms to one side yield the equivalent inequality

$$\left(\frac{1}{x - 1}\right)^2 - \left(\frac{2}{x + 1}\right)^2 > 0$$

Now, if we perform the indicated operations, the last inequality takes the equivalent form

$$\left(\frac{1}{x-1}+\frac{2}{x+1}\right)\left(\frac{1}{x-1}-\frac{2}{x+1}\right)>0$$

or

$$\frac{3x-1}{(x-1)(x+1)}\cdot\frac{-x+3}{(x-1)(x+1)}>0$$

or

$$\frac{(3x-1)(-x+3)}{(x-1)^2(x+1)^2}>0 \qquad \text{or} \qquad (3x-1)(-x+3)>0$$

However, if both sides of the last inequality are multiplied by -1, we obtain

$$(3x-1)(x-3)<0$$

Once again, Table 6.3 establishes that the set

$$S=\{x \mid \tfrac{1}{3}<x<3\}$$

constitutes the solution set of our inequality.

The geometric meaning of inequalities and their solutions provide an intuitive appreciation of the realities involved in many physical problems. Geometric pictures of solutions reinforce the learning and retaining process. Geometrically, since any conditional inequality involving one variable x can be written in the form $f(x) \geq 0$ or $f(x) \leq 0$, the values of x such that $f(x) > 0$ are those values for which the graph of $y = f(x)$ lies above the x axis (why?).

Table 6.3

x axis	$3x-1$	$x-3$	$(3x-1)(x-3)<0$	
	+	+	+	
3		0		
	+	-	-	$\frac{1}{3}<x<3$
$\frac{1}{3}$	0			
	-	-	+	

Example 6.15 Find graphically the values of x for which $x^2 > 2x + 3$.

Solution Consider $x^2 - 2x - 3 > 0$, and let

$$f(x) = x^2 - 2x - 3$$

Then we are looking for the set of all x such that

$$y = f(x) = x^2 - 2x - 3 > 0$$

The graph of

$$y = x^2 - 2x - 3$$

is presented in Figure 6.1.

From the graph we can see that $y = f(x) > 0$ for all $x > 3$ and for all $x < -1$ or that the solution set of $x^2 > 2x + 3$ is

$$S = \{x \mid x > 3\} \cup \{x \mid x < -1\}$$

The graph of S is indicated in Figure 6.1.

Subsequently in working with either single inequalities or systems of inequalities and their solution sets as subsets of the plane we shall recognize that at times these subsets are significantly different, and it is because of these differences that we emphasize Definition 6.2.

DEFINITION 6.2 A set of points S is said to be convex if and only if any two points of the set determine a segment that lies entirely in S.

Figure 6-1

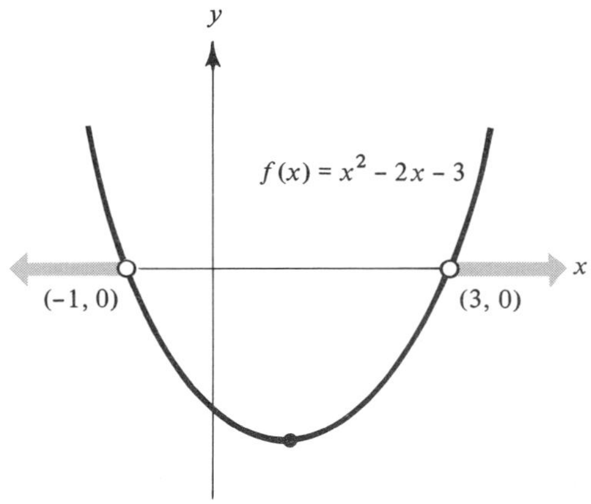

Figure 6-2

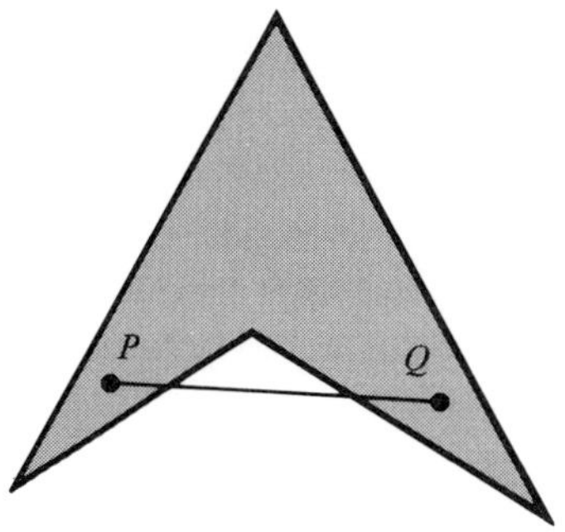

Figure 6-3

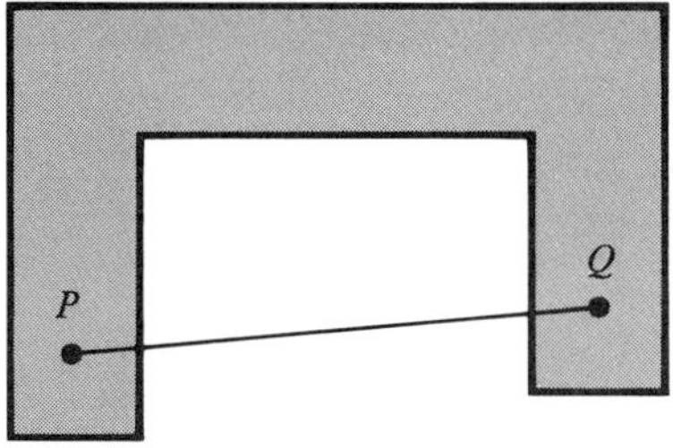

Thus the interiors of circles, triangles, and squares constitute convex sets of points. On the other hand the sets in Figures 6.2 and 6.3 are nonconvex sets, since the segment PQ does not lie in the set.

An inequality involving two variables, x and y, describes a subset of the plane. For example, the line $x + y = 2$ separates the rest of the plane, that is, all the points in the plane not in the line into two convex sets[4] called *half planes*, namely, the half plane H_1 consisting of all points (x, y) such that $x + y > 2$, and the half plane H_2 consisting of all points (x, y) such that $x + y < 2$ (Figure 6.4).

Likewise, the circle $x^2 + y^2 = 9$ separates the rest of the plane, that is, all points in the plane not on the circle, into two sets of points (regions), namely, the set

$$R_1 = \{(x, y) \mid x^2 + y^2 < 9\}$$

and the set

$$R_2 = \{(x, y) \mid x^2 + y^2 > 9\}$$

These two sets are graphically shown in Figure 6.5.

Figure 6–4

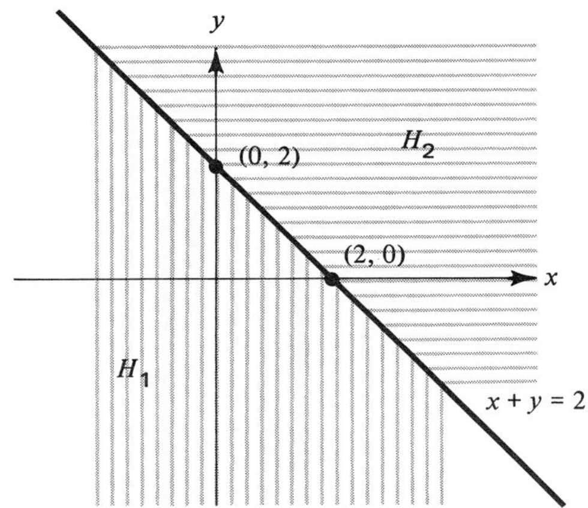

It is clear that the regions R_1 and R_2 of this example are not both convex sets. In fact, R_1 is a convex set, whereas R_2 is not (why?).

Just as a single inequality in two variables describes a subset of the plane, a system of inequalities describes a more restricted subset of the plane. Specifically, in linear systems of inequalities the solution set is always a convex subset of the plane, whereas that of nonlinear systems of inequalities need not be. We illustrate these ideas in the following examples.

Example 6.16 Shade the part of the plane described by the following

Figure 6–5

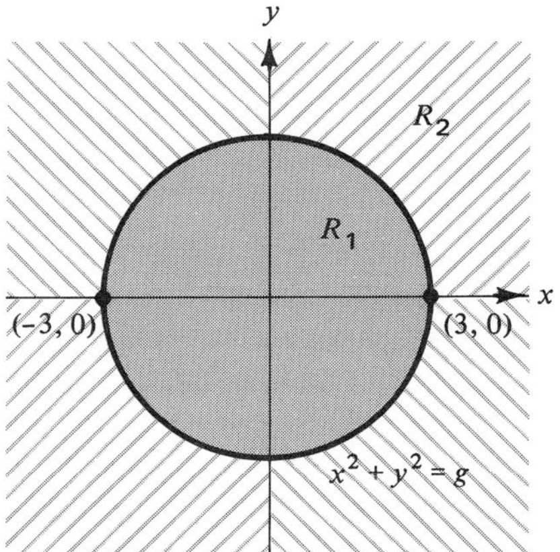

system of inequalities

$$x + y < 1$$
$$x - y > -1$$
$$x - 5y < 7$$

Solution In Figure 6.6 we graph the lines

$$x + y = 1, \qquad x - y = -1, \qquad \text{and} \qquad x - 5y = 7$$

Then we determine the half planes H_1, H_2, and H_3 described by the inequalities

$$x + y < 1, \qquad x - y > 1, \qquad \text{and} \qquad x - 5y < 7,$$

respectively. The coordinates of the points A, B, and C satisfy the simultaneous systems of equations

$$\left.\begin{aligned} x + y &= 1 \\ x - y &= -1 \end{aligned}\right\}, \qquad \left.\begin{aligned} x - y &= -1 \\ x - 5y &= 7 \end{aligned}\right\}, \qquad \text{and} \qquad \left.\begin{aligned} x + y &= 1 \\ x - 5y &= 7 \end{aligned}\right\}$$

respectively. Therefore, the solutions of these systems yield that

$$A = (0, 1), \qquad B = (-3, -2), \qquad \text{and} \qquad C = (2, -1)$$

Hence the solution set of this system of linear inequalities is the set

$$S = H_1 \cap H_2 \cap H_3$$

shown as the interior of the triangle ABC in Figure 6.6.

Figure 6-6

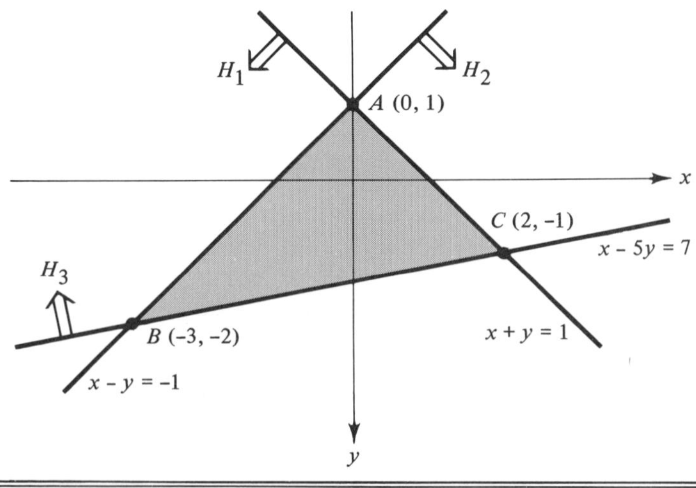

The points $B(-3, -2)$, $C(2, -1)$, and $A(0, 1)$ of the previous example shall be called *corner points*, or *extreme points*, of the triangular solution set S. As we have seen, these kinds of points can always be found by considering the equations corresponding to the inequalities of any system of inequalities and solving these equations in pairs or triples or whatever number of equations is required.

Example 6.17 Shade the part of the plane described by the following system of inequalities.

$$x - y \leq 0$$

$$2x - y \geq -1$$

$$2x + 3y \leq 6$$

Solution We consider first the corresponding equations

$$x - y = 0, \qquad 2x - y = -1, \qquad \text{and} \qquad 2x + 3y = 6$$

and solve them in pairs such as

$$(2x - y = -1, \qquad x - y = 0)$$

$$(x - y = 0, \qquad 2x + 3y = 6)$$

and

$$(2x - y = 1, \qquad 2x + 3y = 6)$$

Figure 6–7

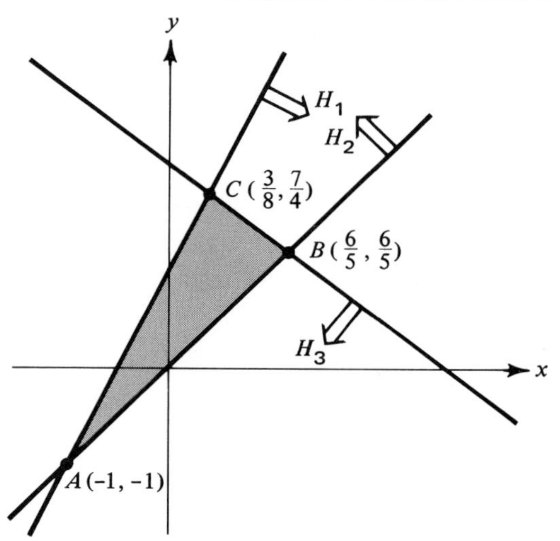

The solutions of these pairs are the so-called *corner* or *extreme points*

$$A(-1, -1), \qquad B\left(\frac{6}{5}, \frac{6}{5}\right), \qquad \text{and} \qquad C\left(\frac{3}{8}, \frac{6}{4}\right)$$

respectively (Figure 6.7).

The solution of this system of inequalities is the shaded interior and perimeter of the triangle ABC, that is, the intersection of the half planes H_1, H_2, and H_3 shown in Figure 6.7.

Exercises

Graphically, find the values of x for which the following inequalities are true, and graph the solution set for each inequality.

1. $x^2 + 3x > 4$.
2. $x^2 + 10 < 7x$.
3. $x^2 > x + 1$.
4. $x^2 + 5x < -6$.

Shade the part of the plane described by the following inequalities or systems of inequalities. What are the *corner* or *extreme* points in Problems 8 through 15?

5. $4y < x^2$.
6. $x^2 + y^2 > 1$.
7. $x^2 + 9y^2 < 1$.
8. $y > x^2, \quad x + y < 1$.
9. $x^2 + y^2 \leq 1, \quad y \geq x, \quad x + y \geq -1$.
10. $y < \sqrt{x}, \quad x^2 + y^2 < 4$.
11. $y < \sqrt{x}, \quad y > x^2$.
12. $-4x + y \leq 1$, $-x + y \geq 2$, $x + y \leq 4$.
13. $-3x + y \leq 3$, $x + y \geq 0$, $y \geq 0$.
14. $3x + 6y \geq 6$, $x \geq 0$, $y \geq 0$.
15. $2x + y \geq -1$, $-x + y \leq 4$, $x \leq 2$.

Find the values of x for which each of the following inequalities is true. Graph the solution for each inequality.

16. $\dfrac{x+3}{2} \geq \dfrac{2-x}{-2}$.
17. $\dfrac{4x-3}{2} \geq \dfrac{2-x}{3}$.

Ans. $x \geq 13/14$

18. $\dfrac{2}{2x - 1} < \dfrac{1}{x + 2}.$

19. $\dfrac{2}{x + 1} > \dfrac{1}{3x - 1}.$

Ans. $x > 3$ and $\frac{1}{3} < x < -1$

20. $\dfrac{x^2 - 1}{4x^2 - 1} > 0$

21. $\dfrac{3x^2 - 4}{x^2 + x + 1} < 0$

Ans. (21) $x > (2\sqrt{3})/3$ and $x < -(2\sqrt{3})/3$

22. $(x - 1)(x + 2)(x + 3)(2x - 1)(3x + 4) > 0$

23. $\left|\dfrac{2}{2x - 1}\right| > \left|\dfrac{3}{x - 3}\right|$

24. $\left|\dfrac{x}{3x + 1}\right| < \left|\dfrac{x}{x - 2}\right|$

Ans. (24) $x > 1/4$ and $x < -3/2$

6.4 Linear Transformations and the Characteristic Value

In Section 6.1 we discussed the theory of linear systems $AX = B$ where A is a constant matrix, B a constant column vector, and X the vector of unknowns. These systems frequently appear as

$$\begin{aligned} a_{11}x_1 + a_{12}x_2 + \cdots + a_{1n}x_n &= y_1 \\ a_{21}x_1 + a_{22}x_2 + \cdots + s_{2n}x_n &= y_2 \\ \cdots\cdots\cdots\cdots\cdots\cdots\cdots\cdots \\ a_{m1}x_1 + a_{m2}x_2 + \cdots + a_{mn}x_n &= y_m \end{aligned} \qquad \textbf{(6.11)}$$

or simply in the matrix form

$$AX = Y$$

where A is a constant matrix and both X and Y are variable (or unknown) vectors. When this is the case, we say that the matrix A transforms or maps the vector X into the vector Y or that the equation $AX = Y$ defines a linear transformation[1] with matrix A. This transformation shall be denoted by T_A, and the vector Y shall be called the image of X under T_A. Clearly, the inverse of a linear transformation determined by a matrix A exists if and only if A^{-1} exists; that is, the vector X can be obtained in terms of its image Y (or $X = A^{-1}Y$) by simply multiplying, from the left, both sides of the equation $AX = Y$ by A^{-1}.

DEFINITION 6.3 The linear transformation $Y = AX$ is said to be nonsingular if and only if its matrix A has an inverse. Otherwise, the transformation is called *singular*.

Now, suppose that T_A and T_B are the linear transformations $Y = AX$ and $Z = BY$, respectively. Then T_A transforms a vector X into a vector Y, and T_B maps or transforms a vector Y into some vector Z. Moreover, if T_A is applied to a vector X, and then T_B is applied to the resulting vector Y, the vector X is transformed into the vector Z. The equation of the resultant transformation T_{BA}, equivalent to following T_A by T_B, is obtained by eliminating Y between $Y = AX$ and $Z = BY$. This is done by substituting AX for Y in the equation $Z = B(AX)$ or $Z = (BA)X$, which establishes the following important theorem.

Theorem 6.10 The result of following a linear transformation T_A: $Y = AX$ with the linear transformation T_B: $Z = BY$ is the resultant transformation T_{BA}: $Z = (BA)X$ whose matrix is the product BA.

In routine business and marketing situations linear transformations are utilized in transferring (or mapping) price vectors into revenue vectors. For instance, if a manufacturer produces the commodities $C_1, C_2, \ldots, C_n$, if he sells them in the markets $M_1, M_2, \ldots, M_n$ with annual sales volumes given by the matrix

$$A = \begin{array}{c} \\ M_1 \\ M_2 \\ \vdots \\ M_k \end{array} \begin{array}{c} \begin{array}{ccc} C_1 & C_2 \cdots & C_n \end{array} \\ \begin{pmatrix} a_{11} & a_{12} \cdots & a_{1n} \\ a_{21} & a_{22} \cdots & a_{2n} \\ \cdots & \cdots & \cdots \\ a_{k1} & a_{k2} \cdots & a_{kn} \end{pmatrix} \end{array}$$

where a_{ij} designates the annual sales volume of commodity C_j in market M_i for each $i = 1, 2, \ldots, k$ and $j = 1, 2, 3, \ldots, n$, if the unit sales prices are given by the price vector

$$\begin{array}{ccc} C_1 & C_2 \cdots & C_n \end{array}$$
$$X = (x_1, x_2, \ldots, x_n)^t$$

and if the total revenue in each market is specified by the revenue vector

$$\begin{array}{ccc} M_1 & M_2 \cdots & M_k \end{array}$$
$$Y = (Y_1 \quad Y_2 \ldots, Y_k)^t$$

then the price vector is always transformed by the matrix A into the revenue vector Y via the linear transformation

$$Y = AX$$

Specifically, if a manufacturer produces three products P_1, P_2, and P_3 which he sells in two markets M_1 and M_2, if the annual sales volumes are indicated by the matrix

$$A = \begin{array}{c} \\ M_1 \\ M_2 \end{array} \begin{array}{c} \begin{array}{ccc} P_1 & P_2 & P_3 \end{array} \\ \begin{pmatrix} 10{,}000 & 3{,}000 & 16{,}000 \\ 6{,}000 & 2{,}000 & 8{,}000 \end{pmatrix} \end{array}$$

and if the unit sales prices are given by the price vector

$$X = \begin{array}{ccc} P_1 & P_2 & P_3 \\ (\$2.00 & \$1.50 & \$1.00)^t \end{array}$$

then the total revenue vector in each market is obtained as the image of X under the linear transformation determined by the matrix A; that is, the revenue vector is given by

$$Y = AX \text{ or } Y = \begin{pmatrix} 10{,}000 & 3{,}000 & 16{,}000 \\ 6{,}000 & 2{,}000 & 8{,}000 \end{pmatrix} \begin{pmatrix} \$2.00 \\ \$1.50 \\ \$1.00 \end{pmatrix}$$

$$= \begin{pmatrix} \$40{,}500 \\ \$23{,}000 \end{pmatrix}$$

Hence the total revenues from markets M_1 and M_2 are \$40,500 and \$23,000, respectively.

In factor analysis the matrix $Y = (y_{ij})_{t\times n}$ of observed scores is obtained as the image of the matrix $X = (x_{ij})_{q\times n}$ of standard scores under the linear transformation defined by the matrix $A = (a_{ij})_{t\times q}$ of factor loadings as follows.

Let each of n subjects be tested on t tests. Let the score y_{ji} for subject i on test j depend on both the degree to which the subject i possesses certain ability factors and the extent to which the test measures these factors. More precisely, let the equation

$$y_{ji} = a_{j1}x_{1i} + a_{j2}x_{2i} + \cdots + a_{jq}x_{qi} \tag{6.12}$$

where a_{jk}, $k = 1, 2, \ldots, q$, denotes the degree to which test j measures factor k, and x_{ki} the standard score of individual i on factor k. In matrix form Equation (6.12) becomes

$$Y = AX \tag{6.13}$$

Now, by scaling the observed scores and standard scores so that

$$\sum_{i=1}^{n} y_{ji} = 0, \qquad \frac{1}{n}\sum_{i=1}^{n} y_{ji}^2 = 1 \qquad \text{for each } \quad = 1, 2, \ldots, t$$

$$\sum_{i=1}^{n} x_{ki} = 0, \qquad \frac{1}{n}\sum_{i=1}^{n} x_{ki}^2 = 1 \qquad \text{for each } k = 1, 2, \ldots, q \tag{6.14}$$

the elements in each row of Y and X add to zero, and the sum of their squares is equal to n.

The correlation r_{kl} between tests k and l is defined to be

$$r_{kl} = \frac{1}{n}\sum_{i=1}^{n} y_{ki}y_{li} \tag{6.15}$$

and the correlation z_{kl} between factors k and l is defined as

$$z_{kl} = \frac{1}{n}\sum_{i=1}^{n} x_{ki}x_{li} \tag{6.16}$$

Thus the matrix forms of Equations (6.15) and (6.16) are

$$R = \frac{1}{n}YY^t \qquad \text{and} \qquad Z = \frac{1}{n}XX^t, \tag{6.17}$$

respectively, where $R = (r_{ij})_{t\times t}$ is the matrix of interest correlations and $Z = (z_{ij})_{q\times q}$ is the matrix of interfactor correlations. Hence, substituting Equation (6.13) into Equation (6.17) we obtain

$$R = \frac{1}{n}YY^t = \frac{1}{n}AXX^tA^t = A\left(\frac{1}{n}XX^t\right)A^t = AZA^t$$

or

$$R = AZA^t$$

However, since the ideal situation occurs when correlations between different factors are zero, that is, when the factors are uncorrelated and $Z = I$, we have

$$R = AZA^t \qquad \text{or} \qquad R = AIA^t$$

or

$$R = AA^t$$

as the basic equation of factor analysis.

Furthermore, in input-output analysis if an economic system

contains n industries each producing a single commodity and if every commodity enters the system as an input, that is, if we assume that the unit output (production) of the jth commodity requires a fixed quantity a_{ij} of the ith commodity as an input then these inputs a_{ij} constitute the $n \times n$ *input-output* matrix

$$A = \underbrace{\begin{bmatrix} a_{11} & a_{12} \cdots a_{1n} \\ a_{21} & a_{22} \cdots a_{2n} \\ \cdots & \cdots \\ a_{n1} & a_{n2} \cdots a_{nn} \end{bmatrix}}_{\substack{1 \quad 2\,3 \cdots n \\ \text{output commodities}}} \left.\begin{matrix} 1 \\ 2 \\ \cdot \\ \cdot \\ \cdot \\ n \end{matrix}\right\} \text{input commodities}$$

The entries in the jth column of A $(j = 1, 2, \ldots, n)$ specify the input requirements for production of one unit of the jth commodity. For example, the second column of the input-output matrix

$$A = \underbrace{\begin{pmatrix} \frac{2}{3} & 0 & \frac{1}{3} \\ \frac{1}{2} & \frac{1}{5} & 0 \\ 0 & \frac{2}{5} & \frac{1}{2} \end{pmatrix}}_{\substack{1 \quad 2 \quad 3 \\ \text{output commodities}}} \left.\begin{matrix} 1 \\ 2 \\ 3 \end{matrix}\right\} \begin{matrix} \text{input} \\ \text{commodities} \end{matrix}$$

shows that the production of one unit of commodity 2 requires no units of commodity 1, $\frac{1}{5}$ units of commodity 2, and $\frac{2}{5}$ units of commodity 3. Similarly, we may determine the requirements for commodities 1 and 3 from the other two columns of A. Since x units $(j = 1, 2, \ldots, n)$ of production (or output) of the jth industry require $a_{ij}x_j$ units of the ith commodity, the total quantity of required output i is the image Y of X under the transformation determined by the input-output matrix A; that is,

$$Y = AX$$

where A is the input-output matrix, and $X = (x_1, x_2, \ldots, x_n)^t$ and $Y = (y_1, y_2, \ldots, y_n)^t$ are the vectors of input requirements and outputs, respectively.

DEFINITION 6.4 An economic system is said to be closed if and only if there are no input services outside the system.

Since in a closed economic system no input sources exist outside the system, it is clear that such a system cannot function when an input requirement y_i is greater than an output x_i. Therefore we say that a closed economic system is viable if and only if

$$Y = AX \leq X$$

where $Y \leq X$ means the components of the vector Y are less or equal to the corresponding components of the vector X. Furthermore, we say that a vector X is an interior equilibrium of the system if and only if it satisfies the equilibrium equation

$$AX = X \quad \text{and} \quad X \geq 0 \tag{6.18}$$

The equilibrium equation $AX = X$ is a special case of the more general equation

$$AX = kX \tag{6.19}$$

which specifies that a vector X maps, under T_A: $Y = AX$, into a scalar multiple of itself, that is, into a vector parallel to itself, or that the direction of the vector X remains the same (invariant) under the linear transformation T_A.

DEFINITION 6.5 The values of k for which there exist nonzero vector(s) X such that $AX = kX$ under T_A are called characteristic or eigenvalues of A (or T_A). The solution vector(s) of $AX = kX$ for each value k_i of k is called the characteristic or eigenvector(s) of A corresponding to k_i. Finally, the equation $|A - kI| = 0$, satisfied by the eigenvalues, is called the characteristic equation of A (or T_A).

Since the linear system of equations $AX = kX$ or equivalently

$$(A - kI)X = 0$$

has nonzero (nontrivial) solutions if and only if $|A - kI| = 0$, the roots of the characteristic equation $|A - kI| = 0$ of a matrix A constitute the characteristic or eigenvalues of the matrix A.

Example 6.18 Given the matrix

$$A = \begin{pmatrix} 2 & -3 \\ 0 & -4 \end{pmatrix}$$

determine the following about the linear transformation, T_A: $Y = AX$, specified by this matrix:

a. The image of the vectors $(1, 2)^t$ and $(x_1, x_2)^t$.
b. The vector X in terms of Y.
c. The characteristic values and the characteristic vectors of A.

Solution

(a) Since the image of any vector X under T_A is the vector Y obtained from $Y = AX$, the images of $(1, 2)^t$ and $(x_1, x_2)^t$ must be

$$\begin{pmatrix} 2 & -3 \\ 0 & -4 \end{pmatrix} \begin{pmatrix} 1 \\ 2 \end{pmatrix} = \begin{pmatrix} -4 \\ -8 \end{pmatrix}$$

and

$$\begin{pmatrix} 2 & -3 \\ 0 & -4 \end{pmatrix} \begin{pmatrix} x_1 \\ x_2 \end{pmatrix} = \begin{pmatrix} 2x_1 - 3x_2 \\ - 4x_2 \end{pmatrix}$$

respectively.

(b) Since the inverse of A is

$$A^{-1} = \begin{pmatrix} \frac{4}{8} & -\frac{3}{8} \\ 0 & -\frac{2}{8} \end{pmatrix} \qquad \text{(Why?)}$$

and since $X = A^{-1}Y$, we have

$$X = \begin{pmatrix} x_1 \\ x_2 \end{pmatrix} = A^{-1}Y = \begin{pmatrix} \frac{4}{8} & -\frac{3}{8} \\ 0 & -\frac{2}{8} \end{pmatrix} \begin{pmatrix} y_1 \\ y_2 \end{pmatrix}$$

$$\begin{pmatrix} x_1 \\ x_2 \end{pmatrix} = \begin{pmatrix} \frac{4}{8}y_1 - \frac{3}{8}y_2 \\ -\frac{2}{8}y_2 \end{pmatrix}$$

or

$$x_1 = \tfrac{4}{8}y_1 - \tfrac{3}{8}y_2 \qquad \text{and} \qquad x_2 = -\tfrac{2}{8}y_2$$

Thus,

$$X = (x_1, x_2) = (\tfrac{4}{8}y_1 - \tfrac{3}{8}y_2, - \tfrac{2}{8}y_2)$$

(c) In order to find both the characteristic values and the characteristic vectors we must solve the equation

$$AX = kX \qquad \text{or} \begin{pmatrix} 2 & -3 \\ 0 & -4 \end{pmatrix} \begin{pmatrix} x_1 \\ x_2 \end{pmatrix} = k \begin{pmatrix} x_1 \\ x_2 \end{pmatrix} \tag{6.20}$$

This equation, $AX = kX$, can be written as $AX - kX = 0$ or

$$(A - kI)X = 0 \tag{6.21}$$

where I designates the identity matrix. Now Equation (6.21) constitutes a homogeneous linear system of two equations in two unknowns which has nontrivial solutions if and only if the determinant $|A - kI|$ of coefficients is zero; that is, Equation (6.21) has nontrivial solution(s) for each of the roots of the characteristic equation

$$|A - kI| = \left| \begin{pmatrix} 2 & -3 \\ 0 & -4 \end{pmatrix} - k \begin{pmatrix} 1 & 0 \\ 0 & 1 \end{pmatrix} \right| = 0$$

or

$$\begin{vmatrix} 2 - k & -3 \\ 0 & -4 - k \end{vmatrix} = 0 \qquad \text{or} \qquad (2 - k)(-4 - k) = 0$$

Therefore the characteristic values of this matrix are $k = 2$ and $k = -4$. Finally, to obtain the corresponding characteristic vectors it becomes necessary to solve Equation (6.21) for each value of k; that is, we must solve

$$(A - 2I)X = 0 \qquad \text{and} \qquad (A + 4I)X = 0$$

or

$$\begin{pmatrix} 2 - 2 & -3 \\ 0 & -4 - 2 \end{pmatrix} \begin{pmatrix} x_1 \\ x_2 \end{pmatrix} = \begin{pmatrix} 0 \\ 0 \end{pmatrix}$$

and

$$\begin{pmatrix} 2 + 4 & -3 \\ 0 & -4 + 4 \end{pmatrix} \begin{pmatrix} x_1 \\ x_2 \end{pmatrix} = \begin{pmatrix} 0 \\ 0 \end{pmatrix}$$

$$0x_1 - 3x_2 = 0$$

$$0x_1 - 6x_2 = 0 \tag{6.22}$$

and

$$6x_1 - 3x_2 = 0$$

$$0x_1 + 0x_2 = 0 \tag{6.23}$$

From the system of Equations (6.22) we obtain $x_2 = 0$ and $x_1 = a$ (any number). On the other hand, the system of Equations (6.23) yields $x_1 = b$ (any number) and $x_2 = 2b$. Thus the characteristic vectors corresponding to $k = 2$ and $k = -4$ are $X = (a, 0)$ and $X = (b, 2b)$, respectively, or simply $X = (1, 0)$ and $X = (1, 2)$ in their representative form.

Example 6.19 Show that the vectors $X = (1, 1, 1)^t$ and $Y = (2, 2, 2)^t$ are both interior equilibrium vectors for the closed economic system with input-output matrix

$$A = \begin{pmatrix} 0.2 & 0.3 & 0.5 \\ 0.4 & 0.4 & 0.2 \\ 0.1 & 0.5 & 0.4 \end{pmatrix}$$

Solution Since $AX = X$ and $AY = Y$, that is, since

$$\begin{pmatrix} 0.2 & 0.3 & 0.5 \\ 0.4 & 0.4 & 0.2 \\ 0.1 & 0.5 & 0.4 \end{pmatrix} \begin{pmatrix} 1 \\ 1 \\ 1 \end{pmatrix} = \begin{pmatrix} 1 \\ 1 \\ 1 \end{pmatrix}$$

and

$$\begin{pmatrix} 0.2 & 0.3 & 0.5 \\ 0.4 & 0.4 & 0.2 \\ 0.1 & 0.5 & 0.4 \end{pmatrix} \begin{pmatrix} 2 \\ 2 \\ 2 \end{pmatrix} = \begin{pmatrix} 2 \\ 2 \\ 2 \end{pmatrix}$$

as matrix multiplication verifies, the vectors X and Y are indeed interior equilibrium vectors for this system. In fact, matrix multiplication alone establishes that every three-dimensional vector whose components are positive and equal constitutes an interior equilibrium for this system.

Although the idea of characteristic values and vectors of a matrix was introduced in connection with problems in input-output analysis and with those of the invariant direction of vectors under a linear transformation, these concepts have many other important applications in the algebra of matrices and its utilizations as suggested by the following partial list of theorems.

Theorem 6.11 A matrix is singular if and only if at least one of its characterisic values is zero.

Theorem 6.12 If the matrices A and B are similar, then they have the same characteristic values.

Theorem 6.13 A characteristic vector of a square matrix cannot correspond to two distinct characteristic values.

Theorem 6.14 The characteristic values of any $n \times n$ real symmetric matrix are real numbers, and the matrix possesses n

linearly independent characteristic vectors. If X_i and X_j are the characteristic vectors corresponding to the distinct characteristic values k_i and k_j of this matrix, then X_i and X_j are perpendicular for all $i \neq j$.

Exercises

1. Find the characteristic values and the corresponding characteristic vectors for each of the following matrices.

a. $\begin{pmatrix} 4 & 6 & 6 \\ 1 & 3 & 2 \\ -1 & -5 & -2 \end{pmatrix}$ b. $\begin{pmatrix} 11 & -4 & -7 \\ 7 & -2 & -5 \\ 10 & -4 & -6 \end{pmatrix}$

c. $\begin{pmatrix} 4 & 6 & 6 \\ 1 & 3 & 2 \\ -1 & -4 & -3 \end{pmatrix}$ d. $\begin{pmatrix} -4 & 5 & 5 \\ -5 & 6 & 5 \\ -5 & 5 & 6 \end{pmatrix}$

e. $\begin{pmatrix} -1 & 4 \\ 1 & 2 \end{pmatrix}$ f. $\begin{pmatrix} -1 & 5 \\ 1 & 3 \end{pmatrix}$

Ans. (a) $k_1 = 1$, $k_2 = 2$ (double), $X_1 = (4, 1, -3)$,
$X_2 = (3, 1, -2)$

(b) $k = 0, 1, 2$ and $X = (1, 1, 1), (1, -1, 2), (2, 1, 2)$

(c) $k = -1, 1, 4$, and $X = (6, 2, -7), (0, 1, -1)$,
$(3, 1, -1)$

(d) $k_1 = 1$ (double), $k_2 = 6$, $X_1 = (1, 1, 0), (1, 0, 1)$,
$X_2 = (1, 1, 1)$

2. Using the matrix of Exercise 1(b), illustrate Theorem 6.11.

3. Using the similar matrices

$$A = \begin{pmatrix} 2 & 0 & 0 \\ 0 & 4 & 0 \\ 0 & 0 & 6 \end{pmatrix} \quad \text{and} \quad B = \begin{pmatrix} 5 & -2 & -1 \\ -1 & 4 & -1 \\ 1 & -2 & 3 \end{pmatrix},$$

illustrate Theorem 6.12.

4. Using the real symmetric matrix

$$A = \begin{pmatrix} 1 & 0 & 0 \\ 0 & 2 & 0 \\ 0 & 0 & 3 \end{pmatrix}$$

illustrate Theorem 6.14.

5. Show that the vector $(1, 2, 3)^t$ is an interior equilibrium vector for the closed economic system with input-output matrix

$$A = \begin{pmatrix} 0.2 & 0.4 & 0 \\ 0.1 & 0.5 & 0.3 \\ 0.3 & 0.3 & 0.7 \end{pmatrix}.$$

6. Given the linear transformation

$$\begin{aligned} y_1 &= 3x_1 - 7x_2 - 5x_3 \\ y_2 &= 2x_1 + 4x_2 + 3x_3 \\ y_3 &= x_1 + 2x_2 + 2x_3 \end{aligned}$$

find

a. The matrix of the transformation.
b. The image of the vector $(-3, 1, 1)^t$.
c. The characteristic values and vectors of this transformation.
d. Solve for x_1, x_2, and x_3 in terms of y_1, y_2, and y_3.

7. Show that the characteristic equation of the matrix

$$A = \begin{pmatrix} 1 & 2 \\ 2 & 1 \end{pmatrix}$$

is $k^2 - 2k - 3 = 0$. Moreover, show that $A^2 - 2A - 3I = 0$, that is, A satisfies its own characteristic equation. This result is true for all square matrices and constitutes the very important theorem known as the "Cayley-Hamilton theorem" which states that "Every matrix satisfies its own characteristic equation."

6.5 Markov Chains

Probabilistic models of independent trials do not always provide sufficient methods to analyze managerial, social, and behavioral processes. In fact, as we shall subsequently demonstrate, there exists a wide variety of applications related to mathematical models of dependent trials. For instance, the so-called *Markov-chain* models are based on dependent trials,[5] and although probabilistic in nature, they turn out to be mathematically very deterministic since mathematics employed in their analysis can also be used to study many deterministic models. Specifically, the following two examples present typical realizations of Markov-chain processes.

Example 6.20 In a T maze a rat may turn right and obtain food or may turn left and receive an electric shock. Naturally we assume that the events "turning right" and "turning left" are equiprobable. Furthermore, we hypothesize that once it obtained food on one trial, the rat's probabilities of turning right and left become .7 and .3, respectively, on the next trial, and if the rat was shocked on a certain trial, its new probabilities of turning right or left become .8 and .2.

The movements of the rat can easily be described by the vector

$$\begin{matrix} & R & L \\ X = & (\frac{1}{2}, & \frac{1}{2}) \end{matrix}$$

and the matrix

$$P = \begin{matrix} & R & L \\ R & .7 & .3 \\ L & .8 & .2 \end{matrix} \quad \text{i.e.} \quad P = \begin{pmatrix} .7 & .3 \\ .8 & .2 \end{pmatrix}$$

where R and L designate right and left turns of the rat, respectively, and the components of the vector X represent the initial probabilities in the rat's movements. The entries .7 and .3 in the first row of P are the probabilities that the rat will turn right and left if it obtained food in the previous trial. Similarly, the entries .8 and .2 in the second row of P represent the probability that the rat will turn right or left if it was shocked in the preceding trial.

Example 6.21 Three toy manufacturers A, B, and C decide to introduce a new toy at a time when they will share the toy market equally, and the following are the first-year projections:

i. A retains 40% of his customers and loses 30% to B and 30% to C.
ii. B retains 30% of his customers and loses 60% to A and 10% to C.
iii. C retains 30% of his customers and loses 60% to A and 10% to B.

If this trend continues, determine what share of the market each manufacturer will have at the end of

a. Two years.
b. Three years.
c. In the long-run.

Clearly, the initial distribution of the market can be described by the vector

$$X = \begin{matrix} A & B & C \\ (\frac{1}{3} & \frac{1}{3} & \frac{1}{3}) \end{matrix}$$

where the component $\frac{1}{3}$ represents the probability that each manufacturer intially has one third of the market. On the other hand the entries of the matrix

$$P = \begin{matrix} & A & B & C \\ A & .4 & .3 & .3 \\ B & .6 & .3 & .1 \\ C & .6 & .1 & .3 \end{matrix}$$

are the following probabilities:

a. The entries .4, .3, and .3 in the first row of P are the probabilities that A retains 40% of the market and loses 30% to each of B and C.
b. The entries .6, .3, and .1 in the second row of P represent the probabilities that B loses 60% of the market to A, retains 30%, and loses 10% to C.
c. The elements .6, .1, and .3 in the last row of P designate the probabilities that C loses 60% to A and 10% to B, while he retains 30% of the market, respectively.

Subsequently we shall demonstrate that XP^2 and XP^3 are the vectors that determine the share of the market possessed by each manufacturer at the end of two and three years, respectively. On the other hand the solution vector of the system $X = PX$ designates their share of the market in the long run. Vectors and matrices such as those encountered in the above two examples are quite useful, and we define them as follows.

DEFINITION 6.6 We say that a row vector

$$X = (x_1, x_2, \ldots, x_n)$$

is a probability vector if and only if $x_i \geq 0$ for all $i = 1, 2, \ldots, n$ and

$$x_1 + x_2 + \cdots + x_n = 1$$

DEFINITION 6.7 We say that a square matrix

$$P = (P_{ij})$$

is a stochastic matrix if and only if each of its rows is a probability vector.

Obviously, the vectors X and the matrices P of the previous two examples are probability vectors and stochastic matrices, respectively. Moreover, any probability vector

$$X = (x_1, x_2, \ldots, x_n)$$

can be represented by the vector $(x_1, x_2, \ldots, x_{n-1}, 1 - x_1 - x_2 - x_{n-1})$ which contains only $n - 1$ unknown components rather than n. Of course this is the case since

$$x_1 + x_2 + \cdots + x_{n-1} + x_n = 1$$

and consequently any of the components can be expressed in terms of the others. In particular, the two- and three-dimensional probability vectors (x, y) and (x, y, z) can be represented by $(x, 1 - x)$ and $(x, y, 1 - x - y)$, respectively. To demonstrate further the concepts of probability vectors and stochastic matrices we consider the following examples.

Example 6.22 Determine whether or not the following are probability vectors, and in each probability vector replace one of its components in terms of the others.

a. $A = (\frac{1}{2}, 0, \frac{1}{4}, \frac{1}{4})$
b. $B = (\frac{2}{3}, \frac{1}{2}, 0, -\frac{1}{3})$
c. $C = (\frac{1}{3}, \frac{1}{3}, \frac{1}{3})$
d. $D = (3/11, 0, 2/11, 1/11, 6/11)$

Solution

a. Since the components of A are nonnegative and add up to one, A is a probability vector. Moreover, A can be written as

$$A = (\tfrac{1}{2}, 0, \tfrac{1}{4}, 1 - \tfrac{1}{2} - \tfrac{1}{4}) \quad \text{or} \quad A = (1 - \tfrac{1}{4} - \tfrac{1}{4}, 0, \tfrac{1}{4}, \tfrac{1}{4}) \quad \text{etc.}$$

b. Since the fourth component of B is negative, B cannot be a probability vector.

c. Since all components of C are nonnegative with sum equal to one, C is a probability vector. In fact, C can be represented by any one of the following vectors:

$$C = (\tfrac{1}{3}, \tfrac{1}{3}, 1 - \tfrac{1}{3} - \tfrac{1}{3})$$

$$C = (\tfrac{1}{3}, 1 - \tfrac{1}{3} - \tfrac{1}{3}, \tfrac{1}{3})$$

or

$$C = (1 - \tfrac{1}{3} - \tfrac{1}{3}, \tfrac{1}{3}, \tfrac{1}{3})$$

d. Although the components of D are nonnegative, D cannot be a probability vector, since the sum $12/11$ of its components is greater than one.

Example 6.23 Are the following matrices stochastic?

$$A = \begin{pmatrix} \frac{1}{3} & \frac{1}{3} & \frac{1}{3} \\ \frac{2}{5} & \frac{1}{5} & \frac{2}{5} \end{pmatrix} \qquad B = \begin{pmatrix} \frac{3}{2} & -\frac{1}{2} \\ \frac{1}{4} & \frac{3}{4} \end{pmatrix} \qquad C = \begin{pmatrix} .2 & .3 & .5 \\ .4 & .4 & .2 \\ .1 & .5 & .4 \end{pmatrix}$$

Solution Since the row vector of A and C are probability vectors, the matrices A and C must be stochastic. However, the matrix B is not stochastic because its first row vector contains a negative component.

In the subsequent section we consider some of the properties and characteristics of stochastic matrices, as well as their classification. To begin with let us consider the following stochastic matrices:

$$A = \begin{pmatrix} 0 & 1 \\ \frac{1}{2} & \frac{1}{2} \end{pmatrix} \quad \text{and} \quad B = \begin{pmatrix} 1 & 0 \\ \frac{2}{3} & \frac{1}{3} \end{pmatrix}$$

Then,

$$AB = \begin{pmatrix} \frac{2}{3} & \frac{1}{3} \\ \frac{5}{6} & \frac{1}{6} \end{pmatrix} \quad \text{and} \quad A^2 = \begin{pmatrix} \frac{1}{2} & \frac{1}{2} \\ \frac{1}{4} & \frac{3}{4} \end{pmatrix}$$

Thus the product AB, of the stochastic matrices A and B is itself a stochastic matrix, and the second power of A, namely, A^2, possesses strictly positive entries. The generalizations of these special cases hold true as Theorem 6.15 indicates.

Theorem 6.15 The product of stochastic matrices is a stochastic matrix. In particular, all powers A^n of a stochastic matrix A are stochastic matrices.

DEFINITION 6.8 A stochastic matrix P is said to be regular if and only if some power of it, p^k, possesses all positive entries.

DEFINITION 6.9 A nonzero vector $X = (x_1, x_2, \ldots, x_n)$ is said to be an invariant or fixed point (or vector) of the square matrix P if and only if

$$XP = X$$

Thus a fixed point of matrix P is a vector whose image under a linear transformation determined by P is the vector itself. In other words the vector is an eigenvector corresponding to the characteristic value $k = 1$ of the matrix P. The interior equilibrium vector in a closed economic system provides a good example of a fixed point for some square input-output matrix. Additionallly, the vector $X = (4, -2)$ is a fixed point of the matrix

$$A = \begin{pmatrix} 2 & 1 \\ 2 & 3 \end{pmatrix}$$

since

$$XA = (4, -2)\begin{pmatrix} 2 & 1 \\ 2 & 3 \end{pmatrix} = (4, -2) = X$$

Theorem 6.16 If X is a fixed vector of a matrix A, then kX is also a fixed vector of A for all real numbers $k \neq 0$.

Proof $(kX)A = k(XA) = kX$, since $XA = X$. Therefore kX is a fixed vector of A.

A useful relationship between regular stochastic matrices and invariant (or fixed) points is contained in the following important theorem. Once again, the proof of this theorem is beyond our scope, and instead we consider its illustration by way of examples.

Theorem 6.17 If P is a regular stochastic matrix, then

a. The matrix P has a unique invariant probability vector X whose components are positive (greater than zero).
b. The sequence $P, P^2, P^3, \ldots, P^n, \ldots$ approaches the matrix M whose rows are each the invariant vector X, that is, every entry of P^n approaches the corresponding entry of M as n increases indefinitely.
c. The sequence of vectors $YP, YP^2, \ldots, YP^n, \ldots$ approaches the invariant (fixed) vector X for any probability vector Y.

Example 6.24 Using the regular matrix

$$P = \begin{pmatrix} 0 & 1 \\ \frac{1}{2} & \frac{1}{2} \end{pmatrix}$$

illustrate parts (a) and (b) of Theorem 6.17.

Solution Since the row vectors of the matrix P are probability vectors, P is stochastic. Moreover, since

$$P^2 = \begin{pmatrix} \frac{1}{2} & \frac{1}{2} \\ \frac{1}{4} & \frac{3}{4} \end{pmatrix}$$

that is, since all entries of P^2 are positive, P must be regular.

(a) We seek a probability vector $X = (x, y)$ or $X = (x, 1 - x)$ such that $XP = X$ or

$$(x, 1 - x)\begin{pmatrix} 0 & 1 \\ \frac{1}{2} & \frac{1}{2} \end{pmatrix} = (x, 1 - x)$$

or

$$\left(\frac{1}{2} - \frac{1}{2}x, x + \frac{1}{2} - \frac{x}{2}\right) = (x, 1 - x)$$

or

$$\frac{1}{2} - \frac{x}{2} = x \qquad \text{and} \qquad x + \frac{1}{2} - \frac{x}{2} = 1 - x$$

Thus

$$x = \tfrac{1}{3} \qquad \text{and} \qquad X = (x, 1 - x) = (\tfrac{1}{3}, \tfrac{2}{3})$$

is the unique invariant probability vector of P.

(b) Since

$$M = \begin{pmatrix} \frac{1}{3} & \frac{2}{3} \\ \frac{1}{3} & \frac{2}{3} \end{pmatrix}$$

or

$$M = \begin{pmatrix} .333 & .666 \\ .333 & .666 \end{pmatrix}$$

and

$$P^2 = \begin{pmatrix} 1/2 & 1/2 \\ 1/4 & 3/4 \end{pmatrix}, \qquad P^3 = \begin{pmatrix} 1/4 & 3/4 \\ 3/8 & 5/8 \end{pmatrix}, \ldots,$$

$$P^6 = \begin{pmatrix} 11/32 & 21/32 \\ 21/64 & 43/64 \end{pmatrix}, \ldots$$

it appears as though

$$P^6 = \begin{pmatrix} .343 & .656 \\ .328 & .671 \end{pmatrix}$$

approaches

$$M = \begin{pmatrix} .333 & .666 \\ .333 & .666 \end{pmatrix}$$

as indicated by Theorem 6.17.

Probability vectors, stochastic matrices, and invariant vectors of square matrices find their main applications in finite sequences of experiments (or trials) called *Markov-chain processes.* These are formally defined in Definition 6.10.

DEFINITION 6.10 Let $x_1, x_2, \ldots, x_n$ be the outcomes of a sequence of experiments (or trials) performed on a system S. And let the following assumptions be made concerning these outcomes:

a. Each outcome belongs to a finite set of outcomes $S_1, S_2, \ldots, S_n$ called "states of the system."
b. If S_i is the outcome on the nth trial, we say the system is in state S_i at time n or at the nth step.
c. The outcome of any trial depends at most on the outcome of the immediately preceding experiment and not on any other previous outcome.
d. The number p_{ij} is the probability that S_j occurs immediately after S_i occurs.

Such a sequence of experiments is called a "Markov-chain process" or simply a Markov chain. The numbers p_{ij} are called the "transition probabilities," and the matrix $P = (p_{ij})_{n\times n}$ is called the "transition matrix" of the Markov chain.

Clearly, each state S_i corresponds to the ith row vector $(p_{i1},$

$p_{i2}, \ldots, p_{in})$ of the transition matrix P. This vector represents the probabilities of all possible outcomes of the trial directly following the state S_i. Therefore each row of P is a probability vector, and consequently the transition matrix P of any Markov chain is a stochastic matrix.

The transition probabilities can be exhibited either by a square matrix called "transition matrix," or by transition diagrams. For instance, in a Markov chain with states S_1, S_2, and S_3 the transition matrix is written as

$$P = \begin{array}{c} \\ S_1 \\ S_2 \\ S_3 \end{array} \begin{array}{c} \begin{array}{ccc} S_1 & S_2 & S_3 \end{array} \\ \begin{pmatrix} p_{11} & p_{12} & p_{13} \\ p_{21} & p_{22} & p_{23} \\ p_{31} & p_{32} & p_{33} \end{pmatrix} \end{array}$$

and the transition diagram is illustrated in Figure 6.8. The arrows from each state indicate the possible states to which a process can move from the given state.

Example 6.25 A man either has lunch out at a restaurant or brings it from home. Suppose he never brings his lunch two days in a row; but

Figure 6-8

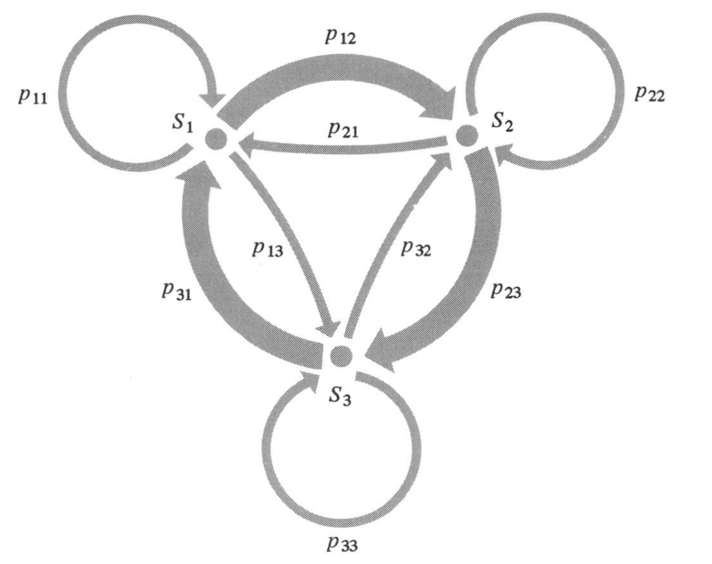

if he eats out, then the next day he is as likely to eat out again as he is to bring his own lunch.

Solution The states of the system are S_1, bring his own lunch, and S_2, eat at a restaurant. This stochastic process is a Markov chain since the outcome of any day depends only on what happened on the preceding day. The transition matrix of the Markov chain is

$$P = \begin{array}{c} \\ S_1 \\ S_2 \end{array} \begin{array}{c} \begin{array}{cc} S_1 & S_2 \end{array} \\ \begin{pmatrix} 0 & 1 \\ \frac{1}{2} & \frac{1}{2} \end{pmatrix} \end{array}$$

The first row $(0, 1)$ corresponds to the fact that he never brings his lunch two days in a row, and so he definitely will eat out the day after he brings his own lunch. The second row $(\frac{1}{2}, \frac{1}{2})$ represents the fact that the day after he eats out he will eat out or bring his lunch with equal probability.

The transition diagram for this Markov chain is shown in Figure 6.9. The entry of 0 indicates that the transition is impossible.

The entry p_{ij} in the transition matrix P of the Markov chain is the probability that the system goes to state S_j immediately after state S_i in one step. It is often necessary to know the probability $p_{ij}^{(n)}$ that the system goes to state S_j in exactly n steps immediately after state S_i. For instance, in a country the Communists, Democrats, Republicans, and Socialists always nominate candidates for the prime minister of the country. The probability of winning depends

Figure 6-9

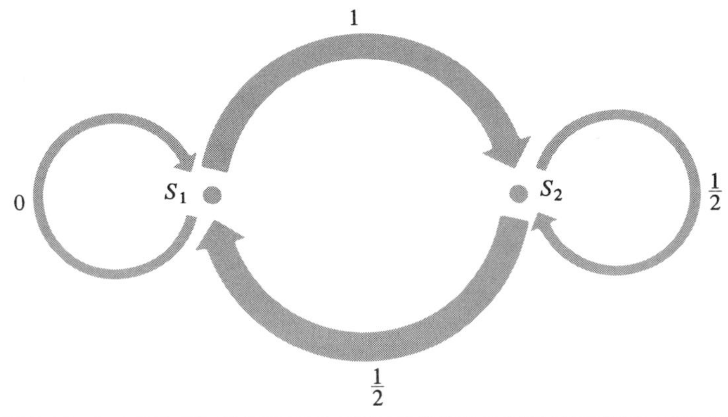

on the party in control and is given by the matrix

$$P = \begin{array}{c} \\ C \\ D \\ R \\ S \end{array} \begin{array}{c} \begin{array}{cccc} C & D & R & S \end{array} \\ \begin{pmatrix} .20 & .35 & .35 & .10 \\ .10 & .25 & .35 & .30 \\ .05 & .35 & .50 & .10 \\ .10 & .40 & .25 & .25 \end{pmatrix} \end{array}$$

where C, D, R, and S are the states Communist, Democrat, Republican, and Socialist, respectively. Now, given that a Republican is the prime minister, find the probability that after two elections (a) a Republican will be prime minister, and (b) a Democrat will be prime minister. The solution of this problem clearly requires the probabilities of going from state R to states R and D in exactly two steps. In general, this type of problem becomes simply a problem in matrix multiplication as Theorem 6.18 indicates.

Theorem 6.18 If P is the transition matrix of a Markov chain process, then the n-step transition matrix, that is, the matrix of the probabilities $p_{ij}^{(n)}$ that the system goes to state S_j immediately after the state S_i in exactly n steps is

$$P^{(n)} = (p_{ij}^{(n)}) = P^n$$

Thus, returning to the previous example on prime ministers, we find that (a) the probability of a Republican being elected is designated by the element $p_{33}^{(2)} = .4150$ of the third row and third column of the matrix

$$(p_{ij}^{(2)}) = P^{(2)} = P^2$$

$$= \begin{array}{c} \\ C \\ D \\ R \\ S \end{array} \begin{array}{c} \begin{array}{cccc} C & D & R & S \end{array} \\ \begin{pmatrix} .20 & .35 & .35 & .10 \\ .10 & .25 & .35 & .30 \\ .05 & .35 & .50 & .10 \\ .10 & .40 & .25 & .25 \end{pmatrix} \end{array} \begin{pmatrix} .20 & .35 & .35 & .10 \\ .10 & .25 & .35 & .30 \\ .05 & .35 & .50 & .10 \\ .10 & .40 & .25 & .25 \end{pmatrix}$$

or

$$P^2 = \begin{pmatrix} .1025 & .3200 & .3925 & .1825 \\ .0925 & .3400 & .3725 & .1950 \\ .0800 & \textcircled{.3200} & \textcircled{.4150} & .1850 \\ .0975 & .3225 & .3625 & .2175 \end{pmatrix}$$

and (b) the probability of a Democrat being elected is the element $p_{32}^{(2)} = .32$ of the third row and second column of the same matrix $P^{(2)} = P^2$.

As in all mathematical problems, the initial conditions of Markov-chain processes are extremely important; that is, it is essential to specify the beginning of the process. Thus if the initial state is chosen by some chance device that selects state S_i with probability $p_i^{(0)}$, $i = 1, 2, \ldots, n$, then these probabilities represented by the vector

$$P^{(0)} = (p_1^{(0)}, p_2^{(0)}, \ldots, p_n^{(0)})$$

specify the beginning of the experiment, and the vector $P^{(0)}$ is referred to as the initial probability distribution vector of the system. Analogously, if $p_i^{(0)}$ is the probability that the process is in state S_i, $i = 1, 2, 3, \ldots, n$, after n steps, then the n-step probability distribution of the system is represented by the vector

$$P^{(n)} = (p_1^{(n)}, p_2^{(n)}, \ldots, p_n^{(n)})$$

Interestingly, these probability vectors $P^{(i)}$, $i = 1, 2, \ldots, n$, relate to each other as Theorem 6.19 specifies.

Theorem 6.19 If P is the transition matrix of a Markov chain, and if $P = (p_1, p_2, \ldots, p_n)$ is the probability distribution of the system at any time, then the probability distributions of the system in the very next step and in the next n steps are pP and pP^n, respectively. Specifically,

$$p^{(1)} = p^{(0)}P, p^{(2)} = p^{(1)}P, \ldots, p^{(n)} = p^{(n-1)}P,$$

and

$$p^{(n)} = p^{(0)}P^n$$

Example 6.26 Let us suppose that three basketball players S_1, S_2, and S_3 are throwing a ball to each other. Let us also assume that S_1 always throws the ball to S_3, and S_2 always throws it to S_1. However, S_3 is just as likely to throw the ball to S_2 as to S_1. If

$$P_i = (p_1{}^i, p_2{}^i, p_3{}^i), \qquad i = 1, 2, 3$$

designates the probability that the ball is thrown to S_i, is this process a Markov chain?

Solution Since the player throwing the ball is not influenced by those previously having the ball, that is, since the outcome of any trial depends at most on the outcome of the immediately preceding experiment only, the process must be a Markov chain with transition matrix

$$P = \begin{matrix} & \begin{matrix} S_1 & S_2 & S_3 \end{matrix} \\ \begin{matrix} S_1 \\ S_2 \\ S_3 \end{matrix} & \begin{pmatrix} 0 & 0 & 1 \\ 1 & 0 & 0 \\ \frac{1}{2} & \frac{1}{2} & 0 \end{pmatrix} \end{matrix}$$

and transition diagram (Figure 6.10).

The rows (0, 0, 1) and (1, 0, 0) represent the fact that S_1 and S_2 always throw the ball to S_3 and S_1, respectively. On the other hand the row $(\frac{1}{2}, \frac{1}{2}, 0)$ of P corresponds to the fact that S_3 is equally likely to throw the ball to S_1 and S_2. Of course, no player throws the ball to himself.

Figure 6–10

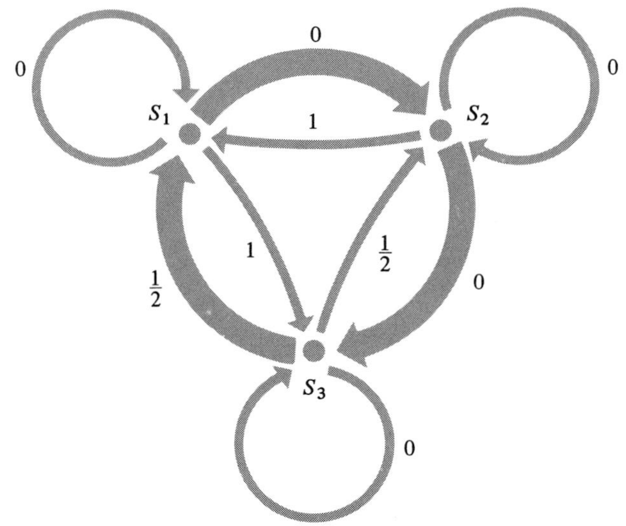

If in this example we assume that S_2 had the ball at the beginning, that is, if the initial probability distribution is $p^{(0)} = (0, 1, 0)$, then by Theorem 6.19

$$p^{(1)} = p^{(0)}P = (0, 1, 0)\begin{pmatrix} 0 & 0 & 1 \\ 1 & 0 & 0 \\ \frac{1}{2} & \frac{1}{2} & 0 \end{pmatrix} = (1, 0, 0)$$

$$p^{(2)} = p^{(1)}P = (1, 0, 0)\begin{pmatrix} 0 & 0 & 1 \\ 1 & 0 & 0 \\ \frac{1}{2} & \frac{1}{2} & 0 \end{pmatrix} = (0, 0, 1)$$

$$p^{(3)} = p^{(2)}P = (0, 0, 1)\begin{pmatrix} 0 & 0 & 1 \\ 1 & 0 & 0 \\ \frac{1}{2} & \frac{1}{2} & 0 \end{pmatrix} = (\tfrac{1}{2}, \tfrac{1}{2}, 0)$$

Thus after three throws the probability of either S_1 or S_2 having the ball is one half, as expected (why?).

Example 6.27 A woman has the following dating habits:

a. If she dates one night, she is 90% sure not to date the next night.
b. The probability that she does not date two nights in a row is 0.7.

Find the transition matrix of the system and the three-step probability distribution vector of the system assuming the initial probability distribution to be

$$p^{(0)} = (.1, .9)$$

How often does she date in the long run?

Solution Since the probability distribution is $p^{(0)} = (.1, .9)$, and the transition matrix is

$$P = \begin{matrix} & \begin{matrix} D & D' \end{matrix} \\ \begin{matrix} D \\ D' \end{matrix} & \begin{pmatrix} .1 & .9 \\ .3 & .7 \end{pmatrix} \end{matrix}$$

where D and D' represent the states "dates" and "does not date," respectively, Theorem 6.19 yields $p^{(3)} = p^{(0)}P^3$ for the three-step probability distribution vector; that is,

$$p^{(3)} = (.1 \quad .9)\begin{pmatrix} .1 & .9 \\ .3 & .7 \end{pmatrix}^3 = (.1, .9)\begin{pmatrix} .244 & .756 \\ .252 & .748 \end{pmatrix} = (.25 \quad .75)$$

To determine what happens in the long run, we must find the fixed vector X such that

$$XP = X$$

or

$$(x, 1 - x)\begin{pmatrix} .1 & .9 \\ .3 & .7 \end{pmatrix} = (.3 - .2x, .2x + .7) = (x, 1 - x)$$

or

$$0.3 - 0.2x = x \qquad \text{and} \qquad 0.2x + 0.7 = 1 - x$$

Thus $x = 3/12$, and the invariant vector X is

$$X = \left(\frac{3}{12}, \frac{9}{12}\right)$$

Hence, the woman dates 3/12 of the time. The transition diagram of this process is shown in Figure 6.11.

Suppose now that the transition matrix P of a Markov chain is regular; that is, P^k possesses positive entries for some value of k. Then Theorem 6.20 holds true.

Figure 6–11

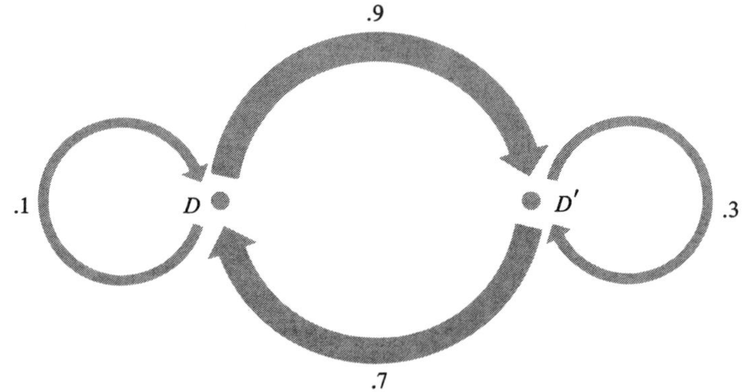

Theorem 6.20 If the transition matrix P of a Markov chain is regular, in the long run the probability that any state S_i occurs is approximately equal to the component x_i of the fixed probability vector X of P.

The above theorem says, in effect, that the initial probability distribution or the initial state of the process wears off as we increase the number of steps in the experiment. Moreover, every sequence of probability distributions approaches the invariant probability vector X of a regular transition matrix P. This invariant vector is called *stationary distribution* of the Markov chain and it describes the probability distribution in the long run. For instance, in Example 6.26

$$P = \begin{matrix} & \begin{matrix} S_1 & S_2 & S_3 \end{matrix} \\ \begin{matrix} S_1 \\ S_2 \\ S_3 \end{matrix} & \begin{pmatrix} 0 & 0 & 1 \\ 1 & 0 & 0 \\ \frac{1}{2} & \frac{1}{2} & 0 \end{pmatrix} \end{matrix}$$

was the transition matrix of the Markov chain described by the throwing of the ball among three basketball players. The invariant vector $X = (x, y, z)$ of P can be obtained from

$$XP = X \qquad \text{or} \qquad (x, y, 1 - x - y) \begin{pmatrix} 0 & 0 & 1 \\ 1 & 0 & 0 \\ \frac{1}{2} & \frac{1}{2} & 0 \end{pmatrix}$$

$$= (x, y, 1 - x - y)$$

or

$$\left(y + \frac{1}{2} - \frac{x}{2} - \frac{y}{2}, \frac{1}{2} - \frac{x}{2} - \frac{y}{2}, x\right) = (x, y, 1 - x - y)$$

Thus equating corresponding components, we obtain

$$\begin{aligned} 3x - y &= 1 \\ x + 3y &= 1 \\ 2x + y &= 1 \end{aligned} \qquad \textbf{(6.24)}$$

Solving the system of Equations (6.24) we obtain $x = \frac{2}{5}$, and $y = \frac{1}{5}$ or

$$X = \left(\tfrac{2}{5}, \tfrac{1}{5}, \tfrac{2}{5}\right)$$

for the invariant vector of P. Therefore in the long run the ball will be thrown to each of the players S_1 and S_3 40% of the time, and to S_2 20% of the time.

Since P^n is the transition matrix of a Markov chain in its n step, the ith row of P remains the same in P^n if and only if it contains one on the main diagonal and zero elsewhere; that is, if the ith row of P has the form $(0, 0, \ldots, p_{ii}, 0, \ldots)$ where $p_{ii} = 1$ for all i, then it remains the same in P^n. When this is true about a given row of P, we say that such a row represents an *absorbing state* S_i of the Markov chain or that S_i is an *absorbing state* of the Markov chain. Thus if

$$P = \begin{matrix} & S_1 & S_2 & S_3 \\ S_1 & \frac{1}{3} & \frac{1}{3} & \frac{1}{3} \\ S_2 & 0 & 1 & 0 \\ S_3 & 0 & 0 & 1 \end{matrix}$$

is the transition matrix of a Markov chain whose transition diagram is shown in Figure 6.12 then the states S_2 and S_3 are absorbing, since the rows $(0, 1, 0)$ and $(0, 0, 1)$ have one on the main diagonal of P and zero elsewhere.

Figure 6-12

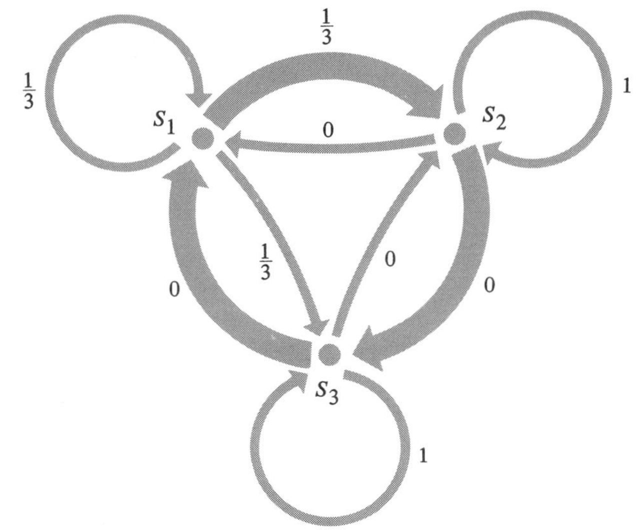

Exercises

1. Are the following vectors probability vectors?

a. $(\frac{2}{5}, 0, \frac{1}{5}, \frac{2}{5})$ b. $(\frac{1}{6}, \frac{1}{2}, \frac{2}{3})$
c. $(\frac{1}{2}, 0, 0, -\frac{1}{2})$ d. $(\frac{3}{7}, \frac{1}{7}, \frac{1}{7}, \frac{2}{7})$
e. $(\frac{1}{4}, \frac{1}{8}, 0, \frac{1}{4}, \frac{3}{8})$

Ans. (b) No; (d) Yes; (e) Yes.

2. In each of the probability vectors of Exercise 1, express the last component in terms of the remaining ones.

3. Write any three stochastic matrices and any three nonstochastic ones.

4. Verify that the following matrices are regular stochastic, draw the associated transition diagrams for each matrix, and find their fixed points.

a. $\begin{pmatrix} \frac{1}{2} & \frac{1}{4} & \frac{1}{4} \\ \frac{1}{2} & 0 & \frac{1}{2} \\ 0 & 1 & 0 \end{pmatrix}$ b. $\begin{pmatrix} 0 & 1 & 0 \\ \frac{1}{6} & \frac{1}{2} & \frac{1}{3} \\ 0 & \frac{2}{3} & \frac{1}{3} \end{pmatrix}$

c. $\begin{pmatrix} \frac{1}{2} & \frac{1}{4} & \frac{1}{4} \\ 0 & \frac{1}{2} & \frac{1}{2} \\ \frac{1}{2} & \frac{1}{2} & 0 \end{pmatrix}$

Ans. (a) (4/11, 4/11, 3/11); (b) (1/10, 6/10, 3/10)

5. Concerning the behavior of rats in connection with a particular feeding, a psychologist makes the following assumptions about their movements in a T maze:

a. For any particular trial 80% of the rats that went right on the previous trial will go right again.
b. 60% of the rats that went left will go right.

If 50% of the rats went right on the first trial, what will the psychologist predict for (i) the second trial, (ii) the third trial, and (iii) in the long run?

Ans. $P = \begin{pmatrix} .8 & .2 \\ .6 & .4 \end{pmatrix}$, (.7, .3), (.74, .26), (.75, .25)

6. Given the transition matrix

$$P = \begin{pmatrix} \frac{1}{2} & \frac{1}{2} \\ 1 & 0 \end{pmatrix}$$

with initial probability distribution $P^{(0)} = (\frac{2}{3}, \frac{1}{3})$, what are the numbers $P^{(3)}$, $P_{21}^{(3)}$, and $P_2^{(3)}$, and what do they mean?

7. A salesman's territory includes the cities C_1, C_2, and C_3. He never sells in the same city on successive days. If he sells in city C_1, then the next day he sells in city C_3. However, if he sells in either C_2 or C_3, then the next day he is twice as likely to sell in city C_1 as in the other city. In the long run, how often does he sell in each of the cities?

Ans. $P = (8/20, 3/20, 9/20)$

8. A man either drives his car or rides a bicycle to work each day. He never rides a bicycle two days in a row; but if he drives to work, then the next day he is just as likely to drive again as he is to ride his bicycle. Find the transition matrix if this process is a Markov chain.

9. A player has \$4. He bets \$2 at a race and wins with a probability of one half. He stops playing if he loses the \$4 or wins \$8. What is the probability that he has lost his money at the end of, at most, five plays? What is the probability that the game lasts more than seven plays?

10. Each year a man trades his car in for a new one. If he has a Ford, he trades it for a Chevrolet. If he has a Chevrolet, he trades it for a Buick. However, if he has a Buick, he is just as likely to trade it for a new Buick as to trade for a new Ford or Chevrolet. In 1956 he bought his first car that was a Ford. Find the probability that he has

(a) A 1958 Buick. (b) A 1958 Ford.
(c) A 1959 Chevrolet. (d) A 1959 Buick.

In the long-run, how often will he have a Buick?

11. A smoker's habit is as follows: If he smokes cigarettes for one week, he switches to a pipe the next week with probability .2. However, the probability that he smokes a pipe two weeks in a row is .7. In the long run, how often does he smoke cigarettes? How often does he smoke a pipe?

Ans. $\frac{3}{5}$, $\frac{2}{5}$

12. Suppose that of the sons of laborers, 80% move upward in the social structure and the rest move downward; of the sons of blue-collar workers, 60% move upward, 20% move downward, and 20% remain blue-collar workers; and of the sons of white-collar workers, 50% move upward, 40% move downward, and 10% remain at the same stability.
    i. What is the probability that the grandson of a blue-collar worker will move upward? Downward? Horizontally?
    ii. In the long run what will be the fraction of men in each of the three social classifications?

References for Supplementary Readings

1. P. Horst, *Matrix Algebra for Social Scientists* (New York: Holt, Rinehart and Winston, 1963).
2. R. I. Levin and R. Lamone, *Linear Programming for Management Decisions* (Homewood, Ill.: Irwin, 1969).
3. M. A. Munem and J. Yizze, *Functional Approach to Precalculus*, Chapters 1 and 8 (New York: Worth Publishers, 1971).
4. P. J. Kelly and N. Ladd, *Fundamental Mathematical Structures—Geometry* (Palo Alto, Calif.: Scott, Foresman, 1965).
5. J. G. Kemeny and J. L. Snell, *Finite Markov Chains* (Princeton, N.J.: Van Nostrand, 1960).

CHAPTER 7

Linear Programming

In the field devoted to resource allocation the fundamental problem is to arrange limited resources in a way that will best meet desired goals. The most fundamental method applied to such problems is based on the theory and solutions of systems of inequalities or equations (or both) called *constraints*. This method constitutes the technology of resource allocation best known as mathematical programming. A special area of mathematical programming deals with idealized situations in which all the constraints (inequalities) on the variables are assumed to be linear. This particular type of mathematical programming is known as *linear programming*. Those areas of mathematical programming for which the constraints on the variables are not linear call for very complicated formulations as well as computations and are widely known as *nonlinear programming*.

Historically, problems of management, operations research, and optimization theory, as well as problems associated with quality control and the theory of the firm, have led to the development of a relatively new branch of applied mathematics known as "operations research." Linear programming is one of the quantitative tools employed in operations research and the decision-making processes.[1] It was originally developed to facilitate the solution of large tactical and strategic military problems during World War II. In fact, in 1947 George B. Dantzig and his associates in the U.S. Department of the Air Force recognized that many military problems involved the task of maximizing and minimizing certain linear functions whose variables were restricted to values satisfying a set of linear inequalities or equations (or constraints). Surprisingly, the Russian mathematician L. V. Kantorovich had been studying similar problems since 1939. However, Dantzig developed independently a method for solving linear programming problems. This method, recognized as the most effective, is known as the *simplex method.*

7.1 Definitions and Examples

In this section we shall be primarily concerned with the basic concepts of linear programming and its fundamental definitions. These we shall illustrate in a sequence of examples.

Example 7.1 Of the 54 employees of a small autoparts plant 36 are assemblers and 18 are packers. The plant produces two parts packages X and Y, each of which requires the labor of both assemblers and packers. The labor requirements per parts package are indicated in Table 7.1.

If each employee works 8 hours a day, there are $288 = 36 \times 8$ assembler man-hours and $144 = 18 \times 8$ packer man-hours available per day, and if x and y are the daily production of packages X and Y, respectively, then clearly

$$x \geq 0 \qquad \text{and} \qquad y \geq 0$$

since there cannot be negative production. Moreover, the capacity restrictions for assemblers and packers dictate that

$$x + 2y \leq 288 \qquad \text{and} \qquad x + \tfrac{1}{2}y \leq 144 \qquad \text{(Table 7.1)}$$

Now suppose that in this example the packages X and Y yield per unit profits of \$1.50 and \$1.00, respectively. How then does one allocate available labor so as to achieve maximum daily profits?

Table 7.1

Labor Classification	Man-hour Required per Unit Packages: X	Y
Assembly	1	2
Packing	1	$\frac{1}{2}$

Solution The profits function P is given by

$$P = 1.50x + 1.00y = \tfrac{3}{2}x + y \qquad (7.1)$$

and the constraints or restrictions on it are included in the following system of inequalities:

$$\begin{aligned} x &\geq 0 \\ y &\geq 0 \\ x + 2y &\leq 288 \\ x + \tfrac{1}{2}y &\leq 144 \end{aligned} \qquad (7.2)$$

These inequalities are satisfied in the convex subset of the plane shaded in Figure 7.1.

Since the maximum or minimum is attained only at corner or extreme points (see Theorem 7.4), we shall be concerned only with such points. The corner points or the extreme points are only the points (0, 0) (144, 0), (96, 96), and (0, 144), not all six points obtained by solving in pairs the four equations that correspond to the four inequalities.

It is easy to see that at each of the four corner points the value of

$$P = \tfrac{3}{2}x + y$$

Figure 7–1

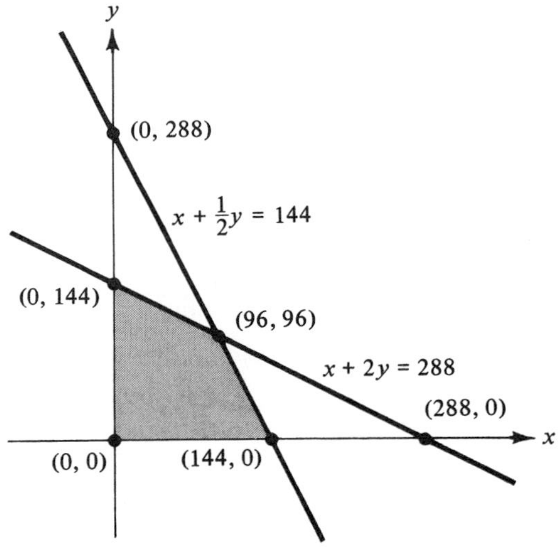

is

a. $P = 0$ at $(0, 0)$.

b. $P = 216$ at $(144, 0)$.

c. $P = 240$ at $(96, 96)$.

d. $P = 144$ at $(0, 144)$.

That is, P is maximum, 240, at the point $(96, 96)$, and therefore the maximum daily profit is $240 achieved when 96 packages of each are produced.

Example 7.2 A dietition must prepare a mixture of foods, X, Y, and Z. Each unit of food X contains 3 ounces protein, 2 ounces carbohydrate, and 6 ounces fat, and costs $0.60. Each unit of food Y contains 5 ounces protein, 6 ounces carbohydrate, and 2 ounces fat, and it costs $1.25. Finally, each unit of food Z contains 3 ounces protein, 4 ounces carbohydrate, and 5 ounces fat, and costs $0.50. The mixture must contain at least 20 ounces protein, 18 ounces carbohydrate, and 30 ounces fat. Find the mixture of minimal total cost subject to these constraints (requirements or specifications).

Solution Let x, y, and z be the amounts of foods X, Y, and Z, respectively. Let w represent the total cost of the mixture, and consider Table 7.2 in which we summarize all the data.

Table 7.2

Food	Amount	Protein	Carbohydrate	Fat	Cost
X	x	3	2	6	$0.60
Y	y	5	6	2	$1.25
Z	z	3	4	5	$0.50
TOTAL		20	18	30	

Now the problem may be expressed as follows:

i. Cost function: $W = 0.60x + 1.25y + 0.50z$

ii. Constraints:

$$3x + 5y + 3z \geq 20$$

$$2x + 6y + 4z \geq 18$$

$$6x + 2y + 5z \geq 30$$

$$x, y, z \geq 0$$

Again, this problem of minimizing (maximizing) a linear function (i) subject to constraints (ii) is another example of linear programming. Such problems are often referred to as linear programs. The solution of this problem we defer for later consideration because the number of variables involved requires geometric considerations in the space of three dimensions or some other algebraic method that we have not yet developed.

DEFINITION 7.1 In general the form of a linear programming problem appears as maximize (or minimize) the function

$$W = a_1x_1 + a_2x_2 + \cdots + a_nx_n \tag{7.3}$$

subject to the constraints (or requirements or specifications)

$$\begin{aligned} a_{11}x_1 + a_{12}x_2 + \cdots + a_{1n}x_n &\leq b_1 \\ a_{21}x_1 + a_{22}x_2 + \cdots + a_{2n}x_n &\leq b_2 \\ a_{m1}x_1 + a_{m2}x_2 + \cdots + a_{mn}x_n &\leq b_m \\ x_1, x_2, \ldots, x_n &\geq 0 \end{aligned} \tag{7.4}$$

The function W is called the *objective function*, and the solution set of the constraints is called the *constraint set*. In general this set is a set of points commonly known or referred to as *corner*, *feasible*, or *extreme points*. If $(\bar{x}_1, \bar{x}_2, \ldots, \bar{x}_n)$ is a solution vector of the constraint set, that is, if $(\bar{x}_1, \bar{x}_2, \ldots, \bar{x}_n)$ satisfies the constraints in (7.4), and if

$$W \leq \sum_{i=1}^{n} a_i\bar{x}_i \qquad \left(\text{or } W \geq \sum_{i=1}^{n} a_i\bar{x}_i\right) \qquad \text{for all } x_1, x_2, \ldots, x_n$$

then we say that $(\bar{x}_1, \bar{x}_2, \ldots, \bar{x}_n)$ is the *solution* of the linear program (7.3) through (7.4) and that

$$W_0 = \sum_{i=1}^{n} a_i\bar{x}_i$$

is the *value* of this linear program.

It must be understood that the constraints of the program may involve inequalities of the opposite sign or that some constraints may even be equations instead of inequalities. Moreover, it may be that some of the variables are not positive. Whatever the case may be, one thing should be clear, which is that the program can always be reduced to an equivalent one of the form (7.3) through (7.4) (why?). Other possibilities and forms of linear programs exist. However, such special forms will not be considered here.

In discussing the second example on linear programming we were faced with geometric considerations beyond the ordinary Euclidean plane and its convex subsets such as the half planes. The standard form of a linear programming problem as it was described above certainly suggests that we need to develop some geometric ideas in higher dimensional geometry if we are to have a geometric picture of the meaning and relevance of this type of problem. Thus in the following discussion we digress slightly either to review or introduce for the first time some fundamental ideas and principles of three-dimensional Euclidean geometry and its generalizations which should be, at least intuitively, comprehensible.

Exercises

1. Find the values of x and y that maximize the function $P = 3x + 4y$ subject to the following constraints:

a. $0 \leq x \leq 10$
 $0 \leq y \leq 8$
 $x + y \geq 6.$

b. $x - y \geq 0$
 $x - 3y \leq 0$
 $x + y \leq 4.$

c. $x \geq 1$
 $x - y \leq 1$
 $x - y \geq -1$
 $x + y \leq 5.$

Ans. (b) (2, 2)

2. Find the values of x and y that minimize the function of Exercise 1 subject to the same constraints.

Ans. (b) (0, 0)

3. A candy manufacturer makes two kinds of candy A and B. Candy A sells at a profit of $0.40 per box, and B sells for a profit of $0.50 a box. The manufacturing process involves the operations of blending, cooking, and packaging as Table 7.3 specifies.

Table 7.3

Candy	Blending (min.)	Cooking (min.)	Packaging (min.)
A	1	5	3
B	2	4	1

For each production run, the blending, cooking, and packaging equipment are available for a maximum of 12, 30, and 15 machine hours, respectively. If this machine time can be allocated to the making of either type of candy at all times that it is available, find the number of boxes of each kind the manufacturer should make in order to maximize his profit.

Ans. 120, 300

4. Work Exercise 3 if the profits are as follows:

   a. \$0.30 and \$0.40 per box of candy A and B, respectively.
   b. \$0.50 and \$0.40 per box of candy A and B, respectively.

7.2 Geometric Topics

In orienting a locus of points and in describing its properties in a three-dimensional space we can employ various systems of coordinates.[2] However, for practical purposes it should be sufficient if we familiarize ourselves only with the so-called rectangular or Cartesian system of coordinates.

The frame of reference for such a system consists of three mutually perpendicular planes having one point in common. The planes of the system are called *coordinate planes,* and their common point is called the *origin.* The lines in which the coordinate planes intersect in pairs are called *coordinate* axes. The position of a point in space, and in this system, is determined by its three directed distances from the point to the coordinate planes. These distances are called *rectangular* or Cartesian coordinates of the point.

In Figure 7.2 the three mutually perpendicular lines XX', YY', and ZZ' through the origin O are called the coordinate axes and shall be referred to as the x axis, y axis, and z axis, respectively. The planes YOZ, XOZ, and XOY are the coordinate planes referred to as the yz plane, xz plane, and xy plane, respectively. Conventionally, we consider the part of the x axis that extends from O in front of the yz plane as positive, and the other part as negative; the part of the y

Figure 7–2

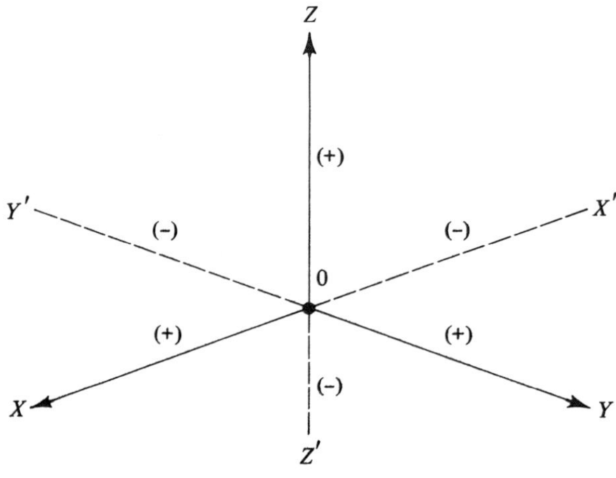

axis extending to the right of O as positive, and the other part as negative; the part of the z axis extending upward from O as positive, and the other part as negative. Now the position of a point P is designated by $P(x, y, z)$ where x, y, and z represent the directed distances of P from the yz, xz, and xy planes, respectively. For example, the points $A(x, y, z)$, $B(1, 1, 1)$, and $C(1, 2, -3)$ are shown in Figure 7.3. Similarly, the points $A(x, y, O)$, $B(x, O, z)$, and $C(O, y, z)$ lie in the xy, xz, and yz planes, respectively, and the points $M(x, O, O)$, $N(O, y, O)$, and $P(O, O, z)$ lie on the axes. About the geometry of space the following theorems are true.

Theorem 7.1 The distance between the points $P(x, y, z)$ and $Q(x', y', z')$ is

$$d = \sqrt{(x - x')^2 + (y - y')^2 + (z - z')^2}$$

and the coordinates of the midpoint M of the segment PQ are $(x + x')/2$, $(y + y')/2$, $(z + z')/2$.

Theorem 7.2 Any linear equation of the form

$$ax + by + cz = d$$

where a, b, c, and d are constants represents a plane that divides the space into two convex subsets called "half spaces."

Theorem 7.3 The planes

$$ax + by + cz = d \qquad \text{and} \qquad a_1x + b_1y + c_1z = d_1$$

Figure 7-3

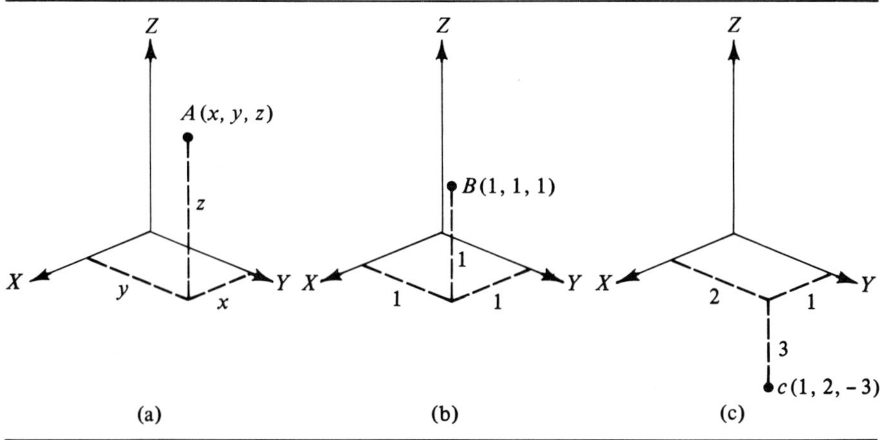

are parallel if and only if their coefficient vectors (a, b, c) and (a_1, b_1, c_1) are parallel, that is,

$$\frac{a}{a_1} = \frac{b}{b_1} = \frac{c}{c_1}$$

Since two planes intersect in a straight line, unless the planes are parallel or coincide, the coordinates of any point on the line of intersection must satisfy the equation of each of the planes, and a line then is a set

$$\{(x, y, z) \mid ax + by + cz = d \quad \text{and} \quad a_1x + b_1y + c_1z = d_1\}$$

Similarly, a point is the intersection of some three planes, or if three planes are not all parallel or parallel in pairs and they do not have a line in common, they then intersect at a point. We shall now illustrate some of these ideas via a sequence of examples.

Example 7.3

a. The distance between the points $A(1, 9, 0)$ and $B = (1, 5, -3)$ is

$$d = \sqrt{(1-1)^2 + (9-5)^2 + (0+3)^2} = \sqrt{16+9} = 5$$

The midpoint of the segment AB is the point

$$P\left(\frac{1+1}{2}, \frac{5+9}{2}, \frac{0-3}{2}\right) \quad \text{or} \quad P\left(1, 7, -\frac{3}{2}\right)$$

b. The planes

$$3x + y + z = 2 \quad \text{and} \quad 6x + 2y + 2z = 21$$

are parallel since $\frac{3}{6} = \frac{1}{2} = \frac{1}{2}$, or the coefficient vectors $(3, 1, 1)$ and $(6, 2, 2)$ of these planes are indeed parallel.

c. Find the equation of the plane determined by the points $(1, 2, 0)$, $(-2, 0, 1)$, and $(1, -3, 1)$.

Solution We need to determine a, b, c, and d so that the plane

$$ax + by + cz + d = 0$$

will contain the three given points. For the points to be on the plane we must have

$$a(1) + b(2) + c(0) + d = 0$$

$$a(-2) + b(0) + c(1) + d = 0$$

$$a(1) + b(-3) + c(1) + d = 0$$

or equivalently

$$\begin{aligned} a + 2b + 0c + d &= 0 \\ -2a + 0b + 1c + d &= 0 \\ a - 3b + 1c + d &= 0 \end{aligned}$$

The solution of this system yields $a = -1/3d$, $b = -1/3d$, and $c = -5/3d$. Thus the equation of the plane becomes

$$-\frac{1}{3}dx - \frac{1}{3}dy - \frac{5}{3}dz + d = 0$$

or

$$x + y + 5z = 3$$

Example 7.4 Shade the part of space described by the inequality

$$x + y + 5z > 3$$

Solution We first graph the plane

$$x + y + 5z = 3$$

(Figure 7.4) as follows:

Let $x = y = 0$, then $z = 3/5$.
Let $x = z = 0$, then $y = 3$.
Let $y = z = 0$, then $x = 3$.

Figure 7-4

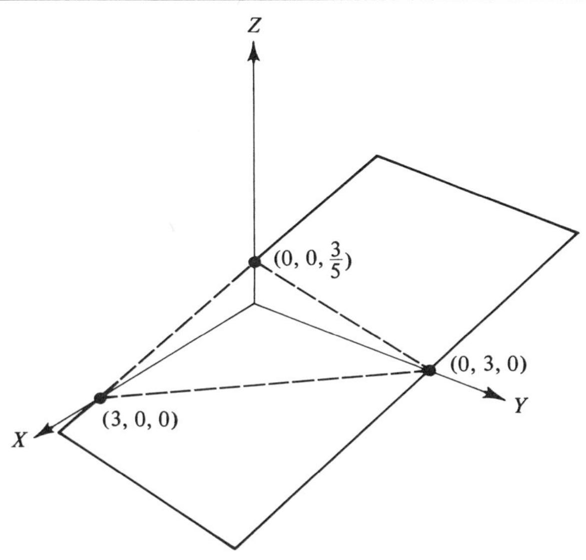

Hence, the plane intersects the axes at the points $(0, 0, \frac{3}{5})$, $(0, 3, 0)$, and $(3, 0, 0)$. This plane then divides space into two half spaces:

$$S_1 = \{(x, y, z) \mid x + y + 5z > 3\}$$

and

$$S_2 = \{(x, y, z) \mid x + y + 5z < 3\}$$

whereby S_2 is in the back of the plane (determined by the triangle) and S_1 is in the front of this same plane. Therefore the solution set to the inequality

$$x + y + 5z > 3$$

is the half plane S_1.

Exercises

1. Locate or graph the points $(-2, -3, 2)$, $(1, -1, 1)$, and $(-1, 1, -1)$.

2. Find the value of c if the point $(2, -1, 1)$ lies in the plane $x - 2y + cz = 4$.

   Ans. 0

3. If the points $(1, 2, 3)$ and $(0, -1, -2)$ lie in the plane $2x + by + cz = 2$, find the values of b and c.

   Ans. 6, -4

4. Find the equation of the plane determined by the following points:

   a. $(0, 3, 0)$, $(1, 3, -1)$, and $(7, 3, 0)$.
   b. $(-2, -3, 2)$, $(1, -1, 1)$, and $(-1, 1, -1)$.

   Ans. $y = 3$, $x - z = 0$

5. Find the lengths of the sides of the triangle with vertices as follows:

   a. $A(0, 1, 3)$, $B(3, 5, 3)$, and $C(2, 1, 1)$.
   b. The midpoints of the segments AB, AC, and BC.

6. Give the coordinates of any point on the line determined by the planes

   $2x - y + 3z = 1$ and $x - 2y + z = 4$.

7. How would you generalize (a) Problem 5? (b) Problem 6?

7.3 The Linear Programming Problem

As an extension of the plane definition

$$ax_1 + bx_2 + cx_3 = d_3$$

in the space of three dimensions, we define a hyperplane of an n-dimensional space S_n to be the set of all n-tuples $(x_1, x_2, \ldots, x_n)$ or points of S_n that satisfy the equation

$$\sum_{i=1}^{n} a_i x_i = d$$

where the a_i's and d are constants. This equation we call the "equation of the hyperplane." It is true, as in the case of the plane, that a hyperplane divides the space S_n into two convex subspaces,

$$S_1 = \{(x_1, x_2, \ldots, x_n) \mid a_1x_1 + a_2x_2 + \cdots + a_nx_n < d\}$$

and

$$S_2 = \{(x_1, x_2, \ldots, x_n) \mid a_1x_1 + a_2x_2 + \cdots + a_nx_n > d\}$$

called "half hyperplanes."

Now the general maximum (minimum) linear programming problem can be restated in a matrix form as maximize

$$W = \sum_{i=1}^{n} a_i x_i \tag{7.5}$$

subject to the constraints $AX \leq B$, where

$$A = \begin{pmatrix} a_{11} & a_{12} \cdots a_{1n} \\ a_{21} & a_{22} \cdots a_{2n} \\ \cdots & \cdots \\ a_{m1} & a_{m2} \cdots a_{mn} \end{pmatrix}, \quad X = \begin{pmatrix} x_1 \\ x_2 \\ \vdots \\ x_n \end{pmatrix}, \quad B = \begin{pmatrix} b_1 \\ b_2 \\ \vdots \\ b_m \end{pmatrix} \tag{7.6}$$

and

$$x_1, x_2, \ldots, x_n \geq 0$$

Thus the constraint set consists of all the points $(x_1, x_2, \ldots, x_n)$ that satisfy the constraints

$$\begin{aligned} a_{11}x_1 + a_{12}x_2 + \cdots + a_{1n}x_n &\leq b_1 \\ a_{21}x_1 + a_{22}x_2 + \cdots + a_{2n}x_n &\leq b_2 \\ \cdots\cdots\cdots\cdots\cdots\cdots\cdots \\ a_{m1}x_1 + a_{m2}x_2 + \cdots + a_{mn}x_n &\leq b_m \\ x_1, x_2, \ldots, x_n &\geq 0 \end{aligned} \tag{7.7}$$

If bounded, this convex set will be bounded by the hyperplanes whose equations are those corresponding to the inequalities above. Its extreme points shall be those solutions of all systems of n equations in n unknowns obtained from the inequalities of (7.7), taken n at a time, which satisfy the remaining inequalities in (7.7). For example, let us maximize

$$W = 2x + y + 3z \tag{7.8}$$

subject to the constraints

$$\begin{aligned}
&1: \quad x + 2y + z \leq 25\\
&2: \quad 3x + 2y + 2z \leq 30\\
&3: \quad x \geq 0\\
&4: \quad y \geq 0\\
&5: \quad z \geq 0
\end{aligned} \tag{7.9}$$

If bounded, the constraint set of this linear programming problem is bounded by the five planes determined by the equations corresponding to the inequalities in (7.9). The extreme points are those of the 10 points of intersections of five planes taken three at a time, as indicated in Table 7.4, that satisfy the remaining constraints.

Table 7.4 indicates that of the 10 points of intersections only 6 are indeed extreme points. Moreover, the values of W, as indicated in Table 7.4, suggest that if there is a maximum for W, it must be $W = 45$ which is obtained at the point $(0, 0, 15)$. This observation is in fact the case since multiplication of the first constraint,

$$x + 2y + z \leq 25$$

by 3 yields

$$3x + 6y + 3z \leq 75$$

Table 7.4

Equations from Constraints (Eqs. 7.9)	Point of Intersection	Satisfies the Remaining Constraints	Value of $W = 2x + y + 3z$
1, 2, 3	(0, 10, 5)	Yes	25
1, 2, 4	(—20, 0, 45)	No	
1, 2, 5	(5/2, 45/4, 0)	Yes	6 5/4
1, 3, 4	(0, 0, 25)	No	
1, 3, 5	(0, 25/2, 0)	Yes	25/2
1, 4, 5	(25, 0, 0)	No	
2, 3, 4	(0, 0, 15)	Yes	45
2, 3, 5	(0, 15, 0)	No	
2, 4, 5	(10, 0, 0)	Yes	20
3, 4, 5	(0, 0, 0)	Yes	0

or

$$W = 2x + y + 3z \leq 3x + 6y + 3z$$

since x, y, and $z \geq 0$. Therefore $W \leq 75$; that is, W is not arbitrarily large. This and the fact that the constraint set contains its boundary points yield that W has a maximum. The last remarks constitute the substance of Theorem 7.4.

Theorem 7.4 The maximum of the objective function, $W = \sum_{i=1}^{n} a_i x_i$, subject to the constraints $AX \leq B$ and $x_i \geq 0$ for all i, where

$$A = \begin{pmatrix} a_{11} & a_{12} \cdots a_{1n} \\ a_{21} & a_{22} \cdots a_{2n} \\ a_{m1} & a_{m2} \cdots a_{mn} \end{pmatrix}, \qquad X = \begin{pmatrix} x_1 \\ x_2 \\ \vdots \\ x_n \end{pmatrix}$$

and

$$B = \begin{pmatrix} b_1 \\ b_2 \\ \vdots \\ b_m \end{pmatrix}$$

if such a maximum exists, will be attained at one of the extreme points of the constraint set.

In analyzing the previous example, in projecting the number of steps that may be involved in working with the general linear programming problem, and in considering the statement of Theorem 7.4 that the objective function attains its maximum at its extreme points, if such a maximum exists, we come to realize that we have a method for solving linear programs in a finite number of steps since the number of extreme points is finite. This method we shall call the *geometric method*. However, there are two major difficulties associated with this method of solving linear programming problems. The first is related to the fact that there is no guarantee of the existence of a maximum, and the second has to do with the fact that although the extreme points are finitely many, their number may be quite large, and finding them all might be beyond the capabilities of even the largest computers. Nevertheless, for the purpose of appreciating the method and its difficulties we shall provide some further examples to illustrate its success in many practical problems.

Example 7.5 An interviewer is hiring for job classifications A and B. He has been instructed to hire between 2 and 4 people for job A and at least 4 people for job B, but no more than 10 people altogether. From experience it is known that 80% of those hired for job A are successful, whereas of those hired for job B 90% are successful. A successful employee in either job is valued at \$10,000, whereas an unsuccessful employee costs the company \$4000 in job A and \$9000 in job B. How many people should he hire for each job in order to maximize the total average return to the company?

Solution Let x be the number of people hired for job A and y for job B. Then, on the average, $0.80x$ and $0.90y$ will be successful in jobs A and B, whereas $0.20x$ and $0.10y$ will be unsuccessful at the same jobs, respectively. Thus the total average return will be

$$R = (10{,}000)(0.80x) - (4000)(0.20x) + (10{,}000)(0.90y) - (9000)(0.10y)$$

or

$$R = 7200x + 8100y$$

Now, we need to maximize R subject to the restrictions

$$\begin{aligned} 1: &\quad x \geq 2 \\ 2: &\quad x \leq 4 \\ 3: &\quad y \geq 4 \\ 4: &\quad x + y \leq 10 \\ 5: &\quad x \geq 0 \\ 6: &\quad y \geq 0 \end{aligned} \tag{7.10}$$

The constraint set of (7.10) is shaded in Figure 7.5.

Of the 15 points of intersection of the six constraints taken two at a time, the extreme points are those that satisfy the remaining constraints. The results of this process we state in Table 7.5.

Thus to maximize the total average return R to the company he must hire two people for job A and eight for job B. Then the total average return will be \$79,200.

Example 7.6 Maximize

$$W = x + 2y \tag{7.11}$$

Figure 7–5

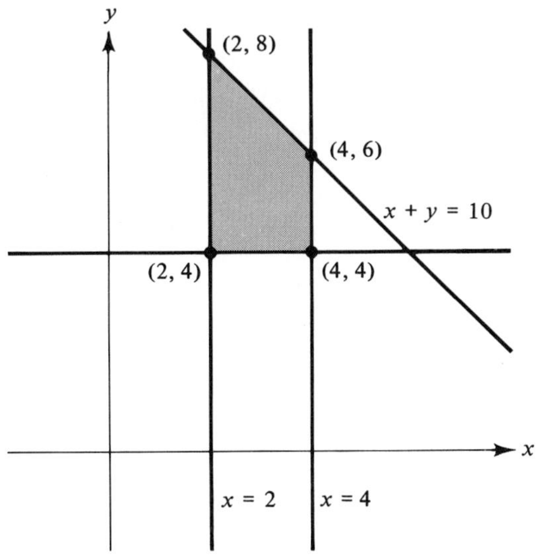

Table 7.5

Equations from Constraints (7.10)	Point of Intersection	Satisfies Other Constraints	Value of $R = 7200x + 8100y$
1, 3	(2, 4)	Yes	\$46,800
1, 4	(2, 8)	Yes	\$79,200
2, 3	(4, 4)	Yes	\$61,200
2, 4	(4, 6)	Yes	\$77,400
1, 2; 1, 5; 2, 5; 3, 6	Undefined, i.e., the lines are parallel	No	
For all other combinations	Exists	No	

subject to the constraints

$$\begin{aligned} 1: &\quad -2x + y \leq 10 \\ 2: &\quad x - 2y \leq 6 \\ 3: &\quad x \geq 0 \\ 4: &\quad y \geq 0 \end{aligned} \tag{7.12}$$

Solution Once again we take the constraints two at a time and solve them as equations to obtain the extreme points (Table 7.6). In

Figure 7-6

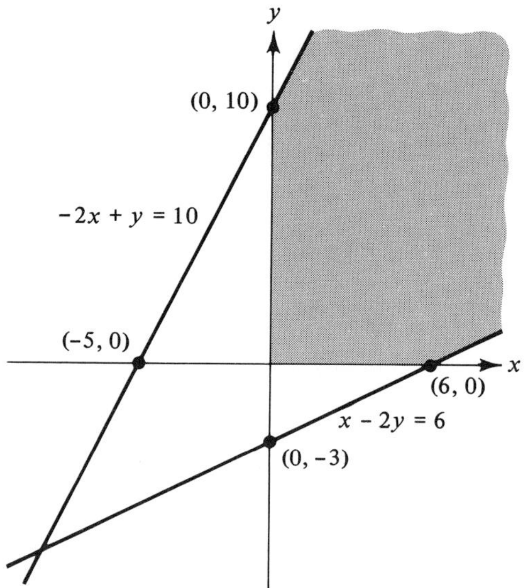

Figure 7.6 we shade the constraint set of this problem, and in Table 7.6 we state the details of the solutions that lead to the extreme points and perhaps the maximum.

Thus from Table 7.6 it follows that if there is a maximum, it should be 20, and it must be attained at the point $(0, 10)$. However, the constraint set indicates that it is unbounded, and the function $W = x + 2y$ becomes arbitrarily large as either both or one of the variables x, y becomes arbitrarily large. Therefore the problem fails to have a solution,[3] and the method of looking at the extreme points fails us.

Table 7.6

Equations from Constraints	Point of Intersection	Satisfies Other Constraints	Value of $W = x + 2y$
1, 2	$(-26/3, -22/3)$	No	
1, 3	$(0, 10)$	Yes	20
1, 4	$(-5, 0)$	No	
2, 3	$(0, -3)$	No	
2, 4	$(6, 0)$	Yes	6
3, 4	$(0, 0)$	Yes	0

Exercises

Solve each of the following linear programs and sketch the constraint set together with the extreme points.

1. Minimize $W = 3x + 2y$ subject to the constraints

$$\begin{aligned} x + 3y &\geq 6 \\ 2x + y &\geq 3 \\ x &\geq 0 \\ y &\geq 0. \end{aligned}$$

2. Maximize $W = 4x + 8y - 12$ subject to the constraints

$$\begin{aligned} -4x + y &\leq 1 \\ -x + y &\geq 2 \\ x + y &\leq 4. \end{aligned}$$

Ans. Maximum, 88/5 at (3/5, 17/5)

3. Maximize and minimize $W = 2x + y$ subject to the constraints

$$\begin{aligned} x + 3y &\leq 12 \\ 2x + y &\leq 10 \\ x + y &\leq 4 \\ x &\geq 0 \\ y &\geq 0. \end{aligned}$$

4. Minimize $W = -5x + 2y$ subject to the constraints

$$\begin{aligned} 2x + y &\geq -1 \\ -x + y &\leq 4 \\ x &\leq 2. \end{aligned}$$

Ans. Minimum, −20 at (2, −5)

5. Maximize and minimize $W = 3x + 5y$ subject to the constraints

$$\begin{aligned} x + 3y &\leq 12 \\ 2x + y &\leq 10 \\ x + y &\leq 4 \\ x &\geq 0 \\ y &\geq 0. \end{aligned}$$

6. Maximize $W = 15x + 25y$ subject to the constraints

$$\begin{aligned} x &\geq 0 \\ y &\geq 0 \\ x + y &\leq 25 \\ 2x + 2y &\geq 10. \end{aligned}$$

Ans. Maximum, 625 at (0, 25)

7. A company has 600 pounds of peanuts and 400 pounds of walnuts. The peanuts alone can be sold for $0.20 per pound. However, the company can mix the peanuts and walnuts in a ratio of three parts peanuts and one part walnuts, or in a ratio of one part peanuts and two parts walnuts. The first mixture sells at $0.35 per pound and the second at $0.50 per pound. How much of each mixture should it produce to maximize sales revenue?
8. A food processing company has 1000 pounds of African coffee, 2000 pounds of Brazilian coffee, and 500 pounds of Colombian coffee. It produces two grades of coffee. Grade A is a mixture of equal parts of African and Brazilian coffees, and sells for $0.60 per pound. Grade B is a mixture of three parts Brazilian and one part Colombian coffee, and sells for $0.85 per pound. How much of each should it produce to maximize sales revenue?

7.4 The Simplex Method

The sequence of examples considered thus far on linear programming clearly indicates the impracticality and uncertainty of the geometric method. The uncertainty stems from the theorem stating that if a maximum exists, it is attained at an extreme point. Its impracticality is the result of having to solve a large number of systems of equations in order to obtain the extreme points and of having to test each of these points. For example, in a program with 6 variables and 15 constraints we will be required to solve $\binom{15}{6} = 12{,}705$ systems of six equations in six unknowns, and check 12,705 solution vectors without guarantee that one of these vectors will yield a maximum for the objective function. All of this implies the need for developing a more efficient and economical method that will lead to a solution if one exists or else will reveal its nonexistence. The so-called *simplex method*, developed by G. B. Dantzig, provides such an alternative, as the following examples demonstrate.

Example 7.7 A company manufactures two types, A and B, of cloth by using three different colors of wool as indicated in Table 7.7.

Table 7.7

For 1-yard Length of Type			
A (ounces)	B (ounces)	Wool Color	Wool Available for Manufacture (ounces)
4	5	Red wool	1000
5	2	Green wool	1000
2	8	Yellow wool	1200

Assuming that the manufacturer wishes to make a profit of \$5 on each yard of type A and \$3 on each yard of type B, how should he use this material to maximize his total profit?

Solution Let x and y be the number of yards produced of the types A and B, respectively. Then we need to maximize the profit function

$$F = 5x + 3y \tag{7.13}$$

subject to the constraints

$4x + 5y \leq 1000$ (only 1000 ounces of red wool are available)

$5x + 2y \leq 1000$ (only 1000 ounces of green wool are available)

$3x + 8y \leq 1200$ (only 1200 ounces of yellow wool are available)

$$x \geq 0$$

$$y \geq 0 \tag{7.14}$$

For the sake of comparison, the solution of this problem shall be obtained in both the geometric method, as in the previous examples, and in the *simplex method* as follows.

GEOMETRIC METHOD: We consider the equations

$$4x + 5y = 1000$$

$$5x + 2y = 1000$$

$$3x + 8y = 1200$$

$$x = 0$$

$$y = 0$$

corresponding to the constraints of the problem, and solve them in pairs to obtain the extreme points (2000/17, 1800/17), (0, 150) (200, 0), and (3000/17, 1000/17) (see Figure 7.7).

In testing the extreme points we now find that the point (3000/17, 1000/17) yields maximum \$18,000/17 for the profit function; that is, the maximum of F is

$$F = 5x + 3y = 5\,\frac{3000}{17} + 3\,\frac{1000}{17} = \frac{\$18{,}000}{17}$$

Figure 7–7

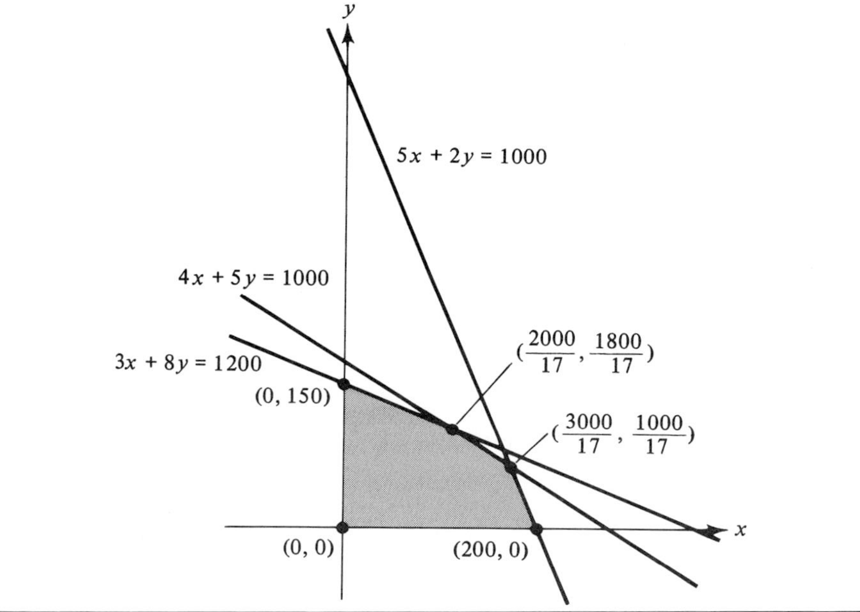

and it will be obtained when $x = 3000/17$ yards of type A and $y = 1000/17$ yards of type B are produced.

SIMPLEX METHOD: Since $a \leq b$ if and only if there is $k \geq 0$ such that $a + k = b$, we may find positive numbers z, v, w called *slack variables* such that

$$\begin{aligned} 4x + 5y + z \qquad\qquad &= 1000 \\ 5x + 2y \quad + v \qquad &= 1000 \\ 3x + 8y \qquad + w \quad &= 1200 \\ -5x - 3y \qquad\qquad + F &= 0 \end{aligned} \tag{7.15}$$

where x, y, z, v, and w are all positive or zero. This system of equations is obviously obtained from the constraints of our problem and the objective function that we shall always write as the last equation of the system. The system (7.15) involves four equations in six unknowns. In such a system we can solve, in general, for four of the unknowns in terms of the other two by using elementary row operations on the augmented matrix of the system just as we did with the Gauss reduction method in solving systems of equations. The augmented matrix of the system in (7.15) is

$$\begin{pmatrix} 4 & 5 & 1 & 0 & 0 & 0 & 1000 \\ \textcircled{5} & 2 & 0 & 1 & 0 & 0 & 1000 \\ 3 & 8 & 0 & 0 & 1 & 0 & 1200 \\ -5 & -3 & 0 & 0 & 0 & 1 & 0 \end{pmatrix}. \tag{7.16}$$

R_i will designate the ith row of a matrix.

The negative number with the greatest absolute value is -5 in the last row, R_4, of the matrix (7.16). In this method such a number will always be an entry of what we shall henceforth call the *pivot column* of this method. From the pivot column we now pick all the positive entries, if there are any, and divide them into the corresponding entries of the last column of the matrix (7.16). The number of the pivot column that yields the smallest quotient shall be called the *pivot number*, and the row containing it will be referred to as the *pivot row*. In this case, the smallest of the quotients $1000/4 = 250$, $1000/5 = 200$, and $1200/3 = 400$ is 200. Therefore the first column is the pivot column, the second row is the pivot row, and the circled number 5 is the pivot number. The elementary row operations $\frac{1}{5}R_2$, first, and then $-4R_2 + R_1$, $-3R_2 + R_3$, $5R_2 + R_4$, in any order, reduce the matrix (7.16) into the equivalent matrix (7.17).

$$\begin{pmatrix} 0 & \textcircled{17/5} & 1 & -4/5 & 0 & 0 & 200 \\ 1 & 2/5 & 0 & 1/5 & 0 & 0 & 200 \\ 0 & 34/5 & 0 & -3/5 & 1 & 0 & 600 \\ 0 & -1 & 0 & 1 & 0 & 1 & 1000 \end{pmatrix} \tag{7.17}$$

Again, since -1 is the negative number with the greatest absolute value in the last row of matrix (7.17), repeating the previous process on matrix (7.17), we find that 17/5 is a new pivot number and that the elementary row operations $5/17R_1$ first, and then $-2/5R_1 + R_2$, $-34/5R_1 + R_3$, $R_1 + R_4$, in any order, reduce matrix (7.17) into the equivalent matrix (7.18)

$$\begin{pmatrix} 0 & 1 & 5/17 & -4/17 & 0 & 0 & 1000/17 \\ 1 & 0 & -2/17 & 5/17 & 0 & 0 & 3000/17 \\ 0 & 0 & -2 & 1 & 1 & 0 & 3400/17 \\ 0 & 0 & 5/17 & 13/17 & 0 & 1 & 18{,}000/17 \end{pmatrix} \tag{7.18}$$

We terminate this process whenever the nonzero elements of the last row are all positive. In this case we stop with matrix (7.18), thus obtaining the equivalent system (system with same solutions) (7.19).

$$
\begin{aligned}
y + \frac{5}{17}z - \frac{4}{17}v \qquad\qquad &= \frac{1000}{17} \\
x - \frac{2}{17}z + \frac{5}{17}v \qquad\qquad &= \frac{3000}{17} \\
-2z + v \quad + w \qquad &= \frac{3400}{17} \\
\frac{5}{17}z + \frac{13}{17}v \qquad + F &= \frac{18{,}000}{17}
\end{aligned}
\tag{7.19}
$$

From the last equation of this system we obtain

$$F = \frac{18{,}000}{17} - \frac{5}{17}z - \frac{13}{17}v$$

which becomes maximum,

$$F = \frac{\$18{,}000}{17},$$

if and only if $z = v = 0$. Thus solving system (7.19) we find that the maximum $F = 18{,}000/17$ is obtained when $x = 3000/17$ yards of type A and $y = 1000/17$ yards of type B are produced.

In the next example we shall once again illustrate the simplex method, and as a final observation we shall briefly discuss the consideration of minimizing a function subject to some constraints.

Example 7.8 Maximize the function

$$N = 5x + 4y + 6z \tag{7.20}$$

subject to the constraints

$$
\begin{aligned}
x + y + z &\leq 100 \\
3x + 2y + 4z &\leq 210 \\
3x + 2y \qquad &\leq 150 \\
x, y, z &\geq 0
\end{aligned}
\tag{7.21}
$$

Solution We first introduce *slack* variables u, v, and w to transform the inequalities (7.21) into equations, and then we consider the equation

$$-5x - 4y - 6z + N = 0$$

as the last equation of the system:

$$
\begin{aligned}
x + y + z + u \qquad\qquad\qquad &= 100\\
3x + 2y + 4z \qquad + v \qquad\qquad &= 210\\
3x + 2y \qquad\qquad\qquad + w \qquad &= 150\\
-5x - 4y - 6z \qquad\qquad\qquad + N &= 0
\end{aligned}
\tag{7.22}
$$

Now the augmented matrix of this system is

$$
\begin{pmatrix}
1 & 1 & 1 & 1 & 0 & 0 & 0 & 100\\
3 & 2 & \textcircled{4} & 0 & 1 & 0 & 0 & 210\\
3 & 2 & 0 & 0 & 0 & 1 & 0 & 150\\
-5 & -4 & -6 & 0 & 0 & 0 & 1 & 0
\end{pmatrix}
\tag{7.23}
$$

Since the negative number of largest absolute value in the last row of the matrix (7.23) is -6, since the *pivot* column is the third one, and since the positive numbers 1 and 4 of this column divided into 100, and 210 yield 100/1, and 210/4 as quotients, the *pivot* row is the row (3, 2, 4, 0, 1, 0, 0, 210), and the *pivot* number is 4 (it yields the smallest quotient). Thus the elementary row operations $R_2/4$ first, and then $-R_2 + R_1$, $6R_2 + R_4$ applied on the matrix (7.23) yield the equivalent matrix (7.24).

$$
\begin{pmatrix}
1/4 & 1/2 & 0 & 1 & -1/4 & 0 & 0 & 47\tfrac{1}{2}\\
3/4 & 1/2 & 1 & 0 & 1/4 & 0 & 0 & 52\tfrac{1}{2}\\
3 & \textcircled{2} & 0 & 0 & 0 & 1 & 0 & 150\\
-1/2 & -1 & 0 & 0 & 3/2 & 0 & 1 & 315
\end{pmatrix}
\tag{7.24}
$$

A similar argument reveals that 2 is the new *pivot* number for the matrix (7.24). Hence the elementary row operations $R_3/2$, first, and then $-\frac{1}{2}R_3 + R_1$, $-\frac{1}{2}R_3 + R_2$, and $R_3 + R_4$ applied to matrix (7.24) yield the equivalent matrix (7.25)

$$
\begin{pmatrix}
-1/2 & 0 & 0 & 1 & -1/4 & -1/4 & 0 & 10\\
0 & 0 & 1 & 0 & 1/4 & -1/4 & 0 & 15\\
3/2 & 1 & 0 & 0 & 0 & 1/2 & 0 & 75\\
1 & 0 & 0 & 0 & 3/2 & 1/2 & 1 & 390
\end{pmatrix}
\tag{7.25}
$$

This matrix has no negative numbers in its last row, a fact that in general terminates this process and establishes the following equivalent system:

$$
\begin{aligned}
-\frac{x}{2} + u - \frac{1}{4}v - \frac{1}{4}w &= 10 \\
z + \frac{1}{4}v - \frac{1}{4}w &= 15 \\
\frac{3}{2}x + y + \frac{1}{2}w &= 75 \\
x + \frac{3}{2}v + \frac{1}{2}w + N &= 390
\end{aligned}
$$

From the last equation of this system we obtain

$$N = 390 - x - \frac{3}{2}v - \frac{1}{2}w \qquad \text{or} \qquad N = 390$$

as the maximum of N attained at

$$x = v = w = 0 \qquad \text{and} \qquad y = 75,\ z = 15,\ u = 10$$

which are found when

$$x = y = w = 0$$

are substituted in the other equations of the system.

In many linear programming problems one needs to minimize rather than maximize a linear expression similar to N above. This is easily accomplished, by maximizing $-N$ using the simplex method as we have already indicated, and then obtaining $-(-N)$ as the required minimum. For example, in Example 7.8 let us minimize

$$N = 5x + 4y + 6z$$

subject to the same constraints. To minimize N is equivalent to maximizing

$$F = (-N) = -5x - 4y - 6z$$

or

$$5x + 4y + 6z + F = 0$$

Thus instead of the system (7.22) we have a system obtained from (7.22) by replacing its last equation with $5x + 4y + 6z + F = 0$.

Then the augmented matrix of the system becomes

$$\begin{pmatrix} 1 & 1 & 1 & 1 & 0 & 0 & 0 & 100 \\ 3 & 2 & 4 & 0 & 1 & 0 & 0 & 210 \\ 3 & 2 & 0 & 0 & 0 & 1 & 0 & 150 \\ 5 & 4 & 6 & 0 & 0 & 0 & 1 & 0 \end{pmatrix}$$

Since the last row contains no negative number, the maximum value of

$$F = 0 - 5x - 4y - 6z = -N$$

is $F = 0$ obtained at $x = 0$, $y = 0$, $z = 0$. Hence, the minimum value of N is $-F$ or $N = 0$.

A SUMMARY OF THE STEPS IN THE SIMPLEX METHOD

In order to solve the general linear programming problem using the simplex method we must perform the following tasks:

Step 1. Transform the constraint inequalities into equations by introducing the so-called *slack* variables. This can always be done since $a \leq b$ if and only if there is $x \geq 0$ such that $a + x = b$ or $a - x = b$ whenever $a \geq b$.

Step 2. In considering the system of equations obtained from the constraint inequalities and the objective function, always write the equation of the objective function as the last equation of the system and equate it to zero.

Step 3. Consider the augmented matrix of the system and select the negative number, of the last row, that has the largest absolute value (if there are more than one just use any one of them). This number identifies the *pivot* column. Of the positive numbers (if there are any) of the pivot column now select the one that yields the smallest quotient when these numbers are divided into the corresponding entries of the last column (if more than one yields the same quotient, select any one). This number, say p_{ij}, is called the "pivot number," and the row containing it the "pivot row."

Step 4. Perform the elementary row operation R_i/p_{ij} to place 1 in the position of the pivot number p_{ij}. Then perform elementary row operations to place zeros above and below 1. If this yields positive numbers in the last row of the matrix, then in general we stop and pick the maximum as in the examples. On the other hand, if there are any negative numbers in the last row, then repeat Steps 3 and 4.

Many books[4] discuss in detail why this method actually works. Thus we shall only make some necessary and useful remarks on the above steps of the simplex method. Hence, Steps 1, 2, and 4 present no particular problems, whereas Step 3 requires that some positive number(s) exists on the pivot column. If all entries of the pivot column are negative, then the objective function has an "infinite" maximum. Moreover, in selecting the pivot number it was required that it yield the smallest ratio when divided into the corresponding entry of the last column. However, in such a division process the following two things may happen:

a. The minimum ratio may happen to be zero.
b. There may be a "tie" between two (or more) of these ratios for a smallest value.

In fact, to pick the pivot number requires the existence of a negative number in the last row, which may not always be the case.

Any one of these cases shall be referred to as a degeneracy of the program, and although disconcerting, such a degeneracy seldom becomes fatal in everyday practice. Any further discussion of degeneracies is beyond the scope of this textbook.[4]

Exercises

Using the simplex method, solve the following linear programming problems:

1. Maximize

$$F = 2x + 3y + z$$

subject to

$$\begin{aligned} x + 4y - 2z &\leq 10 \\ 2x - 3y + 4z &\leq 12 \\ 3y + z &\leq 7 \\ x, y, z &\geq 0. \end{aligned}$$

2. Maximize

$$F = x + 2y$$

subject to

$$\begin{aligned} 3x + 5y &\le 14 \\ x + 7y &\le 10 \\ x, y &\ge 0. \end{aligned}$$

Ans. $F = 5$ at $(3, 1)$

3. Maximize

$$F = 40x + 60y + 50z$$

subject to

$$\begin{aligned} x + y + z &\le 27 \\ -x + y - z &\le 0 \\ -x + 2z &\le 0 \\ x, y, z &\ge 0. \end{aligned}$$

Ans. $F = 1395$, $x = 9$; $y = 27/2$, $z = 9/2$

4. Minimize

$$F = 50x + 30y + 20z$$

subject to

$$\begin{aligned} x + y + z &\le 27 \\ x - 2z &\ge 0 \\ x - y + z &\ge 0 \\ x + y + z &\ge 24 \\ x, y, z &\ge 0. \end{aligned}$$

Ans. $F = -840$, $x = 8$, $y = 12$, $z = 4$

5. Maximize

$$F = 2x + 4y + 7z$$

subject to

$$\begin{aligned} 3x + y + 2z &\le 25 \\ x + 3y - z &\le 10 \\ 4y + z &\le 30 \\ x, y, z &\ge 0. \end{aligned}$$

6. Minimize

$$F = 6x + 2y + 3z$$

subject to

$$\begin{aligned} x + 4y - 3z &\geq 15 \\ 2x + 3y + z &\geq 25 \\ -x + 4y + 2z &\geq 12 \\ x, y, z &\geq 0. \end{aligned}$$

7. A firm manufactures two types of screws A and B and sells them at a profit of \$0.03 on type A and \$0.04 on type B. Each type is processed on two machines, an automatic screw machine and a slotting machine. Type A screws require two minutes of processing on the automatic and five minutes on the slotter machines. On the other hand type B screws require three minutes on the automatic and two minutes on the slotter machines. Each machine is available for not more than 60 hours during any working week. In order to maximize its profits how many of each type of screws should the firm produce each week?

   Ans. 327 of A and 982 of B, \$49.09 profit

8. A company has 500 pounds of peanuts, 150 pounds of cashew nuts, and 75 pounds of brazil nuts on hand. It packages and sells in 8-ounce cans three basic mixes of nuts. Type A, consisting of 50% peanuts, 40% cashews, and 10% brazil nuts, sells for \$0.39 per can; type B, consisting of cashew nuts only, sells for \$0.80 per can; and type C, consisting of 40% cashews and 60% brazil nuts, sells for \$0.65 per can. What production schedule will provide maximum returns to the company?

References for Supplementary Readings

1. A. Vazsonyi, *Scientific Programming in Business and Industry* (New York: Wiley, 1958).
2. W. K. Morrill, *Analytic Geometry*, Chapters 2 and 9 (New York: Intext Educational Publishers, 1968).
3. J. G. Kemeny, *Finite Mathematics*, Chapter 5 (Englewood Cliffs, N.J.: Prentice-Hall, 1972).
4. G. Owen, *Finite Mathematics*, Chapter 3 (Philadelphia: Saunders 1970).

CHAPTER 8

Game Theory

8.1 Introduction

Although new applications of linear programming are continuously coming to light, one of its most interesting ramifications was the early discovery of the close relationship between linear programming and game theory. Originally the theory of games was proposed by the great French mathematician Emile Borel about the year 1921. This theory had its origin in an attempt to describe rational behavior in economic competition. Game theory received a strong impetus in 1928 when a paper published by John von Neumann presented an entirely new approach to the treatment of some fundamental problems in economic theory.[1]

The rapid development of linear programming has brought considerable simplification to an otherwise difficult and sophisticated theory of games and its applications. On the other hand the persistent research of many people such as G. B. Dantzig, L. S. Shapley, A. J. Hoffman, P. Wolfe, S. Vajda, and others has resulted in a high state of perfection for both game theory and linear programming.

In a game of strategy between two players, a player's behavior depends on his evaluation of the other player's strategies. In such games each player tries to attain some objective, such as minimizing losses or maximizing winnings. In this connection one of the most useful and striking applications of linear programming is found in the analysis of competitive situations in economics, business, warfare, and all other areas of conflicting interest or competitive behavior.[2]

A typical yet very simple problem in game theory is the game of matching pennies. This game consists of two players R and C who toss a penny independently of each other, and in turn compare the outcomes. If the outcomes match, that is, if they are both heads or both tails, then player R wins a dollar from player C. If the outcomes do not match, that is, if they are heads and tails or tails and heads, then player C wins a dollar from player R. This game can neatly be represented by the 2×2 matrix

$$\begin{pmatrix} 1 & -1 \\ -1 & 1 \end{pmatrix}$$

called the *payoff* matrix of the game. In this matrix game player R plays the rows R_1 and R_2 as heads and tails, respectively, while player C plays the columns C_1 and C_2 in a similar fashion. The elements of the payoff matrix are called "payoffs" and represent the amounts paid to R for each of the outcomes HH, HT, TH, and TT. A negative entry in the payoff matrix represents actual loss of R or the actual pay to C by R. Since each player wins the same amount that the other player loses, we call this game a two-person zero-sum matrix game.

8.2 Matrix Games

In general, a two-person zero-sum matrix game can best be represented by an $m \times n$ matrix whose essential characteristics are the following:

1. It involves two players R and C who play the rows and columns of the so-called *payoff* matrix, respectively. Thus in playing the game, R may choose any one of the rows R_i, $i = 1, 2, \ldots, m$, of the payoff matrix. Likewise C may choose any one of the columns C_i, $i = 1, 2, \ldots, n$, of the same matrix.

2. Once R and C have made a choice, the winner is declared according to the rule of the game. The *payoff* p_{ij} of C to R is the element of the ith row R_i and jth column C_j, where R_i and C_j are the choices made by R and C, respectively. If p_{ij} is positive, R wins the amount p_{ij} from C; if p_{ij} is negative, R loses the amount p_{ij} to C. In the game of matching pennies the rows headed by H (heads) and T (tails) were selected by player R, while the columns, again, headed

by H and T were selected by player C as we indicate in the following table.

$$\begin{array}{cc} & C: \\ & \begin{array}{cc} H & T \end{array} \\ R: & \begin{array}{c} H \\ T \end{array} \begin{pmatrix} 1 & -1 \\ -1 & 1 \end{pmatrix} \end{array}$$

The payoff p_{ij} for this game is \$1 or $-$\$1. Another example of a two-person zero-sum matrix game is provided by two players R and C each holding a three, four, five, and six from a deck of cards. In this game each player selects one card and simultaneously the players place the cards face up on a table. If the cards match, then the payoff to each other is zero. If the cards do not match, then the player with the smaller face value card wins in dollars the sum of the face values of the two cards. This game can also be nicely represented by the payoff matrix.

$$P = (p_{ij}) = \begin{pmatrix} 0 & 7 & 8 & 9 \\ -7 & 0 & 9 & 10 \\ -8 & -9 & 0 & 11 \\ -9 & -10 & -11 & 0 \end{pmatrix}$$

In this game, R plays the rows R_i, $i = 1, 2, 3, 4$ of P, and C plays the columns C_i, $i = 1, 2, 3, 4$ of P, as follows.

The selection of the R_i row by R implies the playing of the card $i + 2$, $i = 1, 2, 3, 4$. Similarly, the selection of the C_j column by C implies the playing of the card $j + 2$, $j = 1, 2, 3, 4$. The payoff

$$p_{ij} = (i + 2) + (j + 2)$$

is zero if $i = j$, and there is no payoff to either player, or it is positive for $i < j$ and R wins the amount p_{ij}, or it is negative for $i > j$ in which case R loses to C the amount $|p_{ij}|$.

As one would expect, the question raised in this type of game is what row should R select or what kind of move should he make (if there is any) to maximize his winnings no matter what C does? Likewise what should C do to minimize his losses?

DEFINITION 8.1 We say that a two-person zero-sum game is *fair* if and only if the payoff is zero whenever both players use their best selections, that is, whenever both players make their best moves.

In the last game, in trying to maximize his winnings R may be inclined to select the third row, thus winning \$11. However, C is attempting to minimize his losses, so he might pick the second column and thus inflict a \$9 loss to R. In view of what may happen, R should reconsider selecting a row that may lead to a minimum loss, or no loss, even though C might be clever enough to pick his best choice. Hence R must examine the set consisting of all the entries of the payoff matrix, each of which is minimum in its row, and select the row that contains the maximum element of this set.

For example, in the game of matching pennies R could select either one of the two rows since there is only one negative entry in the payoff matrix. On the other hand, in the game of selecting cards R should play the first row that provides a gain of either \$9 or \$8 or \$7 or at the worst a zero loss. Thinking similarly, player C can see that in matching pennies it makes no difference whether he selects the first or the second column. However, in playing the game of cards C recognizes that if he should select the third column for a gain of \$11, R might come up with the selection of either the second row or the first row, thus inflicting a loss to C of either \$9 or \$8, respectively. Consequently, C is forced to select the column of the payoff matrix containing the least positive entry of all entries in the payoff matrix. In this case C must select the first column that provides for him a gain of either \$7 or \$8 or \$9 or, at worst, zero losses if it should happen that R selects row number one. Thus it becomes evident that the players must be careful in selecting the rows or the columns on which they play.

DEFINITION 8.2 By a strategy of a player R we shall mean the probability vector

$$P = (p_1, p_2, \ldots, p_m)$$

where p_i, for any i, is the probability with which R chooses the ith row R_i of the payoff matrix. Similarly, by a strategy of a player C we shall mean the probability vector

$$Q = (q_1, q_2, \ldots, q_n)$$

where q_j, for any j, is the probability with which C chooses the column C_j of the payoff matrix. Moreover, we say that a strategy vector is a pure strategy if and only if all the components of the vector are zero except one that is the number 1. A mixed strategy is any strategy that is not pure.

From elementary concepts of probability theory it follows that for any strategy $(p_1, p_2, \ldots, p_n)$, $p_i \geq 0$ for all i, and

$$\sum_{i=1}^{n} p_i = 1$$

As an illustration of the above definitions we may once again consider the penny-matching game in the following three versions:

a. Playing heads or tails with equal frequency, that is, selecting the first row (or column) one-half of the time. In such a game, the strategy is $(\frac{1}{2}, \frac{1}{2})$ for both players R and C.
b. Playing heads and tails with different frequencies, that is, player R may play heads two thirds of the time and tails only one third of the time. Similarly, player C may select to play heads one fourth of the time and therefore tails three fourths of the time. In this version of the game, the strategy of R is $(\frac{2}{3}, \frac{1}{3})$ and that of C is $(\frac{1}{4}, \frac{3}{4})$.
c. Player R always plays the same row, say the first, and player C always plays the same column, say the second. Then the strategy of R is the pure strategy $(1, 0)$, and that of C is again the pure strategy $(0, 1)$.

More generally, if in the penny-matching game player R should play strategy $(x, 1 - x)$ and player C adopts the strategy $(y, 1 - y)$, then in repeated plays of this version of the game R can expect an average gain Z, per play, given by

$$Z = (x, 1 - x)\begin{pmatrix} 1 & -1 \\ -1 & 1 \end{pmatrix}(y, 1 - y)^t \qquad \text{(Why?)}$$

or

$$Z = 4xy - 2x - 2y + 1 \tag{8.1}$$

Now the expected gain, per play, for the C player is, of course,

$$-Z = -4xy + 2x + 2y - 1 \tag{8.2}$$

Consequently, R desires to maximize Z, and C desires to maximize $-Z$ or equivalently to minimize Z.

To further explain the expected gain (or value) of a matrix game, let us consider a game defined by the 2×2 matrix

$$A = \begin{pmatrix} a_{11} & a_{12} \\ a_{21} & a_{22} \end{pmatrix}$$

and let $P = (p_1, p_2)$ and $Q = (q_1, q_2)$ be the strategies of R and C, respectively. Then the probability that R wins the amount a_{11} (or loses the amount $|a_{11}|$ if $a_{11} < 0$) is p_1q_1. Similarly, the probabilities that R wins the amounts a_{12}, a_{21}, and a_{22} are p_1q_2, p_2q_1, and p_2q_2, respectively. Thus the expectation, $E(P, Q)$, of R (or the expected value of the amount R wins) when R uses the strategy P and C the strategy Q is

$$E(P, Q) = p_1q_1a_{11} + p_1q_2a_{12} + p_2q_1a_{21} + p_2q_2a_{22}$$

or

$$E(P, Q) = PAQ^t$$

In the general case of a two-person zero-sum matrix game defined by the matrix

$$A = \begin{pmatrix} a_{11} & a_{12} & \cdots & a_{1n} \\ a_{21} & a_{22} & \cdots & a_{2n} \\ \cdots & \cdots & \cdots & \cdots \\ a_{m1} & a_{m2} & \cdots & a_{mn} \end{pmatrix}$$

if $P = (p_1, p_2, \ldots, p_m)$ and $Q = (q_1, q_2, \ldots, q_n)$ are the strategies of the players R and C, respectively, then the probability that R wins the amount a_{ij} (or loses $|a_{ij}|$ if $a_{ij} \leq 0$) is p_iq_j for all $i = 1, 2, \ldots, m$ $j = 1, 2, \ldots, n$. Therefore the expected value of the amount R wins whenever R and C use the strategies P and Q, respectively, is

$$E(P, Q) = (p_1, p_2, \ldots, p_m) \begin{pmatrix} a_{11} & a_{12} & \cdots & a_{1n} \\ a_{21} & a_{22} & \cdots & a_{2n} \\ \cdots & \cdots & \cdots & \cdots \\ a_{m1} & a_{m2} & \cdots & a_{mn} \end{pmatrix} \begin{pmatrix} q_1 \\ q_2 \\ \vdots \\ q_n \end{pmatrix} \tag{8.3}$$

or in a short matrix form

$$E(P, Q) = PAQ^t \tag{8.4}$$

Therefore player R wishes to maximize $E(P, Q) = PAQ^t$, and player C wishes to maximize $-E(P, Q)$ or equivalently to minimize $E(P, Q) = PAQ^t$.

DEFINITION 8.3 By an optimum strategy for player R in a two-person zero-sum game we shall mean a strategy that maximizes his expected gain $E(P, Q) = PAQ^t$, regardless of what the other player, C, does. Similarly, by an optimum strategy for C we shall mean a strategy that minimizes C's payments to R, regardless of what R does.

Returning once again to the penny-matching game and the version of it in which player R adopts the strategy $(y, 1 - y)$, we recall that the expected gain was

$$Z = 4xy - 2x - 2y + 1 = 4(x - \tfrac{1}{2})(y - \tfrac{1}{2})$$

Thus the maximum long-run gain that R can hope to achieve against a clever opponent C is $Z = 0$, and this is achieved when R adopts his optimum strategy $(\frac{1}{2}, \frac{1}{2})$. Similarly, player C minimizes Z by adopting his optimum strategy $(\frac{1}{2}, \frac{1}{2})$. To determine the optimum strategies of a game is somewhat complicated even in the case of a game determined by a 2×2 matrix. However, for the 2×2 case we shall state a simple rule which when applied will, in general, provide us with such strategies.

OPTIMUM STATEGIES RULE

a. Preliminaries: Let us first define a mapping called "differential operator," D_x, which maps a function $f(x)$ of the independent variable x into a new function as follows:

$$D_x(x^n) = nx^{n-1} \qquad \text{for all } n \tag{8.5}$$

$$D_x\left(\sum_{i=1}^{n} a_i f_i(x)\right) = a_i\left(\sum_{i=1}^{n} D_x f_i(x)\right) \tag{8.6}$$

where the a_i's are constants or variables different from x.

It clearly follows from Equation (8.5) that D_x maps every constant and every variable other than x into zero (why?).
For example,

$$\begin{aligned} D_x(4y^2x^3 - 3yx^2 + 4x + 3) &= 4y^2D_x(x^3) - 3yD_x(x^2) \\ &\quad + 4D_x(x) + 3D_x(x^0) \\ &= 12y^2x^2 - 6yx + 4 + 0 \end{aligned}$$

However,

$$\begin{aligned} D_y(4y^2x^3 - 3yx^2 + 4x + 3) &= 4x^3D_y(y^2) - 3x^2D_y(y) \\ &\quad (4x + 3)\, D_y(y^0) \\ &= 8x^3y - 3x^2 + 0 \end{aligned}$$

It can easily be seen that in the first case every symbol other than x was treated as a constant, and in the second example every symbol other than y was treated as a constant. For a better understanding of this mapping we suggest that the reader verify the following mappings and work on some examples of his own.

1. $D_x(5x^4y^3 - 3x^2y^2 + 4x - y) = 20x^3y^3 - 6xy^2 + 4$
2. $D_y(5x^4y^3 - 3x^2y^2 + 4x - y) = 15x^4y^2 - 6x^2y - 1$
3. $D_x(x^3y^3 - 5x + 4y + 6) = 3x^2y^3 - 5$
4. $D_y(x^3y^3 - 5x + 4y + 6) = 3x^3y^2 + 4$

b. Rule: In general, a nonlinear function F of the variables x and y possesses a maximum or a minimum for the values of x and y for which both $Dx(F)$ and $Dy(F)$ are zero.[3] The maximum or minimum of a linear function F of x and y can be determined geometrically since such a function represents a plane in the space of three dimensions. The function

$$Z = 4xy - 2x - 2y + 1$$

becomes maximum for the values of x and y for which

$$D_x(Z) = 4y - 2 = 0 \qquad \text{and} \qquad D_y(Z) = 4x - 2 = 0$$

that is, Z is maximum if $x = \frac{1}{2}$ and $y = \frac{1}{2}$. These were the values of x and y that led us to the optimum strategies $(\frac{1}{2}, \frac{1}{2})$ and $(\frac{1}{2}, \frac{1}{2})$ for both players in the version of the penny-matching game with strategies $(x, 1 - x)$ and $(y, 1 - y)$.

Let us now return to the general case of a matrix game with payoff matrix the $m \times n$ matrix A and expected gain $E(P, Q) = PAQ^t$, and let us assume that V_R and V_C are the maximum numbers for which there exist strategies P_0 and Q_0 for the players R and C, respectively, such that the expected gains satisfy the following inequalities:

$$E(P_0, Q) \geq V_R, \qquad \text{for every strategy } Q \text{ of } C \tag{8.7}$$

$$E(P, Q_0) \leq V_C, \qquad \text{for every strategy } P \text{ of } R \tag{8.8}$$

Clearly, the strategies P_0 and Q_0 must be optimum strategies of R and C, respectively, since the numbers V_R and V_C were the maximum such numbers, and the inequalities of Equations (8.7) and (8.8) were satisfied for any strategy Q adopted by C in the first case and any strategy P adopted by R in the second case.

Theorem 8.1 For any matrix game with payoff matrix the $m \times n$ matrix A we can find optimum strategies P_0 and Q_0 and numbers V_R and V_C such that $E(P_0, Q) \geq V_R$ for any strategy Q of C and $E(P, Q_0) \leq V_C$ for any strategy P of R. There also exists a number V, called the "value" of the game, such that $V = V_R = V_C$.

Moreover, if P_0 and Q_0 are optimum strategies of R and C and

$$\vec{V} = (V, V, \ldots, V)_{1\times m} \qquad \vec{V}_1 = (V, \ldots, V)^t_{1\times n}$$

then $P_0A \geq \vec{V}$ and $AQ_0{}^t \leq \vec{V}_1$ if and only if V is the value of the game.

DEFINITION 8.4 In a matrix game with payoff matrix A, a strategy P_0 is an optimum strategy for player R, a strategy Q_0 is an optimum strategy for player C, and a number V is the value of the game if and only if P_0, Q_0, and V satisfy the inequalities:

$$P_0AQ^t \geq V \qquad \text{for every strategy } Q \text{ of } C \tag{8.9}$$

$$PAQ_0{}^t \leq V \qquad \text{for every strategy } P \text{ of } R \tag{8.10}$$

The game is said to be fair if and only if its value $V = 0$

Although the optimum strategies of a matrix game are not always unique, its value is unique, and without any proof we formalize this result in Theorem 8.2.

Theorem 8.2 If P_0 and Q_0 are optimum strategies of R and C, respectively, in a matrix game with matrix A and value V, then

$$E(P_0, Q_0) = P_0AQ_0{}^t = V$$

and V is unique. Moreover, V is positive if all the entries of A are positive.

For example, the game of Problem 2a in Exercises 8.3 is determined by the matrix

$$A = \begin{pmatrix} 0 & 3 \\ 2 & 1 \end{pmatrix}$$

and its optimum strategies are

$$P_0 = (\tfrac{1}{4}, \tfrac{3}{4}) \qquad \text{and} \qquad Q_0 = (\tfrac{1}{2}, \tfrac{1}{2})$$

Thus the value of the game $V = 3/2$ is given by

$$V = P_0AQ_0{}^t = \left(\frac{1}{4}, \frac{3}{4}\right)\begin{pmatrix} 0 & 3 \\ 2 & 1 \end{pmatrix}\begin{pmatrix} \frac{1}{2} \\ \frac{1}{2} \end{pmatrix} = \frac{3}{2}$$

Theorem 8.3 If every entry in a matrix A of a game is increased by an amount k, then the value of the game is also increased by k, but the optimum strategies remain the same.

As an illustration of Theorem 8.3, let us consider the matrix

$$A = \begin{pmatrix} -1 & 2 \\ 1 & 0 \end{pmatrix}$$

of Problem 2c in Exercises 8.3. Then the optimum strategies of A are

$$P_0 = (\tfrac{1}{4}, \tfrac{3}{4}) \qquad \text{and} \qquad Q_0 = (\tfrac{1}{2}, \tfrac{1}{2})$$

On the other hand the optimum strategies of

$$A = \begin{pmatrix} -1+k & 2+k \\ 1+k & 0+k \end{pmatrix}$$

are obtained as follows.

Let

$$P = (x, 1-x) \qquad \text{and} \qquad Q = (y, 1-y)$$

be the required strategies, then

$$E(P, Q) = PAQ^t \qquad \text{or} \qquad E(P, Q) = k + 2x + y - 4xy$$

Thus

$$D_xE(P, Q) = 2 - 4y \qquad \text{and} \qquad D_yE(P, Q) = 1 - 4x$$

Consequently, $y = \frac{1}{2}$ and $x = \frac{1}{4}$ when

$$D_xE = D_yE = 0$$

and the optimum strategies remain

$$P = (x, 1-x) = (\tfrac{1}{4}, \tfrac{3}{4}) \qquad \text{and} \qquad Q = (y, 1-y) = (\tfrac{1}{2}, \tfrac{1}{2})$$

However, the value of the game, $V = \frac{1}{2}$, becomes

$$V' = (\tfrac{1}{4}, \tfrac{3}{4}) \begin{pmatrix} -1+k & 2+k \\ 1+k & 0+k \end{pmatrix} \begin{pmatrix} \frac{1}{2} \\ \frac{1}{2} \end{pmatrix} = \tfrac{1}{2} + k = V + k$$

8.3 Strictly and Nonstrictly Determined Games

The optimum strategies of a game determined by a matrix A can often be obtained by finding the entry of A that is both a minimum in its row and a maximum in its column. The existence or nonexistence of such entries classifies games into "strictly" and "nonstrictly" determined. To clarify the difference between the two types of games and motivate their solutions, we present the following examples.

Example 8.1 Let

$$A = \begin{pmatrix} \boxed{3} & 0 & \textcircled{-4} & -3 \\ 2 & \boxed{3} & \boxed{\textcircled{1}} & 2 \\ \textcircled{-4} & 2 & -1 & 3 \end{pmatrix}$$

be the matrix of a game. Let us circle the minimum entry of each row of A and put a box about the maximum entry of each column. This process yields that the entry 1 of the second row and third column is both circled and put in a box. We call the entry 1 the "minimax" element of A, and the point (2, 3) (second row and third column) a saddle point of A.

Example 8.2 Let

$$A = \begin{pmatrix} \boxed{6} & 2 & \boxed{4} & 1 \\ 3 & \textcircled{-1} & 3 & 0 \\ 3 & \boxed{5} & \textcircled{-2} & \boxed{3} \end{pmatrix}$$

be the matrix of a matrix game, and let us follow the procedure of Example 8.1. Then we find that this matrix does not have any entry that is both circled and boxed. Thus this matrix has neither a "minimax" element nor a saddle point.

Example 8.3 Let

$$B = \begin{pmatrix} \boxed{5} & 3 & \boxed{4} & \boxed{\textcircled{2}} \\ 3 & \textcircled{-2} & 3 & 1 \\ 2 & \boxed{4} & \textcircled{-3} & 0 \end{pmatrix}$$

be the matrix of a matrix game, and again let us circle the minimum entry of each row of B, and place a box about the maximum entry of each column of B. This procedure leads to the entry 2, of the first row and the fourth column, which is both circled and placed in a box. Thus, as in Example 8.1, we have 2 as a "minimax" element in B, and (1, 4) (first row and fourth column) as a saddle point of B.

DEFINITION 8.5 If A is the matrix of a matrix game, we say that A has a minimax element a_{ij} if and only if a_{ij} is both the maximum of all row minima and the minimum of all column maxima, that is, if and only if a_{ij} is both the maximum of all entries of A each of which is minimum in its row and the minimum of all entries of A each of which is maximum in its column. The point (i, j)—ith row and jth column—is called a "saddle point" of A. Any game possessing a saddle point (or a minimax element) is said to be strictly determined.

Theorem 8.4 If S_{ij} is a minimax element of the matrix A of a matrix game, that is, if (i, j) is a saddle point of A, then an optimum strategy for R is to always play the ith row; an optimum strategy for C is always to play the jth column; and the value of the game is always the value of the minimax element S_{ij}.

In view of Theorem 8.4 the games determined by the matrices

$$A = \begin{pmatrix} 3 & 0 & -4 & -3 \\ 2 & 3 & \boxed{①} & 2 \\ -4 & 2 & -1 & 3 \end{pmatrix}$$

and

$$B = \begin{pmatrix} 5 & 3 & 4 & \boxed{②} \\ 3 & -2 & 3 & 1 \\ 2 & 4 & -3 & 0 \end{pmatrix}$$

of Examples 8.1 and 8.3 are strictly determined; the values of these games are given by the minimax elements 1 and 2, respectively; and the optimum strategies for players R and C in A are

$$P_0 = (0, 1, 0) \qquad \text{and} \qquad Q_0 = (0, 0, 1, 0)$$

respectively; and in B

$$P_0 = (1, 0, 0) \qquad \text{and} \qquad Q_0 = (0, 0, 0, 1) \qquad \text{(Why?)}$$

On the other hand the game determined by the matrix

$$A = \begin{pmatrix} 6 & 2 & 4 & 1 \\ 3 & -1 & 3 & 0 \\ 3 & 5 & -2 & 3 \end{pmatrix}$$

of Example 8.2 is nonstrictly determined (there is no saddle point in A) and consequently much more difficult to solve.

DEFINITION 8.6 By a solution of a matrix game we shall mean the finding of optimum strategies for both players R and C, and the finding of the value of the game.

As we observed in the previous examples, the solutions of strictly determined games are quite simple to obtain. Similarly, with the use of the differential operator or because of geometric considerations we can easily solve games determined by 2×2 matrices. For all other games the simplex method will provide the solution, as we shall demonstrate in the subsequent section.

Exercises

1. If A defines a two-person zero-sum game and if P and Q are the strategies of the player R and C, then compute the expectation of R in each of the following.

a. $A = \begin{pmatrix} 3 & -2 \\ 1 & -2 \end{pmatrix}$, $P = (\frac{1}{3}, \frac{2}{3})$, $Q = (\frac{1}{2}, \frac{1}{2})$.

b. $B = \begin{pmatrix} 3 & 1 & -4 \\ -2 & 3 & 2 \end{pmatrix}$, $P = (\frac{2}{3}, \frac{1}{3})$, $Q = (\frac{1}{5}, \frac{3}{5}, \frac{1}{5})$.

c. $A = \begin{pmatrix} 3 & -2 & -1 \\ 2 & -3 & 1 \\ -2 & 1 & 3 \end{pmatrix}$, $P = (\frac{2}{5}, \frac{2}{5}, \frac{1}{5})$, $Q = (\frac{3}{9}, \frac{5}{9}, \frac{1}{9})$.

d. $A = \begin{pmatrix} 0 & 5 & 6 \\ -5 & 0 & 7 \\ -6 & -7 & 0 \end{pmatrix}$, $P = (\frac{1}{3}, \frac{1}{3}, \frac{1}{3})$, $Q = (\frac{2}{5}, \frac{1}{5}, \frac{2}{5})$.

Ans. (a) $-\frac{1}{6}$; (c) $-\frac{2}{5}$

2. If A defines a two-person zero-sum game, then determine optimum strategies for both players R and C, and the value of the game for each of the following games. Is the game fair?

a. $\begin{pmatrix} 0 & 3 \\ 2 & 1 \end{pmatrix}$.
b. $\begin{pmatrix} -4 & 3 \\ 6 & 0 \end{pmatrix}$.

c. $\begin{pmatrix} -1 & 2 \\ 1 & 0 \end{pmatrix}$.
d. $\begin{pmatrix} 4 & -3 \\ 0 & 2 \end{pmatrix}$.

e. $\begin{pmatrix} -2 & 3 \\ 3 & -4 \end{pmatrix}$.
f. $\begin{pmatrix} 0 & -3 \\ 3 & 0 \end{pmatrix}$.

Ans. (a) $(\frac{1}{4}, \frac{3}{4})$, $(\frac{1}{2}, \frac{1}{2})$, $\frac{3}{2}$, No; (c) $(\frac{1}{4}, \frac{3}{4})$, $(\frac{1}{2}, \frac{1}{2})$, $\frac{1}{2}$, No;
(e) $(7/12, 5/12)$, $(7/12, 5/12)$, $1/12$, No.

3. Determine optimum strategies and the value of the game defined by

$$A = \begin{pmatrix} a & 0 \\ 0 & b \end{pmatrix}.$$

Ans. $P = Q = \left(\frac{b}{a+b}, \frac{a}{a+b}\right), \; v = \frac{ab}{a+b}$

4. Determine optimum strategies and the value of the game defined by

$$A = \begin{pmatrix} a & -b \\ -c & d \end{pmatrix}, \qquad a, b, c, d \geq 0$$

When would this game be fair?

Ans. $\left(\frac{c+d}{a+b+c+d}, \frac{a+b}{a+b+c+d}\right), \; \left(\frac{b+d}{a+b+c+d}, \frac{a+c}{a+b+c+d}\right), \quad v = \frac{ad-bc}{a+b+c+d}, \; ad - bc = 0$

5. Represent each of the following games by a two-person zero-sum matrix game.
   a. The players R and C choose, independently, one of the numbers, 1, 3, 5, and 7. If both choose the same number, then C pays to R the amount of the chosen number; otherwise R pays to C the amount of his own number.
   b. The players R and C choose, independently, one of the numbers 3, 0, or -2. Let R's choice be denoted by x and that of C by y. Then C pays R the amount $(x - y)(x + y)$.

6. Two stores, R and C, are to be located in one of two towns, T_1 and T_2. Town T_1 has 60% of the population and town T_2 has 40%. If both stores locate in the same town, they split all the business equally. However, if they locate in different towns, each will get all the business of that town. Where should each store be located?

Ans. Both in T_1 or $P = (1, 0)$ and $Q = (1, 0)$

7. Two stores, R and C, are to be located in one of three towns, T_1, T_2, and T_3. The distances from T_1 to T_2, T_1 to T_3, and T_2 to T_3 are 9 miles, 15 miles, and 12 miles, respectively. Each of the towns T_1, T_2, and T_3 has 50%, 30%, and 20% of the popula-

tion, respectively. Now if both stores locate in the same town, they split the business equally, but if they locate in different towns, all the business in the town that does not have a store will go to the closer of the two stores. Where should each store be located?

Ans. R in either T_1 or T_2, $P = (1, 0, 0)$ or $P = (0, 1, 0)$
C in either T_1 or T_2, $Q = (1, 0, 0)$ or $Q = (0, 1, 0)$

8. If

$$A = \begin{pmatrix} a & b \\ c & d \end{pmatrix}$$

is the payoff matrix of a matrix game, then determine the optimum strategies of the players R and C and the value of the game. When would this game be fair?
Question: Can the answer of this problem be used to solve any 2×2 game? Please explain.

9. Find the value and optimum strategies of the matrix games determined by the following matrices:

a. $$A = \begin{pmatrix} 15 & 2 & -3 \\ 8 & 5 & 6 \\ -7 & 3 & 0 \end{pmatrix}.$$

b. $$A = \begin{pmatrix} 1 & -11 & 5 \\ 2 & -4 & 1 \\ 3 & -7 & 2 \\ 3 & -4 & 2 \\ -6 & -4 & 7 \end{pmatrix}.$$

Ans. (a) $v = 5$, $(0, 1, 0)$, $(0, 1, 0)$;
(b) $v = -4$, $(0, a, 0, 1 - a, 0)$, $(0, 1, 0)$

10. In the children's game "stone or scissors or paper," if one says "stone" and the other "scissors," then the former wins a penny. Similarly, scissors beats paper, and paper beats stone. If the two players name the same item, then the game is a tie.
   a. Find the payoff matrix of this game.
   b. Can your solve the game?

11. If every entry in the matrix A of a game is increased by an amount k, then prove that the value of the game is also increased by k, but the optimum strategies remain the same.

8.4 The Simplex Method in Solving Matrix Games

In considering solutions of matrix games we have restricted ourselves thus far to those matrix games that were either strictly determined and therefore easily solved or else were defined by a 2×2 matrix. However, realistic games frequently are nonstrictly determined, and they lead to very large matrices for which the techniques previously developed are inadequate and insufficient.

Since in any matrix game player R is maximizing and player C is minimizing, we may infer equivalence of matrix games and linear programs. Thus if a matrix game can indeed be formulated as a linear programming problem, then the simplex method will provide the solution of such a game. That we can in fact transform a matrix game into a linear programming problem will be illustrated in the following. Before we undertake such a transformation, however, we shall assume that the matrix game is nonstrictly determined and that it contains neither a "recessive row" nor a "recessive column" defined as follows.

DEFINITION 8.7 We say that the matrix A of a matrix game contains a recessive row r_i if and only if there exists another row r_j in A such that $r_i \leq r_j$ (meaning each element of r_i is less or equal to the corresponding element of r_j). Similarly, we say that A contains a recessive column C_i if and only if there exists another column C_j such that $C_i \geq C_j$. The row r_j is said to dominate the row r_i, and the column C_j is said to dominate the column C_i.

It is quite clear that if the matrix A of a game contains a recessive row r_i, then player R would rather play row r_j $(r_i \leq r_j)$ since he can win the same or a larger amount in every play of the game. Likewise, if A should contain a recessive column C_i, then player C would rather play column $C_j(C_i \geq C_j)$. In either case a recessive row or column can always be deleted from the matrix of the game without in any way altering the solution of the game. As an illustration of these ideas let us now consider the following example.

Example 8.4 Let

$$A = \begin{pmatrix} -6 & -5 & -1 \\ 2 & -1 & 3 \\ -2 & 3 & 5 \end{pmatrix}$$

be the matrix of a matrix game, then the row $R_i = (-6, -5, -1)$ is less than the row $R_2 = (2, -1, 3)$ in A since every entry in R_1 is less than the corresponding entry in R_2. Thus R_1 is a recessive row, and it can therefore be omitted from the matrix A of the game. Hence the matrix A can be reduced to the matrix

$$A_1 = \begin{pmatrix} 2 & -1 & 3 \\ -2 & 3 & 5 \end{pmatrix}$$

which defines a game equivalent to that of A, that is, a game with the same solution as that of A. Furthermore, the column

$$C_3 = \begin{pmatrix} 3 \\ 5 \end{pmatrix}$$

is greater than the column

$$C_2 = \begin{pmatrix} -1 \\ 3 \end{pmatrix}$$

since each element of C_3 is greater than the corresponding element of C_2. Thus C_3 is a recessive column of A, and it can therefore be omitted, thus reducing the matrix A_1 to the matrix

$$A_2 = \begin{pmatrix} 2 & -1 \\ -2 & 3 \end{pmatrix}$$

which again defines a game equivalent to that of A_1 and A. Now the solution of the game can be obtained as the solution of that game which A_2 determines. Hence, applying the techniques of the previous section, we obtain the expectation or expected gain of R as follows:

$$E(P, Q) = PA_2Q^t = (x, 1 - x)\begin{pmatrix} 2 & -1 \\ -2 & 3 \end{pmatrix}\begin{pmatrix} y \\ 1 - y \end{pmatrix}$$

or

$$E(P, Q) = 8xy - 4x - 5y + 3$$

Thus to obtain optimum strategies for R and C we need to find x and y such that the function $E(P, Q)$ is a maximum. This we do by applying the *optimum strategies rule* of the differential operators D_x and D_y as follows.

$$D_xE(P, Q) = D_x(8xy - 4x - 5y + 3) = 8y - 4 = 0$$

$$D_yE(P, Q) = D_y(8xy - 4x - 5y + 3) = 8x - 5 = 0$$

or

$$y = \tfrac{1}{2} \qquad \text{and} \qquad x = \tfrac{5}{8}$$

Therefore $(\frac{5}{8}, 1 - \frac{5}{8})$ and $(\frac{1}{2}, 1 - \frac{1}{2})$ or

$$P_0 = (\tfrac{5}{8}, \tfrac{3}{8}) \qquad \text{and} \qquad Q_0 = (\tfrac{1}{2}, \tfrac{1}{2})$$

are optimum strategies for R and C in the game A_2, respectively, and the value of the game A_2 is

$$V = E(P_0, Q_0) = P_0 A_2 Q_0{}^t = (\tfrac{5}{8} \quad \tfrac{3}{8}) \begin{pmatrix} 2 & -1 \\ -2 & 3 \end{pmatrix} \begin{pmatrix} \frac{1}{2} \\ \frac{1}{2} \end{pmatrix} = \tfrac{1}{2}$$

Now the solution of the original game defined by A can easily be obtained as

$$V = \tfrac{1}{2},\ P_0 = (0, \tfrac{5}{8}, \tfrac{3}{8}) \qquad \text{and} \qquad Q_0 = (\tfrac{1}{2}, \tfrac{1}{2}, 0)$$

since R never plays the first row of A and C never plays the last column. They were both recessive in A.

Another way to solve this game is provided by Problem 4 of Exercises 8.3 in which $a = 2$, $b = 1$, $c = 2$, and $d = 3$. Thus

$$P_0 = \left(0, \frac{2 + 3}{2 + 1 + 2 + 3}, \frac{2 + 1}{8}\right), \qquad Q_0 = \left(\frac{1 + 3}{8}, \frac{2 + 2}{8}, 0\right),$$

and

$$V = \frac{2 \times 3 - 1 \times 2}{8}$$

or

$$P_0 = (0, \tfrac{5}{8}, \tfrac{3}{8}), \qquad Q_0 = (\tfrac{1}{2}, \tfrac{1}{2}, 0), \qquad \text{and} \qquad V = \tfrac{1}{2}$$

Explain the zeros in P_0 and Q_0.

With the assumption that the matrix game defined by a matrix A is nonstrictly determined, and that A contains neither a recessive row nor a recessive column, we shall proceed to show that a matrix game can indeed be formulated as a linear program to which the simplex method is applicable. To demonstrate this formulation we consider Example 8.5.

Example 8.5 Demonstrate that the matrix game determined by the matrix

$$A = \begin{pmatrix} 2 & 5 \\ 3 & 1 \\ 0 & 3 \end{pmatrix}$$

is equivalent to a linear programming problem, and use the simplex method to find a solution of this game.

Discussion Let V be the value of this game. Clearly V is positive since every entry of A is positive or zero. In general, we can make the entries of the payoff matrix A of any game positive or zero by adding the same suitable number k to each entry of A. As we have seen in Theorem 8.3 of the last section, this modification of A merely increases the value of the game by k, and it does not in any way alter any optimum strategies. Thus we may always make the entries of A positive thereby obtaining a positive value of the game. Let $Q_0 = (q_1, q_2)$ be an optimum strategy for C, then by Theorem 8.1 of the last section

$$AQ_0{}^t \leq \vec{V}_1 = (V, V)^t$$

or

$$\begin{pmatrix} 2 & 5 \\ 3 & 1 \\ 0 & 3 \end{pmatrix} \begin{pmatrix} q_1 \\ q_2 \end{pmatrix} \leq \begin{pmatrix} V \\ V \end{pmatrix}$$

or equivalently

$$\begin{aligned} 2q_1 + 5q_2 &\leq V \\ 3q_1 + q_2 &\leq V \\ 0q_1 + 3q_2 &\leq V \end{aligned} \tag{8.11}$$

$$q_1, \quad q_2 \geq 0, \quad q_1 + q_2 = 1 \tag{8.12}$$

since $Q_0 = (q_1, q_2)$ is a strategy. Now, since $V > 0$, if we divide by V and if we let

$$x = \frac{q_1}{V} \qquad \text{and} \qquad x_2 = \frac{q_2}{V}$$

the relations of Equations (8.11) and (8.12) become

$$\begin{aligned} 2x_1 + 5x_2 &\leq 1 \\ 3x_1 + x_2 &\leq 1 \\ 0x_1 + 3x_2 &\leq 1 \end{aligned} \tag{8.13}$$

$$x_1, \quad x_2 \geq 0, \quad x_1 + x_2 = \frac{1}{V} \tag{8.14}$$

Thus to find an optimum strategy, $Q_0 = (q_1, q_2)$, for C, we must find x_1 and x_2 subject to the constraints of (8.13) such that V is a minimum. This, of course, is the case since a strategy Q of C is optimum if it makes V a minimum. However, to minimize V is the same as to maximize $1/V$, which leads to the following linear program.

LP: Maximize

$$M = \frac{1}{V} = x_1 + x_2 \tag{8.15}$$

subject to the constraints

$$\begin{aligned} 2x_1 + 5x_2 &\leq 1 \\ 3x_1 + x_2 &\leq 1 \\ 0x_1 + 3x_2 &\leq 1 \\ x_1, \quad x_2 &\geq 0 \end{aligned} \tag{8.16}$$

Thus far we have shown that the matrix game is equivalent to a linear program or that it can be formulated as a linear program. Therefore the simplex method can be applied.

Solution of the Game Introducing the slack variables x_3, x_4, and x_5 we can write the system of inequalities given in (8.16) as a system of equations. If to that system we add the equation, $-x_1 - x_2 + M = 0$, as the last equation of the system we obtain the system:

$$\begin{aligned} 2x_1 + 5x_2 + x_3 &= 1 \\ 3x_1 + x_2 + x_4 &= 1 \\ 0x_1 + x_2 + x_5 &= 1 \\ -x_1 - x_2 + M &= 0 \end{aligned} \tag{8.17}$$

The augmented matrix of Equations (8.17)

$$\begin{pmatrix} 2 & \textcircled{5} & 1 & 0 & 0 & 0 & 1 \\ 3 & 1 & 0 & 1 & 0 & 0 & 1 \\ 0 & 1 & 0 & 0 & 1 & 0 & 1 \\ -1 & -1 & 0 & 0 & 0 & 1 & 0 \end{pmatrix}$$

The pivot number is 5. Thus dividing the first row by 5 we obtain

$$\begin{pmatrix} 2/5 & \textcircled{1} & 1/5 & 0 & 0 & 0 & 1/5 \\ 3 & 1 & 0 & 1 & 0 & 0 & 1 \\ 0 & 1 & 0 & 0 & 1 & 0 & 1 \\ -1 & -1 & 0 & 0 & 0 & 1 & 0 \end{pmatrix}$$

Now we may perform elementary row operations to place zeros below the 1 that occupies the place of the pivot number. These elementary row operations yield the matrix

$$\begin{pmatrix} 2/5 & 1 & 1/5 & 0 & 0 & 0 & 1/5 \\ \boxed{13/5} & 0 & -1/5 & 1 & 0 & 0 & 4/5 \\ -2/5 & 0 & -1/5 & 0 & 1 & 0 & 4/5 \\ -3/5 & 0 & 1/5 & 0 & 0 & 1 & 1/5 \end{pmatrix}$$

The new pivot number is 13/5. Thus dividing the second row by 13/5 we obtain the matrix

$$\begin{pmatrix} 2/5 & 1 & 1/5 & 0 & 0 & 0 & 1/5 \\ \boxed{1} & 0 & -1/13 & 5/13 & 0 & 0 & 4/13 \\ -2/5 & 0 & -1/5 & 0 & 1 & 0 & 4/5 \\ -3/5 & 0 & 1/5 & 0 & 0 & 1 & 1/5 \end{pmatrix}.$$

Let us now perform elementary row operations again to place zeros above and below the 1, of the first column, that occupies the place of the pivot number. These elementary row operations yield the matrix

$$\begin{pmatrix} 0 & 1 & 15/65 & -2/13 & 0 & 0 & 5/65 \\ 1 & 0 & -1/13 & 5/13 & 0 & 0 & 4/13 \\ 0 & 0 & -15/65 & 2/13 & 1 & 0 & 60/65 \\ 0 & 0 & 10/65 & 3/13 & 0 & 1 & 25/65 \end{pmatrix}$$

Hence

$$0x_1 + 0x_2 + \frac{10}{65}x_3 + \frac{15}{65}x_4 + M = \frac{25}{65}$$

or

$$M = \frac{25}{65} - \frac{10}{65}x_3 - \frac{15}{65}x_4$$

and M becomes maximum 25/65 for $x_3 = x_4 = 0$, and $x_2 = 5/65$, $x_1 = 4/13$ as clearly shown in the last matrix. Thus the value of the game is

$$V = \frac{1}{M} \qquad \text{or} \qquad V = \frac{13}{5}$$

and an optimum strategy, $Q_0 = (q_1, q_2)$, for C is the strategy

$$(Vx_1, Vx_2) = (\tfrac{4}{5}, \tfrac{1}{5})$$

To complete the solution of this game we must find an optimum strategy for player R. Thus with V as the value of the game, let

$$P_0 = (p_1, p_2, p_3)$$

be an optimum strategy for R. Once again, by Theorem 8.1

$$P_0A \geq \vec{V} \qquad \text{or} \qquad (p_1, p_2, p_3)\begin{pmatrix} 2 & 5 \\ 3 & 1 \\ 0 & 3 \end{pmatrix} \geq (V, V)$$

or equivalently

$$\begin{aligned} 2p_1 + 3p_2 + 0p_3 &\geq V \\ 5p_1 + p_2 + 3p_3 &\geq V \end{aligned} \tag{8.18}$$

$$p_1, p_2, p_3 \geq 0, \qquad p_1 + p_2 + p_3 = 1 \tag{8.19}$$

since

$$P_0 = (p_1, p_2, p_3)$$

is a strategy.

Now, since $V > 0$, if we divide by V and let

$$x_1 = \frac{p_1}{V}, \qquad x_2 = \frac{p_2}{V}, \qquad \text{and} \qquad x_3 = \frac{p_3}{V}$$

the relations of Equations (8.18) and (8.19) become

$$\begin{aligned} 2x_1 + 3x_2 + 0x_3 &\geq 1 \\ 5x_1 + x_2 + 3x_3 &\geq 1 \end{aligned} \tag{8.20}$$

$$x_1, x_2, x_3 \geq 0, \qquad x_1 + x_2 + x_3 = \frac{1}{V} \tag{8.21}$$

Consequently, finding an optimum strategy $P_0 = (p_1, p_2, p_3)$ for R means finding x_1, x_2, and x_3 subject to the constraints of Equation (8.20) such that V is a maximum. This, of course, is the case since a strategy P_0 of R is optimum if it makes V a maximum. Now, to maximize V is the same as to minimize $1/V$, which leads to the following linear program:

LP: Minimize

$$m = \frac{1}{V} = x_1 + x_2 + x_3 \tag{8.22}$$

subject to the constraints

$$\begin{aligned} 2x_1 + 3x_2 + 0x_3 &\geq 1 \\ 5x_1 + x_2 + 3x_3 &\geq 1 \\ x_1, x_2, x_3 &\geq 0 \end{aligned} \tag{8.23}$$

However, to minimize

$$m = \frac{1}{V} = x_1 + x_2 + x_3$$

is equivalent to maximizing

$$f = -m = \frac{-1}{V} = -x_1 - x_2 - x_3$$

This and multiplication by -1 of the first two inequalities in (8.23) change this linear program into the following equivalent one:

LP: Maximize

$$f = -x_1 - x_2 - x_3 \tag{8.24}$$

subject to the constraints

$$\begin{aligned} -2x_1 - 3x_2 - 0x_3 &\leq -1 \\ -5x_1 - x_2 - 3x_3 &\leq -1 \\ x_1, x_2, x_3 &\geq 0 \end{aligned} \tag{8.25}$$

We shall now apply the simplex method to the above linear program by introducing the slack variables x_4 and x_5 thus obtaining the system

$$\begin{aligned} -2x_1 - 3x_2 - 0x_3 + x_4 \qquad\qquad &= -1 \\ -5x_1 - x_2 - 3x_3 \qquad + x_5 \qquad &= -1 \\ x_1 + x_2 + x_3 \qquad\qquad + f &= 0 \end{aligned} \tag{8.26}$$

The augmented matrix of (8.26) is

$$\begin{pmatrix} -2 & -3 & 0 & 1 & 0 & 0 & -1 \\ -5 & -1 & -3 & 0 & 1 & 0 & -1 \\ 1 & 1 & 1 & 0 & 0 & 1 & 0 \end{pmatrix} \tag{8.27}$$

This matrix has no negative entry in the last row, and it may very easily be assumed that $f = 0$ is the maximum obtained when $x_1 = x_2 = x_3 = 0$. However, this is not a feasible solution since for these values of x_1, x_2, and x_3 the first equation of (8.26) yields $x_4 = -1$, which is absurd, since the slack variable x_4 must be positive or zero. However, if we modify the last row of the matrix (8.27) by adding to it the second row of the same matrix, and if we also modify the first row of (8.27) by adding to it -3 times the second row, we obtain the equivalent matrix

$$\begin{pmatrix} \textcircled{13} & 0 & 9 & 1 & -3 & 0 & 2 \\ -5 & -1 & -3 & 0 & 1 & 0 & -1 \\ -4 & 0 & -2 & 0 & 1 & 1 & -1 \end{pmatrix} \tag{8.28}$$

In the new matrix (8.28) 13 is a pivot number. Therefore let us divide the first row of the matrix by 13, and then via elementary row operations let us place zeros below the position of the pivot number. This sequence of elementary row operations on the matrix (8.28) yields the equivalent matrix

$$\begin{pmatrix} 1 & 0 & 9/13 & 1/13 & -3/13 & 0 & 2/13 \\ 0 & -1 & 6/13 & 5/13 & -2/13 & 0 & -3/13 \\ 0 & 0 & 10/13 & 4/13 & 1/13 & 1 & -5/13 \end{pmatrix}$$

Therefore

$$f = -\frac{5}{13} - \frac{10}{13}x_3 - \frac{4}{13}x_4 - \frac{1}{13}x_5$$

or

$$f = -\frac{5}{13} \qquad \text{(maximum)}$$

for $x_3 = x_4 = x_5 = 0$, and consequently

$$x_1 = \frac{2}{13} \qquad \text{and} \qquad x_x = \frac{3}{13}$$

as can be seen from the last matrix. Since $f = -m = -1/V$, we obtain $V = 13/5$ for the value of the game, as we knew it from the first part of the solution, and

$$P = (p_1, p_2, p_3) = (Vx_1, Vx_2, Vx_3) = \left(\frac{2}{13}\cdot\frac{13}{5}, \frac{3}{13}\cdot\frac{13}{5}, 0\right)$$

or

$$P = \left(\frac{2}{5}, \frac{3}{5}, 0\right)$$

for an optimum strategy of the player R.

The completion of Example 8.5 clearly indicates the transformation of a matrix game into a linear program and the subsequent application of the simplex method.

In general, if $A = (a_{ij})_{m\times n}$ is the matrix of a nonstrictly determined matrix game, if A contains neither a recessive row nor a recessive column, and if all the entries of A are positive, then the value V of the game is positive (see Theorem 8.2). Furthermore, for optimum strategies P_0 and Q_0 of the players R and C, respectively, Theorem 8.1 implies

$$P_0A \geq \vec{V} \quad \text{or} \quad (p_1, p_2, \ldots, p_m)\begin{pmatrix} a_{11} & a_{12} & \cdots & a_{1n} \\ a_{21} & a_{22} & \cdots & a_{2n} \\ a_{m1} & a_{m2} & \cdots & a_{mn} \end{pmatrix} \geq (V, V, \ldots, V)$$

or

$$\begin{aligned} a_{11}p_1 + a_{21}p_2 + \cdots + a_{m1}p_m &\geq V \\ a_{12}p_1 + a_{22}p_2 + \cdots + a_{m2}p_m &\geq V \\ \cdots\cdots\cdots\cdots\cdots\cdots\cdots\cdots \\ a_{1n}p_1 + a_{2n}p_2 + \cdots + a_{mn}p_m &\geq V \end{aligned} \tag{8.29}$$

where $p_i \geq 0$ for all $i = 1, 2, \ldots, m$, and

$$\sum_{i=1}^{m} p_i = 1$$

since $P = (p_1, p_2, \ldots, p_m)$ is a strategy of the game, and

$$AQ_0^t \leq \vec{V} \quad \text{or} \quad \begin{pmatrix} a_{11} & a_{12} & \cdots & a_{1n} \\ a_{21} & a_{22} & \cdots & a_{2n} \\ \cdots & \cdots & \cdots & \cdots \\ a_{m1} & a_{m2} & \cdots & a_{mn} \end{pmatrix}\begin{pmatrix} q_1 \\ q_2 \\ \cdot \\ q_n \end{pmatrix} \leq \begin{pmatrix} V \\ V \\ \cdot \\ V \end{pmatrix}$$

or

$$\begin{aligned} a_{11}q_1 + a_{12}q_2 + \cdots + a_{1n}q_n &\leq V \\ a_{21}q_1 + a_{22}q_2 + \cdots + a_{2n}q_n &\leq V \\ \cdots\cdots\cdots\cdots\cdots\cdots\cdots\cdots \\ a_{m1}q_1 + a_{m2}q_2 + \cdots + a_{mn}q_n &\leq V \end{aligned} \tag{8.30}$$

where $q_i \geq 0$ for all i, and

$$\sum_{i=1}^{m} q_i = 1$$

since $Q_0 = (q_1, q_2, \ldots, q_n)$ is a strategy of the game.

Since V is positive, if we divide both sides of (8.29) and (8.30) by V, if we let $x_i = p_i/V$ for $i = 1, 2, \ldots, m$, in case (8.29), and if we let $y_i = q_i/V$ for $i = 1, 2, \ldots, n$, in case (8.30), we obtain the following two systems of inequalities:

$$\begin{array}{l} a_{11}x_1 + a_{21}x_2 + \cdots + a_{m1}x_m \geq 1 \\ a_{12}x_1 + a_{22}x_2 + \cdots + a_{m2}x_m \geq 1 \\ \cdots\cdots\cdots\cdots\cdots\cdots\cdots\cdots \\ a_{1n}x_1 + a_{2n}x_2 + \cdots + a_{mn}x_m \geq 1 \\ x_1, x_2, \ldots, x_m \geq 0 \quad \text{and} \quad x_1 + x_2 + \cdots + x_m = \dfrac{1}{V} \end{array} \tag{8.31}$$

$$\begin{array}{l} a_{11}y_1 + a_{12}y_2 + \cdots + a_{1n}y_n \leq 1 \\ a_{21}y_1 + a_{22}y_2 + \cdots + a_{2n}y_n \leq 1 \\ \cdots\cdots\cdots\cdots\cdots\cdots\cdots\cdots \\ a_{m1}y_1 + a_{m2}y_2 + \cdots + a_{mn}y_n \leq 1 \\ y_1, y_2, \ldots, y_n \geq 0 \quad \text{and} \quad y_1 + y_2 + \cdots + y_n = \dfrac{1}{V} \end{array} \tag{8.32}$$

Now, since we are interested in a strategy $P_0 = (p_1, p_2, \ldots, p_m)$ of R that maximizes the value V of the game or equivalently minimizes $m = 1/V$, and since we are also interested in a strategy $Q_0 = (q_1, q_2, \ldots, q_n)$ of C that minimizes the value V of the game or equivalently maximizes $M = 1/V$, we have in fact reduced the game into the following two linear programs:

LP_1: Minimize

$$m = \frac{1}{V} = x_1 + x_2 + \cdots + x_m \tag{8.33}$$

subject to the constraints

$$\begin{array}{l} a_{11}x_1 + a_{21}x_2 + \cdots + a_{m1}x_m \geq 1 \\ a_{12}x_1 + a_{22}x_2 + \cdots + a_{m2}x_m \geq 1 \\ \cdots\cdots\cdots\cdots\cdots\cdots\cdots\cdots \\ a_{1n}x_1 + a_{2n}x_2 + \cdots + a_{mn}x_m \geq 1 \\ \qquad x_1, x_2, \ldots, x_m \geq 0 \end{array} \tag{8.34}$$

LP_2: Maximize

$$M = \frac{1}{V} = y_1 + y_2 + \cdots + y_n \tag{8.35}$$

Subject to the constraints

$$\begin{aligned} a_{11}y_1 + a_{12}y_2 + \cdots + a_{1n}y_n &\leq 1 \\ a_{21}y_1 + a_{22}y_2 + \cdots + a_{2n}y_n &\leq 1 \\ \cdots\cdots\cdots\cdots\cdots\cdots\cdots\cdots \\ a_{m1}y_1 + a_{m2}y_2 + \cdots + a_{mn}y_n &\leq 1 \\ y_1, y_2, \ldots, y_n &\geq 0 \end{aligned} \tag{8.36}$$

The simplex method will solve these two linear programs[4] nicely in terms of x_i's, y_i's, m, and M, which in turn will yield the strategies P_0, Q_0, and the value of the game V since

$$p_i = Vx_i,\ q_i = Vy_i \qquad \text{for all } i$$

and

$$V = \frac{1}{m} \qquad \text{or} \qquad V = \frac{1}{M}$$

In applying the simplex method to obtain the solution of a game determined by an $m \times n$ matrix A we required that A has the following characteristics:

1. Its entries are all positive.
2. It contains neither recessive rows nor recessive columns.

Quite frequently this will not be the case, however, and a transformation of the matrix A to a matrix A_0 that possesses the above characteristics will be necessary. This, of course, is possible since by Theorem 8.3 we may modify the matrix A of a game by adding the same number k to each of its entries so that they may become positive or zero thus altering the value V of the game to $V + k$, yet without in any way altering the optimum strategies of R and C. Thus in the final analysis we must subtract k from the value of the game determined by the suitable matrix A_0 in order to obtain the value of the game A. Inasmuch as the deletion of the recessive rows and columns is concerned, we have already demonstrated that such deletion may be performed without any consequences other than replacing by zero the entry or entries corresponding to the deleted rows or columns in the probability vectors that constitute the strategies $P = (p_1, p_2, \ldots, p_m)$ and $Q = (q_1, q_2, \ldots, q_n)$.

As an illustration of the above concepts let us consider the following examples.

Example 8.6

Show that the matrix game determined by the matrix

$$A = \begin{pmatrix} -4 & -1 & 1 \\ -2 & 0 & 2 \\ 2 & -4 & 3 \\ -4 & 1 & 4 \end{pmatrix}$$

is equivalent to a linear program, and use the simplex method to solve the matrix game.

Solution The first row $R_1 = (-4, -1, 1)$ is less than the second row $R_2 = (-2, 0, 2)$ since every entry in R_1 is less than every entry in R_2. Therefore R_1 is a recessive row, and it can be deleted from the matrix A of the game. The matrix obtained after the deletion of R_1 is

$$A_1 = \begin{pmatrix} -2 & 0 & 2 \\ 2 & -4 & 3 \\ -4 & 1 & 4 \end{pmatrix}$$

and it defines a game equivalent to that of A. Furthermore, the column $C_3 = (2, 3, 4)^t$ is greater than the column $C_2 = (0, -4, 1)^t$, again since each entry of C_3 is greater than the corresponding entry of C_2. Thus C_3 is a recessive column in A_1 and can therefore be deleted without significantly altering the solution of the game. The new matrix

$$A_2 = \begin{pmatrix} -2 & 0 \\ 2 & -4 \\ -4 & 1 \end{pmatrix}$$

defines a game whose solution, with minor modification, provides the solution of the original game. The matrix A_2 contains no recessive rows or columns. However, it contains negative entries that can be transformed into positive or zero ones by adding six to each of them.

This transforms the matrix A_2 into the matrix

$$A_3 = \begin{pmatrix} 4 & 6 \\ 8 & 2 \\ 2 & 7 \end{pmatrix}$$

We shall now derive the solution of the game defined by A_3; and from that solution by incorporating the appropriate zeros as entries of the optimum strategies P and Q and by subtracting six from the value of the game of A_3, we shall obtain the required optimum strategies and the value of the game defined by A. Thus let V be the value of the game

$$A_3 = \begin{pmatrix} 4 & 6 \\ 8 & 2 \\ 2 & 7 \end{pmatrix}$$

Clearly, V is positive since all the entries of A_3 are positive. Since A_3 contains no minimax elements this game is nonstrictly determined. If

$$Q_0 = (q_1, q_2) \qquad \text{and} \qquad P_0 = (p_1, p_2, p_3)$$

are optimum strategies for the players C and R, respectively, then by Theorem 8.1

$$A_3Q_0^t \leq (V, V)^t \qquad \text{and} \qquad P_0A_3 \geq (V, V)$$

or

$$\begin{pmatrix} 4 & 6 \\ 8 & 2 \\ 2 & 7 \end{pmatrix} \begin{pmatrix} q_1 \\ q_2 \end{pmatrix} \leq \begin{pmatrix} V \\ V \end{pmatrix}$$

and

$$(p_1, p_2, p_3) \begin{pmatrix} 4 & 6 \\ 8 & 2 \\ 2 & 7 \end{pmatrix} \geq (V, V)$$

or

$$4q_1 + 6q_2 \leq V$$
$$8q_1 + 2q_2 \leq V$$
$$2q_1 + 7q_2 \leq V$$
$$q_1, q_2 \geq 0 \qquad \text{and} \qquad q_1 + q_2 = 1 \tag{8.37}$$

since $Q_0 = (q_1, q_2)$ is a strategy;

$$4p_1 + 8p_2 + 2p_3 \geq V$$
$$6p_1 + 2p_2 + 7p_3 \geq V$$
$$p_1, p_2, p_3 \geq 0 \qquad \text{and} \qquad p_1 + p_2 + p_3 = 1 \tag{8.38}$$

since $P_0 = (p_1, p_2, p_3)$ is a strategy. If we now divide by V the inequalities in (8.37) and (8.38), and if we make the substitutions $x_i = q_i/V$, $i = 1, 2$, in (8.37), and $y_i = p_i/V$, $i = 1, 2, 3$, in (8.38), we obtain the following two linear programs:

LP_1: Minimize V or maximize

$$M = \frac{1}{V} = \frac{q_1}{V} + \frac{q_2}{V} = x_1 + x_2 \tag{8.39}$$

subject to the constraints

$$4x_1 + 6x_2 \leq 1$$
$$8x_1 + 2x_2 \leq 1 \tag{8.40}$$
$$2x_1 + 7x_2 \leq 1$$
$$x_1, x_2 \geq 0$$

LP_2: Maximize V or minimize

$$m = \frac{1}{V} = \frac{p_1}{V} + \frac{p_2}{V} + \frac{p_3}{V} = y_1 + y_2 + y_3$$

or maximize

$$f = -m = \frac{-1}{V} = -y_1 - y_2 - y_3 \tag{8.41}$$

subject to the constraints

$$4y_1 + 8y_2 + 2y_3 \geq 1$$
$$6y_1 + 2y_2 + 7y_3 \geq 1 \tag{8.42}$$
$$y_1, y_2, y_3 \geq 0$$

Thus far we have formulated the game as equivalent to the linear programs LP_1 and LP_2. The solutions of these two linear programs will provide, with very little modification, the solution to our original game A. In the following we shall use the simplex method to solve LP_1, and we shall only give the answer for LP_2, thus leaving the details of its solution as an exercise for the reader.

Solution of LP_1 If we should introduce the slack variables x_3, x_4, and x_5 in the first three inequalities (8.40) of LP_1, we obtain the system

$$\begin{aligned} 4x_1 + 6x_2 + x_3 \qquad\qquad &= 1 \\ 8x_1 + 2x_2 \qquad + x_4 \qquad &= 1 \\ 2x_1 + 7x_2 \qquad\qquad + x_5 &= 1 \end{aligned}$$

The addition of the equation, $-x_1 - x_2 + M = 0$, as the last equation of the above system, yields the following system:

$$\begin{aligned} 4x_1 + 6x_2 + x_3 \qquad\qquad\qquad &= 1 \\ 8x_1 + 2x_2 \qquad + x_4 \qquad\qquad &= 1 \\ 2x_1 + 7x_2 \qquad\qquad + x_5 \qquad &= 2 \\ -x_1 - x_2 \qquad\qquad\qquad + M &= 0 \end{aligned} \tag{8.43}$$

The augmented matrix of (8.43) is

$$\begin{pmatrix} 4 & 6 & 1 & 0 & 0 & 0 & 1 \\ 8 & 2 & 0 & 1 & 0 & 0 & 1 \\ 2 & \textcircled{7} & 0 & 0 & 1 & 0 & 1 \\ -1 & -1 & 0 & 0 & 0 & 1 & 0 \end{pmatrix} \tag{8.44}$$

The pivot number of matrix (8.44) is either 8 or 7 since there is a tie on the negative entries of the last row. Without altering the solution of the game we may pick 7 as the pivot number. On division of the third row of matrix (8.44) by seven we obtain

$$\begin{pmatrix} 4 & 6 & 1 & 0 & 0 & 0 & 1 \\ 8 & 2 & 0 & 1 & 0 & 0 & 1 \\ 2/7 & \textcircled{1} & 0 & 0 & 1/7 & 0 & 1/7 \\ -1 & -1 & 0 & 0 & 0 & 1 & 0 \end{pmatrix} \tag{8.45}$$

If we now perform suitable elementary row operations on matrix (8.45), we may place zeros above and below the place of the pivot number thus reducing (8.45) to the following matrix:

$$\begin{pmatrix} \textcircled{16/7} & 0 & 1 & 0 & -6/7 & 0 & 1/7 \\ 52/7 & 0 & 0 & 1 & -2/7 & 0 & 5/7 \\ 2/7 & 1 & 0 & 0 & 1/7 & 0 & 1/7 \\ -5/7 & 0 & 0 & 0 & 1/7 & 1 & 1/7 \end{pmatrix}$$

Now, with pivot number 16/7 and again with appropriate row operations we reduce the above matrix to the following matrix:

$$\begin{pmatrix} 1 & 0 & 7/16 & 0 & -3/8 & 0 & 1/16 \\ 0 & 0 & -13/4 & 1 & \textcircled{5/2} & 0 & 1/4 \\ 0 & 1 & -1/8 & 0 & 1/4 & 0 & 1/8 \\ 0 & 0 & 5/16 & 0 & -1/8 & 1 & 3/16 \end{pmatrix}$$

Once again, with pivot number 5/2 and appropriate elementary row operations we can put the number 1 in the place of the pivot number and zeros above and below it, thus obtaining the matrix

$$\begin{pmatrix} 1 & 0 & -1/20 & 3/20 & 0 & 0 & 1/10 \\ 0 & 0 & -13/10 & 2/5 & 1 & 0 & 1/10 \\ 0 & 1 & 1/5 & -1/10 & 0 & 0 & 1/10 \\ 0 & 0 & 3/20 & 1/20 & 0 & 1 & 1/5 \end{pmatrix} \tag{8.46}$$

From the last matrix we obtain

$$M = \frac{1}{5} - \frac{3}{20}x_3 - \frac{1}{20}x_4$$

or

$$M = \frac{1}{5}$$

as maximum whenever

$$x_3 = x_4 = 0 \qquad \text{and} \qquad x_1 = x_2 = x_5 = \frac{1}{10}$$

as can easily be seen in matrix (8.46). Therefore, the value of this game is

$$V = \frac{1}{M} = 5$$

$$q_1 = Vx_1 = 5\frac{1}{10} = \frac{1}{2}$$

and

$$q_2 = Vx_2 = 5\frac{1}{10} = \frac{1}{2}$$

However, the value of the original game must be $5 - 6 = -1$ since we added the number 6 to each entry of A to make positive or zero every entry of the matrix. Moreover, an optimum strategy of C must have zero for its third entry since the third column was recessive and thus deleted. Finally, the value of this game is $V = -1$, and an optimum strategy for C is $Q_0 = (\frac{1}{2}, \frac{1}{2}, 0)$. Similarly, the reader will find the same value for the game and an optimum strategy $P_0 = (0, \frac{3}{4}, \frac{1}{4}, 0)$ for player R by solving the linear program LP_2.

Exercises

1. Complete the solution of the game defined in Example 8.6.
2. For the games determined by the following matrices determine (i) whether or not the game is strictly determined, (ii) whether or not there are recessive rows or columns, and (iii) solve each game.

a. $\begin{pmatrix} 0 & -3 \\ 3 & 0 \end{pmatrix}$.

b. $\begin{bmatrix} 5 & 0 & -4 & -3 \\ 2 & 3 & 1 & 2 \\ 3 & -1 & -1 & -4 \\ -4 & 2 & -1 & 3 \end{bmatrix}$.

c. $\begin{pmatrix} 5 & 3 & 6 & 4 & 2 \\ 3 & -2 & 5 & 3 & 1 \\ 2 & 4 & 4 & -3 & 0 \end{pmatrix}$

d. $\begin{pmatrix} 3 & 0 & -4 & -3 \\ 2 & 3 & 1 & 2 \\ -4 & 2 & -1 & 3 \end{pmatrix}$.

Ans. (b) Yes, second column, $V = 1$, $P = (0, 1, 0, 0)$, and $Q = (0, 0, 1, 0)$; (d) Yes; 2 and 4 columns, $V = 1$

3. Solve the games determined by the following matrices:

a. $\begin{pmatrix} 3 & -1 & 0 & 2 \\ -2 & 1 & -1 & 3 \end{pmatrix}$. b. $\begin{pmatrix} 2 & 1 & -1 \\ -1 & 2 & -1 \\ -2 & -3 & -1 \\ -1 & -1 & 0 \end{pmatrix}$.

c. $\begin{pmatrix} -2 & 1 & 2 \\ 3 & -1 & -2 \\ -1 & 1 & 3 \end{pmatrix}$. d. $\begin{pmatrix} -1 & -3 & 2 \\ 1 & -2 & 2 \\ -3 & 0 & -1 \\ 1 & -1 & 0 \end{pmatrix}$.

e. $\begin{pmatrix} 2 & -4 & -3 \\ -1 & -2 & 1 \\ 0 & 1 & 1 \end{pmatrix}$. f. $\begin{pmatrix} -5 & -3 & 1 \\ 2 & -1 & 2 \\ -2 & 3 & 4 \end{pmatrix}$.

Ans. (b) $V = -4/13$, $P_0 = (3/13, 1/13, 0, 9/13)$, $Q_0 = (1/13, 3/13, 9/13)$; (d) $V = -3/5$, $P_0 = (0, 0, 2/5, 3/5)$, $Q_0 = (1/5, 4/5, 0)$; (f) $V = \frac{1}{2}$, $P_0 = (0, 5/8, 3/8)$, $Q_0 = (\frac{1}{2}, \frac{1}{2}, 0)$

References for Supplementary Readings

1. John von Neumann and Oskar Morgenstern, *Theory of Games and Economic Behavior* (Princeton, N.J.: Princeton University Press, 1947).
2. R. D. Luce and H. Raiffa, *Games and Decisions* (New York: Wiley, 1957).
3. G. B. Thomas, *Calculus and Analytic Geometry*, Chapter 15 (Reading, Mass.: Addison-Wesley, 1968).
4. G. Owen, *Finite Mathematics* (Philadelphia: Saunders, 1970).

PART TWO

Calculus

CHAPTER 9

Functions, Limits, and Continuity

9.1 More on Functions and Graphs

In Chapter 2 we discussed in some detail the concept of the function $y = f(x)$ where x is the domain or independent variable and y is the range or dependent variable. Such a function, $y = f(x)$, is generally known as a function of one variable x and is represented graphically in the Cartesian xy plane. Since adequate description and study of many practical problems require functions of more than one variable, our exclusive concern up to this point with functions of a single variable severely restricts our flexibility. Thus we must broaden our point of view by examining the concept of multivariable functions,[1] that is, functions of more than one variable. The existence and usefulness of such functions is suggested by the following examples.

1. The final grade of a student in a mathematics class is usually the average of four or five test scores, his class participation, and homework assignment scores. In other words, the student's final grade depends on more than one variable or simply is a function of two or more variables.

2. A multivariable function is provided by a company's sales S as a function of the price x it charges for its products, the amount

of expenses for advertising y, the consumer population z, and the per capita income t. This function of the independent variables x, y, z, and t is conventionally written as

$$S = f(x, y, z, t)$$

3. The height of a person depends on more than his age, and intellectual achievement certainly depends on more than IQ.

4. In business and economics, costs depend on more than production, and sales on many factors not excluding location and advertising.

5. The total profit per acre on a wheat farm is related to the expenditure per acre for (a) labor, and (b) soil conditioners and fertilizers. If x represents the dollars per acre spent on labor and y the dollars per acre spent on soil improvement, then the function

$$P(x, y) = 40x + 65y + 12xy - 8x^2 - 7y^2$$

represents the profit that a farmer might use to determine the optimum expenditure levels.

DEFINITION 9.1 A function

$$z = f(x, y)$$

of two variables is a correspondence associating with each pair of possible values of the independent variables x and y one and only one value of the dependent variable z. More generally, a function

$$y = f(x_1, x_2, \ldots, x_n)$$

of n variables is a correspondence associating with each n-tuple $(x_1, x_2, \ldots, x_n)$ of values one and only one value of the dependent variable y.

Obviously, regardless of the number of independent variables there is always one and only one dependent variable whose value can be obtained once arbitrary values are assigned to the independent variables. For example, if

$$z = f(x, y) = x^2 + y^2 - 5$$

is a function of two variables, then the values of the function $f(x, y)$ are obtained by assigning arbitrary values to x and y as the following table indicates.

x	y	$z = f(x, y) = x^2 + y^2 - 5$
0	1	$f(0, 1) = -4$
-2	3	$f(-2, 3) = 8$
-1	-2	$(f(-1, -2) = 0$
0	0	$f(0, 0) = -5$
-10	12	$f(-10, 12) = 239$
.	.	.
.	.	.
.	.	.

Now, multivariable functions can be manipulated just as those of one variable. For instance, if

$$f(x, y) = x^2 + 3y^2 - 1$$

and

$$g(x, y) = 2x^2 - 3y^2 + 4$$

then $f + g$, $f - g$, $f \cdot g$, and f/g are specified as follows:

$$(f + g)(x, y) = f(x, y) + g(x, y) = (x^2 + 3y^2 - 1) + (2x^2 - 3y^2 + 4) = 3x^2 + 3$$

$$(f - g)(x, y) = f(x, y) - g(x, y) = (x^2 + 3y^2 - 1) - (2x^2 - 3y^2 + 4) = -x^2 + 6y^2 - 5$$

$$(f \cdot g)(x, y) = f(x, y)g(x, y) = (x^2 + 3y^2 - 1) \times (2x^2 - 3y^2 + 4) = 2x^4 - 9y^4 + 3x^2y^2 + 2x^2 + 15y^2 - 4$$

and

$$(f/g)(x, y) = \frac{f(x, y)}{g(x, y)} = \frac{x^2 + 3y^2 - 1}{2x^2 - 3y^2 + 4} \quad \text{if } g(x, y) \neq 0$$

Specifically, if $x = -1$ and $y = 2$ (two arbitrary values), then

$$(f + g)(-1, 2) = 3(-1)^2 + 3 = 6$$

$$(f - g)(-1, 2) = -(-1)^2 + 6(2)^2 - 5 = 18$$

$$(f \cdot g)(-1, 2) = 2(-1)^4 - 9(2)^4 + 3(-1)^2(2)^2 + 2(-1)^2 + 15(2)^2 - 4 = -72$$

and

$$(f/g)(-1, 2) = \frac{(-1)^2 + 3(2)^2 - 1}{2(-1)^2 - 3(2)^2 + 4} = \frac{12}{-6} = -2$$

Since graphing multivariable functions requires three or higher dimensional geometry,[1] the graphs of such functions usually present difficulties to students with little or no background in such geometries. Therefore in the subsequent sections we will concentrate primarily on the algebra rather than the geometry of multivariable functions. As a minimum exposure and review, however, we present here a sequence of simple examples in both functions of one and more than one variable, together with their graphs.

Example 9.1 On the same axes, graph a firm's

a. Total cost function $C(x) = x^2 - 4x + 100$.
b. Short-run demand function $d(x) = 56 - 2x$.
c. Marginal cost function $c(x) = 2x - 4$.

Solution Since the demand and marginal cost functions are linear, the respective graphs are straight lines (Figure 9.1) determined by the pair of points (0, 56) and (28, 0) on the demand function, and (0, −4) and (2, 0) on the marginal cost function, respectively. On the other hand the graph of the total cost function is a "parabola" as indicated by its table of values and the graph shown in Figure 9.1.

Example 9.2 If a department store has two departments, if the earn-

Figure 9–1

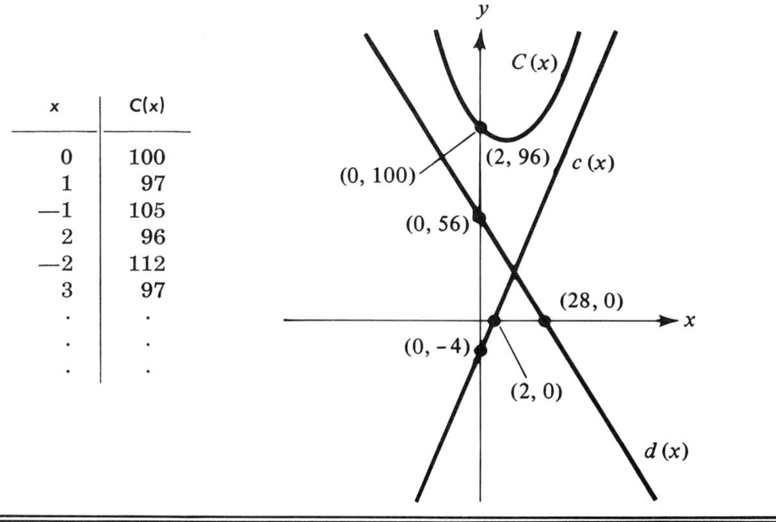

x	C(x)
0	100
1	97
—1	105
2	96
—2	112
3	97
.	.
.	.
.	.

ings (gross revenue minus cost of goods sold) are given by

$$E_1 = 5x + 4y + 3xy - x^2 - 2y^2$$

for one department and

$$E_2 = 4z + 3t + 2zt - 2z^2 - 3t^2 - 10$$

for the other, where x and z are investments in inventory and y and t are floor space used by each department, express the total earnings as a function of x, y, z, and t.

Solution Since the total earnings are the sum of the earnings of each department we have total earnings

$$E = E_1 + E_2$$

or

$$E = 5x + 4y + 3xy - x^2 - 2y^2 + 4z + 3t + 3zt - 2z^2 - 3t^2 - 10$$

An interesting question related to this problem is to determine the allocation of floor space to each department so that the total earnings will be optimum. This and other similar problems are considered in Chapter 11.

Example 9.3 Discuss the domain and graph of the following functions.

a. $f(x) = x^3 + 1.$

b. $g(x) = \dfrac{1}{x^3 + 1}.$

c. $h(x) = \begin{cases} x & \text{if } x < 2. \\ 2 & \text{if } x > 2. \end{cases}$

Solution a. Since the value of the function can be found for all x, the domain of the function is the set of all real numbers. The graph of the function (Figure 9.2) contains neither "holes" nor "breaks or jumps," and indicates that the function is increasing for all x.

b. Clearly the function

$$g(x) = \frac{1}{x^3 + 1}$$

is not defined (has no finite value) at $x = -1$. Therefore the domain of the function contains all real numbers other than -1. Thus the

Figure 9-2

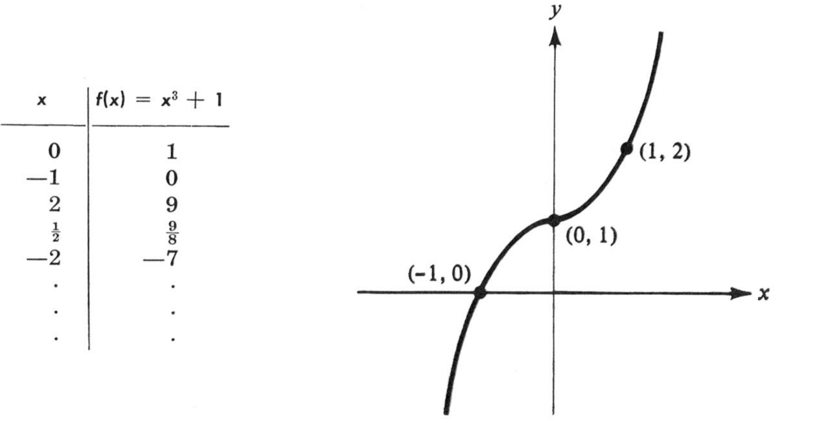

x	$f(x) = x^3 + 1$
0	1
—1	0
2	9
$\frac{1}{2}$	$\frac{9}{8}$
—2	—7
.	.
.	.
.	.

graph (Figure 9.3) of the function possesses a "break" at $x = -1$. Moreover it shows that the function is decreasing for all x and possesses the "asymptote" $x = -1$.

c. The function $h(x)$ is defined as equal to x for all $x < 2$ and equal to 2 for $x > 2$. Since the function is not defined for $x = 2$, its domain consists of all real numbers $x \neq 2$. Thus the graph of the function (Figure 9.4) contains a "hole" at $x = 2$.

Figure 9-3

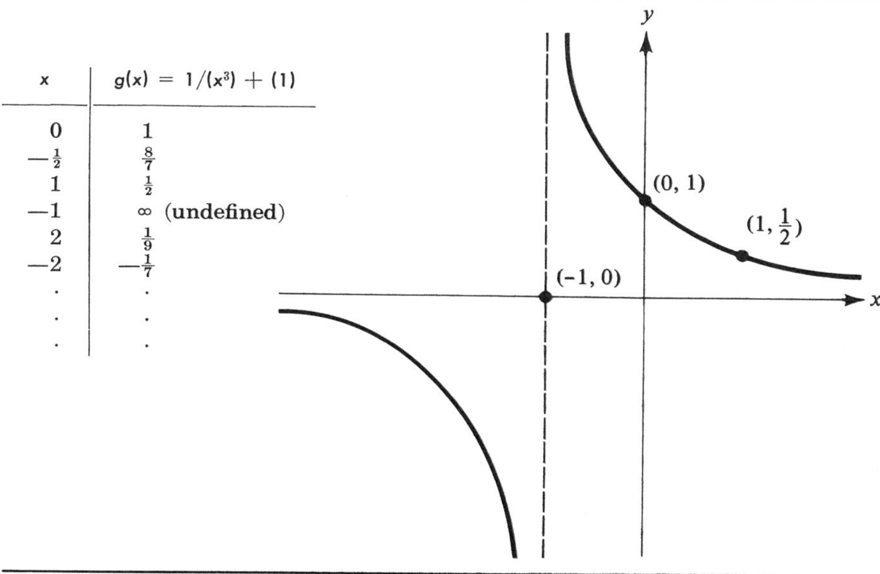

x	$g(x) = 1/(x^3) + (1)$
0	1
$-\frac{1}{2}$	$\frac{8}{7}$
1	$\frac{1}{2}$
—1	∞ (undefined)
2	$\frac{1}{9}$
—2	$-\frac{1}{7}$
.	.
.	.
.	.

Figure 9-4

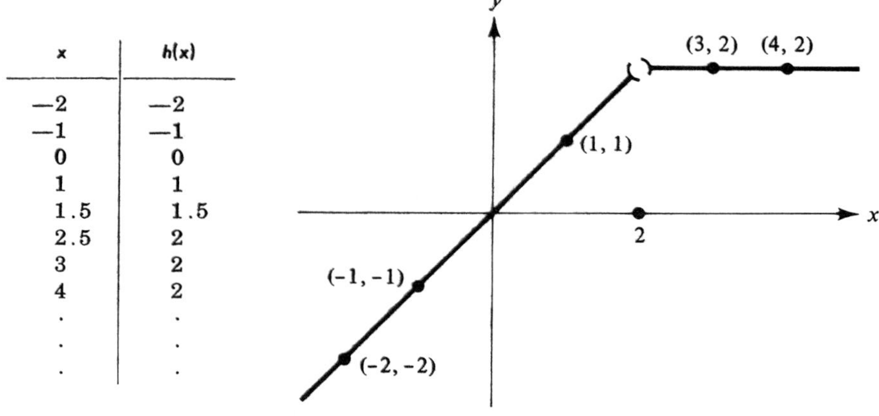

x	h(x)
−2	−2
−1	−1
0	0
1	1
1.5	1.5
2.5	2
3	2
4	2
.	.
.	.
.	.

Example 9.4 Graph the following functions of two variables.

a. $z = f(x, y) = -2x - y + 1.$
b. $z = g(x, y) = x^2 + y^2.$

Solution a. To graph $z = -2x - y + 1$ we first assign arbitrary values to the independent variables x and y, and for each pair of such values we obtain the corresponding value of z. This familiar process yields all ordered triples or points (x, y, z) whose coordinates satisfy the given equation. Since it can be shown that all these points lie in a plane, the function $z = -2x - y + 1$ is referred to as the equation of the plane it describes. In fact, every linear function

$$ax + by + cz = d$$

of two variables represents a plane in the space of three dimensions.[2] The intersections of a plane with the x, y, and z axis are called x, y, and z intercepts, respectively. The intercepts are found if we let two of the three variables be zero and solve for the third. Consequently, the intercepts of the plane $z = -2x - y + 1$ are the points $(\frac{1}{2}, 0, 0)$, $(0, 1, 0)$, and $(0, 0, 1)$. Since three noncollinear points determine one and only one plane and since three points (the intercepts) of the given plane (or function) $z = -2x - y + 1$ have been obtained, the graph of the function and its table of values are shown in Figure 9.5.

b. If we let $z = 1, 4, 9, \ldots$, that is, if we assign certain arbitrary values to z, the function $z = x^2 + y^2$ yields $1 = x^2 + y^2$, $4 = x^2 + y^2, \ldots$, respectively. These equations, however, describe circles[2] of radii 1, 2, 3, . . . in the planes $z = 1$, $z = 2$, $z = 3, \ldots$ of the space

Figure 9-5

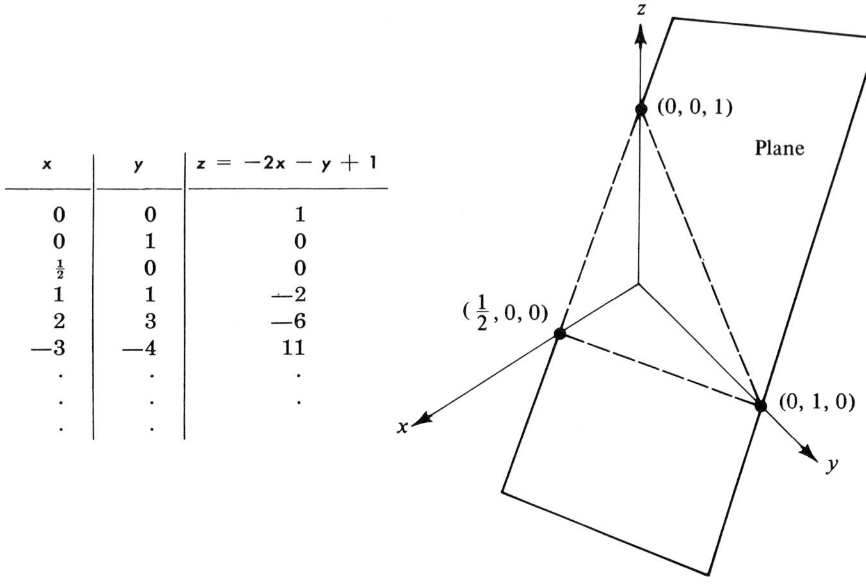

x	y	$z = -2x - y + 1$
0	0	1
0	1	0
$\frac{1}{2}$	0	0
1	1	−2
2	3	−6
−3	−4	11
.	.	.
.	.	.
.	.	

Figure 9-6

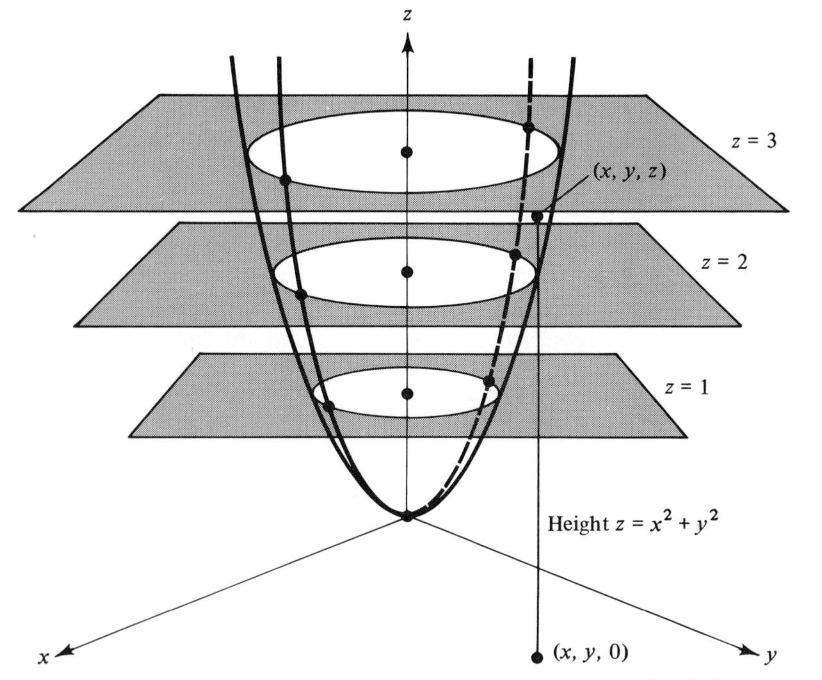

of three dimensions (Figure 9.6). Similarly, the values $y = 1, 2, 3, \ldots$ yield the parabolas[2] $z = x^2 + 1$, $z = x^2 + 4$, $z = x^2 + 9, \ldots$ of the planes $y = 1$, $y = 2$, $y = 3, \ldots$, respectively. Finally, for $x = 1, 2, 3, \ldots$ we obtain the parabolas $z = 1 + y^2$, $z = 4 + y^2$, $z = 9 + y^2, \ldots$ in the planes $x = 1, 2, 3, \ldots$ Summarizing the above information, we obtain the graph of $z = x^2 + y^2$ as the bullet-shaped surface (Figure 9.6) whose height above a point $P(x, y, 0)$ in the xy plane is the value of the function at that point.

Example 9.5 Are the functions

$$f(x, y) = \frac{2}{x^2 + y^2} \quad \text{and} \quad g(x, y) = \frac{x - y}{x^3 - y^3}$$

defined for all the values of x and y?

Solution Since for $x = 0$ and $y = 0$ the function $f(x, y)$ becomes

$$f(0, 0) = \frac{2}{0} = \infty \qquad \text{(undefined)}$$

and since $x = 1$ and $y = 1$ yield

$$g(1, 1) = \frac{0}{0} \qquad \text{(indeterminant)}$$

the functions $f(x, y)$ and $g(x, y)$ are not defined at the points

$$(x, y) = (0, 0) \qquad \text{and} \qquad (x, y) = (1, 1)$$

respectively.

The observations in the previous examples that some functions are or are not defined for all values of the variables involved and that there do or do not exist "breaks or jumps" or "holes" on the graphs of certain functions lead to the very important concepts of "limit" and "continuity" of functions.[3] The study of these concepts we will undertake in Section 9.2.

Exercises

1. If $f(x) = x^2 - 1$, then evaluate the following:

   a. $f(2)$. b. $f(m + 1)$. c. $\dfrac{f(m + h) - f(m)}{h}$.

   Ans. (a) 3; (c) $2m + h$

2. Graph the following functions, and discuss their domain, graphs,

and the means by which existing breaks, jumps, or holes can be eliminated from their graphs:

a. $f(x) = (x^2 - 4)/x + 2.$

b. $g(x) = \begin{cases} x & \text{if } x > 1. \\ x^2 & \text{if } x < 1. \end{cases}$

c. $h(x) = 1/(2x - 4).$

d. $C(x) = \begin{cases} x^3 & \text{if } x < 2. \\ -4x + 16 & \text{if } x > 2. \end{cases}$

e. $m(x) = x/(x^2 + 1).$

f. $T(x) = \begin{cases} x^2 - 1 & \text{for } x < 2. \\ 3 & \text{for } x > 2. \end{cases}$

3. If $f(x, y) = x^2 - xy + y^2$ and $g(x, y) = x^3 - 2y$, then evaluate the following:

a. $f(2, -3)$. b. $g(-2, 0)$. c. $f(1, 3) - g(2, 3)$

d. $(f + g)\ (2, 3)$. e. $\dfrac{f(2, -1)}{g(2, 4)}$. f. $(f \cdot g)\ (0, 1)$.

Ans. (a) 19; (c) 5; (f) −2

4. Graph the following functions of two variables:

a. $f(x, y) = x + 3y - 1$. b. $g(x, y) = 2x - 3y + 4$.
c. $F(x, y) = x^2 - y^2$.

5. If the cost of repairs for inspections at two points in a certain manufacturing process is given by

$$C(x, y) = 4x^2 + 2y^2 - 20x - 12y + 30$$

where x and y are the number of inspections at the two points, find the cost whenever the inspections at the two points are

a. $x = 5$ and $y = 3$. b. $x = 8$ and $y = 10$.

Ans. (a) 24

6. A sociological analysis indicates that a parameter measuring the crime rate z depends on the amount spent on welfare, as measured in x hundreds of millions of dollars, and on the amount spent on prisons as measured by y in hundreds of millions of dollars as follows:

$$z = x^3 + 2y^3 - 6x + 6y - 6xy$$

Find z for the following values of x and y.

a. $x = 5, y = 3$. b. $x = 8, y = 6$.

Ans. (a) 77

9.2 Limits and Continuity

The sequence $f(n) = 1/n$ contains the terms $1, \frac{1}{2}, \frac{1}{3}, \frac{1}{4}, \frac{1}{5}, \ldots, 1/n, 1/n+1, \ldots$ which constitute the points shown in Figure 9.7. Similarly, the sequence $g(n) = (3n+1)/n$ contains the terms $4, 3\frac{1}{2}, 3\frac{1}{3}, 3\frac{1}{4}, \ldots, 3\,1/n, \ldots$ shown as points of the line x (Figure 9.8). Clearly, the terms of the sequences and their graphs suggest that as n becomes larger and larger, the terms of the sequences "cluster" (or "pile up") near a particular number (or point). In the case of $f(n) = 1/n$, that number is 0, and that in $g(n) = (3n+1)/n$ is 3. Accordingly, we write

$$f(n) = \frac{1}{n} \rightarrow 0 \qquad \text{and} \qquad g(n) = \frac{3n+1}{n} \rightarrow 3$$

and say that the sequence $f(n) = 1/n$ approaches 0, whereas $g(n) = (3n+1)/n$ approaches 3 or that their limits are 0 and 3, respectively, as n gets larger and larger or simply as n approaches infinity ($n \rightarrow \infty$). In general, if the terms of a sequence $f(n)$ cluster near a number L as $n \rightarrow \infty$, we say that $f(n)$ has limit L or that $f(n)$ converges[3] to L. Conventionally this is written as

$$\lim_{n \to \infty} f(n) = L$$

and reads "the limit of $f(n)$ as n goes to infinity is the number L."

DEFINITION 9.2 By an h neighborhood of a point L we mean the open interval

$$(L-h, L+h) = \{x \mid L-h < x < L+h\}$$

where L and h are real numbers and $h > 0$.

Geometrically, an h neighborhood of L is a line segment with its endpoints $L-h$ and $L+h$ missing and with midpoint the point

Figure 9-7

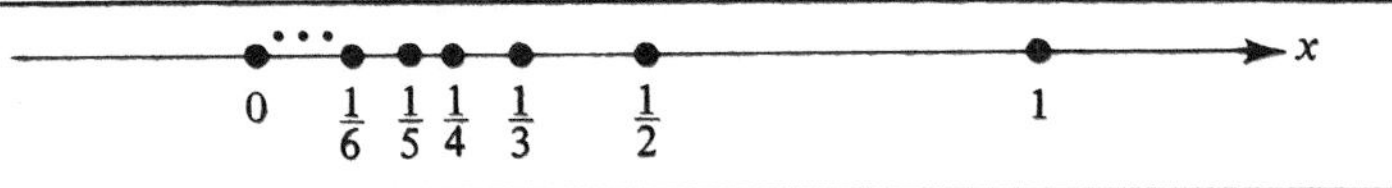

Figure 9-8

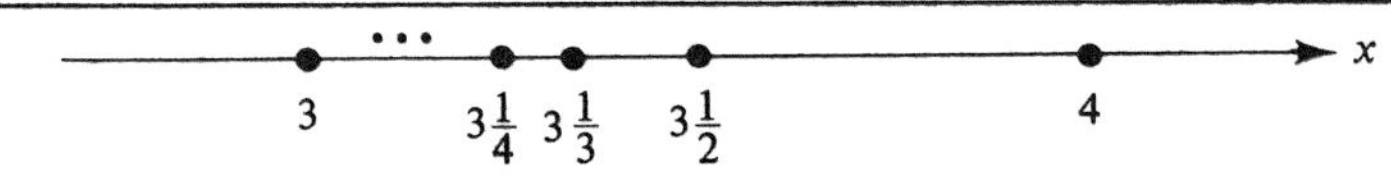

Figure 9-9

L (Figure 9.9). A 2 neighborhood of five and a $\frac{1}{3}$ neighborhood of zero are shown in Figure 9.10. When an h neighborhood contains its endpoints, it is called a "closed interval."

The history of mathematics reveals that infinite or limit processes were known to the ancient Greeks. In fact, Archimedes (287–212 B.C.) used such a process to determine areas bounded by parabolas and circles. The so-called method of exhaustion employed by Archimedes is nicely illustrated in the technique applied to calculate an area bounded by a parabola. The shaded area A under the parabola $y = x^2$ and inside the unit square shown in Figure 9.11(a) is approximated by rectangles under the curve and over the curve as indicated in Figure 9.11(b) and (c).

Since $y = x^2$, the altitude of each rectangle, namely, the value of y, is the square of the corresponding value of x. Thus if s is the sum of the areas of the rectangles under the curve, and if S is that of those over the curve, then

$$s = \frac{1}{4}\left(\frac{1}{4}\right)^2 + \frac{1}{4}\left(\frac{2}{4}\right)^2 + \frac{1}{4}\left(\frac{3}{4}\right)^2 = \frac{1^2 + 2^2 + 3^2}{4^3}$$

and

$$S = \frac{1}{4}\left(\frac{1}{4}\right)^2 + \frac{1}{4}\left(\frac{2}{4}\right)^2 + \frac{1}{4}\left(\frac{3}{4}\right)^2 + \frac{1}{4}\left(\frac{4}{4}\right)^2 = \frac{1^2 + 2^2 + 3^2 + 4^2}{4^3}$$

The geometry of the problem reveals that $s < A < S$ or

$$\frac{7}{32} = \frac{1^2 + 2^2 + 3^2}{4^3} < A < \frac{1^2 + 2^2 + 3^2 + 4^2}{4^3} = \frac{15}{32}$$

Now, if the above process is repeated by doubling the number of rectangles or equivalently by doubling the number of subdivisions

Figure 9-10

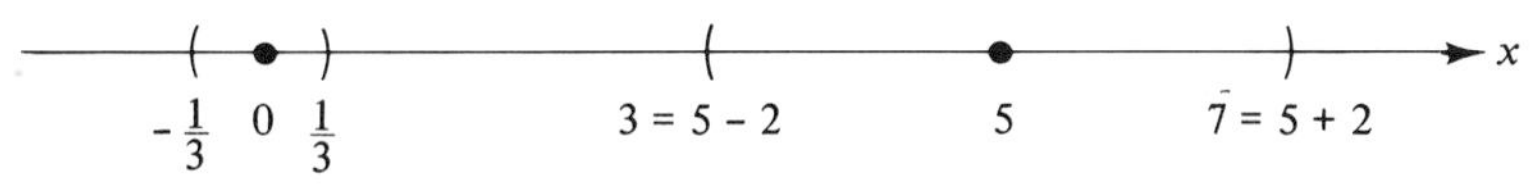

Figure 9–11

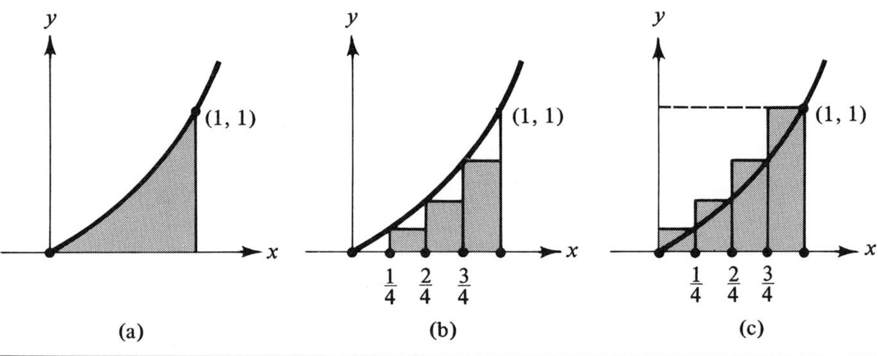

of the interval [0, 1], we obtain

$$\frac{1^2 + 2^2 + 3^2 + 4^2 + 5^2 + 6^2 + 7^2}{8^3} < A < \frac{1^2 + 2^2 + 3^2 + 4^2 + 5^2 + 6^2 + 7^2 + 8^2}{8^3}$$

or

$$\frac{35}{128} < A < \frac{51}{128}$$

The indefinite repetition of the above operations generally becomes an infinite process called the "method of exhaustion." The small areas between the rectangles and the curve are being used up or exhausted as the doubling operation is repeated. As the result of this process, we obtain the following two sets of numbers:

$$U = \left\{\frac{1^2 + 2^2 + 3^2}{4^3}, \frac{1^2 + 2^2 + \cdots + 7^2}{8^3}, \frac{1^2 + 2^2 + \cdots + 15^2}{16^3}, \ldots\right\}$$

$$O = \left\{\frac{1^2 + 2^2 + 3^2 + 4^2}{4^3}, \frac{1^2 + 2^2 + \cdots + 8^2}{8^3}, \frac{1^2 + 2^2 + \cdots + 16^2}{16^3}, \ldots\right\}$$

Finally, Archimedes showed that each of the sets U and O have the same limit, $\frac{1}{3}$; that is, the members of U and O cluster around the number $\frac{1}{3}$. During all this time, the area A has been "squeezed" between the values closer and closer to the limit of each set; that is

$$\tfrac{1}{3} \leq A \leq \tfrac{1}{3} \qquad \text{or} \qquad A = \tfrac{1}{3}$$

As we shall see in subsequent chapters, when the exhaustion operation is generalized and its properties abstracted as a mathematical system, the system is called "integral calculus" and the operation is called "integration."

Generally, the following theorem is quite useful in both determining and verifying limits of sequences.[3]

Theorem 9.1 The limit of a sequence $f(n)$ as $n \to \infty$ is L if and only if every neighborhood of L contains all but finitely many terms of the sequence.

Example 9.6 Find the limit of the sequence $f(n) = (4n + 2)/n$ as $n \to \infty$.

Solution Since $(4n + 2)/n = 4 + 2/n$, and since $2/n \to 0$ as $n \to \infty$, we obtain

$$\lim_{n \to \infty} \frac{4n + 2}{n} = 4$$

Clearly, any neighborhood of four contains all but finitely many terms of the sequence (why?)

The terms of a sequence $f(n)$ often become larger and larger as n approaches infinity (or $n \to \infty$). When this is the case, we say that the sequence does not have a finite limit or that the limit does not exist. This we indicate as

$$\lim_{n \to \infty} f(n) = \infty$$

For instance, the limit of $f(n) = n^2$ as $n \to \infty$ does not exist or it is infinity (∞), that is,

$$\lim_{n \to \infty} (n^2) = \infty$$

Since a sequence is a function whose domain consists only of the natural numbers, the concept of the limit can be extended to functions. Thus to introduce the limit concept about functions, let us consider the following specific cases.

1. The function

$$f(x) = \frac{x^2 - 9}{x - 3}$$

Figure 9-12

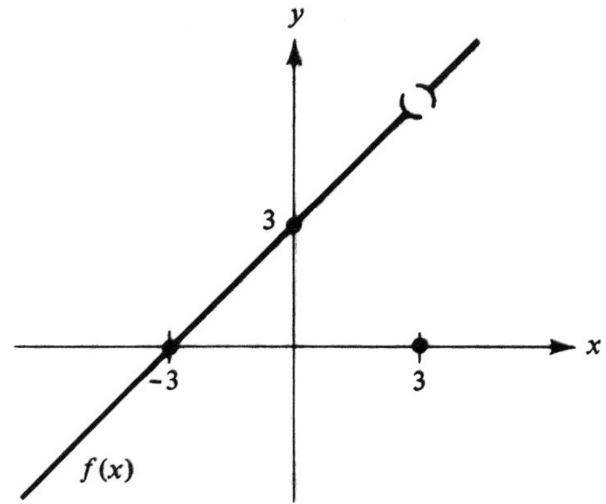

becomes $f(x) = x + 3$ for $x \neq 3$, and it is not defined (its graph contains a hole) if $x = 3$. On the other hand the graph of the function suggests that $f(x)$ is near six or approaches six as x approaches three (Figure 9.12).

2. The graph of the function

$$g(x) = \frac{2}{x^2}$$

indicates that it becomes larger and larger as x gets closer to zero and contains a "break" at $x = 0$; that is, it appears that the closer x gets to zero the closer $g(x)$ gets to infinity (∞) as Figure 9.13 indicates.

Figure 9-13

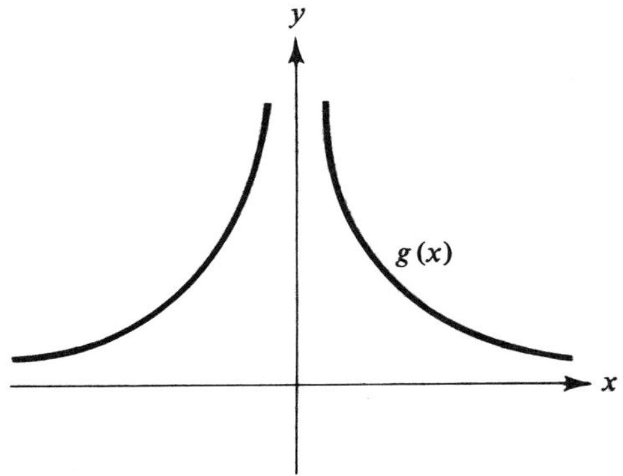

3. Finally, the function

$$h(x) = \begin{cases} 1 & \text{if } x < 0 \\ 3 & \text{if } x > 0 \end{cases}$$

not only is not defined at $x = 0$, but, as its graph suggests (Figure 9.14), the function does not approach a single value as x approaches zero. Instead it contains a "jump" at $x = 0$; that is, the function approaches one as x approaches zero from the left and three as x approaches zero from the right.

The concept underlined in the three examples above and in Example 9.3 is that of the limit extended to all functions of one variable. Accordingly,

$$\lim_{x \to 3} \left(\frac{x^2 - 9}{x - 3}\right) = 6 \qquad \lim_{x \to 0} \left(\frac{2}{x^2}\right) = \infty \qquad \text{(undefined)}$$

and the $\lim_{x \to 0} h(x)$ does not exist. Similarly,

a. $\lim_{x \to -2} \dfrac{x^2 - 4}{x + 2} = -4.$

b. $\lim_{x \to 4} (x^2 + 1) = 17.$

c. $\lim_{x \to \infty} \dfrac{3}{x} = 0.$

d. If $h(x) = \begin{cases} x^2 & \text{for } x > 1 \\ x & \text{for } x < 1 \end{cases}$ then the $\lim_{x \to 1} h(x)$ does not exist.

In general, the limit of a function $f(x)$ as x approaches the number a is the value $f(a)$ of the function at $x = a$. In fact this is the

Figure 9–14

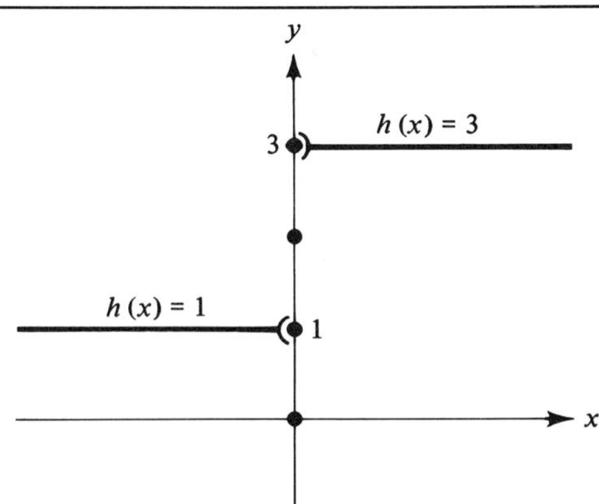

case whenever $f(a)$ exists. Otherwise the limit is the number that "best describes" the value of the function at $x = a$ [see Examples 9.6 and 9.7(b)]. Geometrically, if the graph of a function contains neither "holes" nor "breaks" nor "jumps," the limit of the function as $x \to a$ is always the value of the function[3] at $x = a$. If the graph of the function contains a "hole" at $x = a$, however, then its limit is that number which is needed to "plug" the hole. On the other hand functions whose graphs possess breaks or jumps at $x = a$ have no limits as $x \to a$. The following examples illustrate quite nicely the above observations on limits of functions.

Example 9.7 Evaluate the following limits

a. $\lim_{x \to 1} (x^2 + 1)$. b. $\lim_{x \to -1} \left(\dfrac{x^2 - 1}{x + 1}\right)$. c. $\lim_{x \to 0} \left(\dfrac{1}{1 + 4^{1/x}}\right)$.

Solution a. Since the function

$$f(x) = x^2 + 1$$

is defined for all values of x,

$$\lim_{x \to 1} f(x) = \lim_{x \to 1} (x^2 + 1) = f(1) = (1^2 + 1) = 2$$

The graph of the function (Figure 9.15) contains neither holes nor breaks nor jumps.

b. Obviously, the function

$$g(x) = \frac{x^2 - 1}{x + 1}$$

Figure 9-15

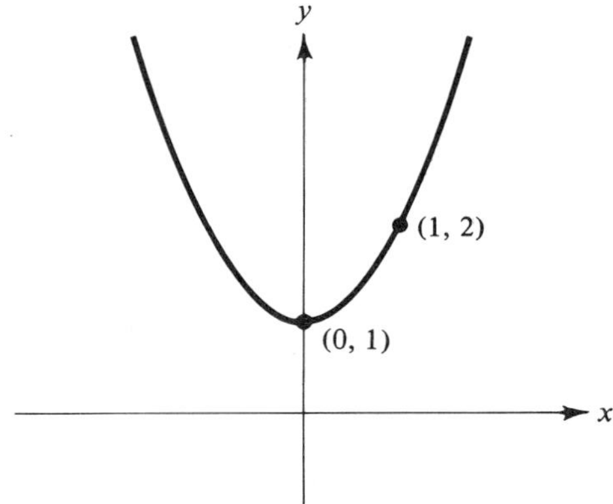

Figure 9–16

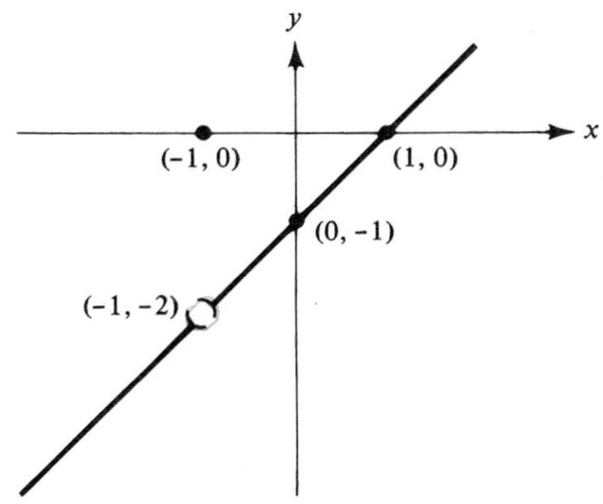

is not defined at $x = -1 (g(-1) = 0/0)$. However, $g(x) = x - 1$ if $x + 1 \neq 0$ since

$$g(x) = \frac{(x+1)(x-1)}{(x+1)}$$

and division by a nonzero number $(x + 1 \neq 0)$ is permissible. The graph of the function (Figure 9.16) suggests that its value at $x = -1$ ought to be -2. Therefore

$$\lim_{x \to -1} g(x) = \lim_{x \to -1} \left(\frac{x^2 - 1}{x + 1}\right) = \lim_{x \to -1} (x - 1) = -2$$

c. Clearly, if x approaches zero from the right, that is, through values for which $x > 0$, the exponent $1/x$ is positive and increases without limit. Moreover, raising four to higher and higher positive powers makes the entire denominator of the function

$$h(x) = \frac{1}{1 + 4^{1/x}}$$

become larger and larger again without limit. This, of course, makes the function approach zero. On the other hand if x approaches zero from the left, so that $x < 0$, the exponent $1/x$ is negative and increases without limit. Furthermore, the higher and higher negative powers of four force $4^{1/x}$ to approach zero. Thus the entire function

$$h(x) = \frac{1}{1 + 4^{1/x}}$$

approaches one as a limit (Figure 9.17). We conclude, then, that the function has a jump at $x = 0$ (Figure 9.17), and therefore $\lim_{x \to 0} 1/(1 + 4^{1/x})$ does not exist.

Figure 9-17

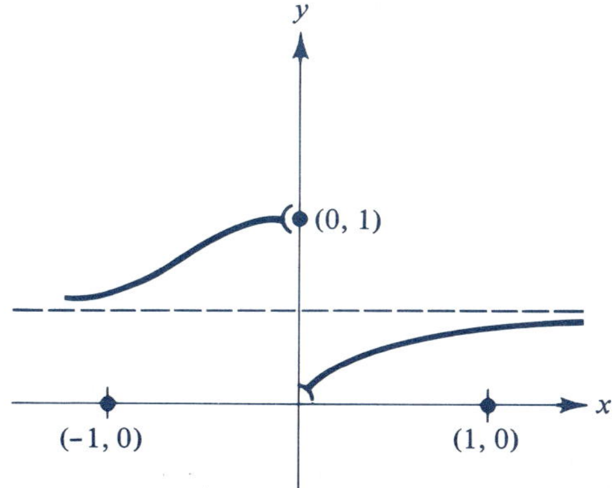

The concept of a limit extends easily to a function of two (or more) variables.[1] Thus

$$\lim_{x \to x_0;\, y \to y_0} f(x, y) = L$$

means that the value of $f(x, y)$ approaches or is L as the point corresponding to (x, y) comes nearer and nearer to that associated with (x_0, y_0). For example,

$$\lim_{x \to 0;\, y \to 0} \left(\frac{9 - x^2}{y + 3}\right) = 3$$

In general, a function of two variables will possess a limit at a point (x_0, y_0) only if the limiting behavior of the function is independent of the way in which the point (x, y) approaches (x_0, y_0). Hence if the function approaches different values when (x, y) approaches (x_0, y_0) along different paths, the limit of the function does not exist (see Example 9.8). For instance,

$$\lim_{x \to 0;\, y \to 0} \left(\frac{x^2}{x^2 + y^2}\right) = \lim_{x \to 0} \left(\frac{x^2}{x^2 + (mx)^2}\right) = \frac{1}{1 + m^2}$$

if (x, y) approaches $(0, 0)$ along the straight line (or path) $y = mx$. Since the limit $1/(1 + m^2)$ of this function depends on the slope m of the path, that is, $1/(1 + m^2)$ is different for different paths (lines) of approach,

$$\lim_{x \to 0;\, y \to 0} \left(\frac{x^2}{x^2 + y^2}\right)$$

does not exist.

In evaluating limits, as we shall have to do over and over again in our subsequent work, much time and labor can be saved by using a few general theorems[3] instead of relying on formal or intuitive definitions. The following theorems will prove particularly useful.

Theorem 9.2 The limit of the sum or difference of two (or more) functions is equal to the sum or difference of their limits, that is,

$$\lim_{x\to a} (f(x) \pm g(x)) = \lim_{x\to a} f(x) \pm \lim_{x\to a} g(x)$$

For instance,

$$\lim_{x\to 2} (x^2 + x) = \lim_{x\to 2} x^2 + \lim_{x\to 2} x = 2^2 + 2 = 6$$

Theorem 9.3 The limit of a constant c times a function $f(x)$ is equal to the constant times the limit of the function, that is,

$$\lim_{x\to a} (cf(x)) = c \lim_{x\to a} f(x)$$

For example,

$$\lim_{x\to -2} 5x^3 = 5 \lim_{x\to -2} x^3 = 5(-2)^3 = -40$$

Theorem 9.4 The limit of the product of two (or more) functions is equal to the product of their limits, that is,

$$\lim_{x\to a} (f(x)g(x)) = (\lim_{x\to a} f(x))(\lim_{x\to a} g(x))$$

For example,

$$\lim_{x\to -1} x^5 = (\lim_{x\to -1} x^3)(\lim_{x\to -1} x^2) = (-1)^3(-1)^2 = -1$$

Theorem 9.5 The limit of a quotient of two functions is the quotient of their limits provided that the limit of the denominator is not zero, that is,

$$\lim_{x\to a} \left(\frac{f(x)}{g(x)}\right) = \frac{\lim_{x\to a} f(x)}{\lim_{x\to a} g(x)} \qquad \text{if} \quad \lim_{x\to a} g(x) \neq 0$$

For instance,

$$\lim_{x\to 3} \left(\frac{x-1}{x+3}\right) = \frac{\lim_{x\to 3} (x-1)}{\lim_{x\to 3} (x+3)} = \frac{3-1}{3+3} = \frac{2}{6} = \frac{1}{3}$$

The existence of the value of a function at $x = a$ and that of its limit as $x \to a$ classify functions as defined below.

DEFINITION 9.3 A function $f(x)$ is said to be continuous at $x = a$ if and only if the following are true:

a. $f(a)$ exists as a finite number.
b. $\lim_{x \to a} f(x) = f(a)$.

Furthermore, we say $f(x)$ is continuous in the open interval

$$(a, b) = \{x \mid a < x < b\}$$

or in the closed interval

$$[a, b] = \{x \mid a \leq x \leq b\}$$

if and only if it is continuous at every point of the respective interval.

To establish the continuity of a function at a point $x = a$, we must demonstrate that both parts (a) and (b) of Definition 9.3 are satisfied. On the other hand we say the function is discontinuous (not continuous) at a point $x = a$ if at least one of the requirements of Definition 9.3 fails to be true.

For example, any polynomial function

$$f(x) = a_0x^n + a_1x^{n-1} + a_2x^{n-2} + \cdots + a_n$$

is continuous for all x and all real numbers $a_0, a_1, a_2, \ldots, a_n$. Likewise, every function of the form

$$h(x) = \frac{f(x)}{g(x)}$$

where $f(x)$ and $g(x)$ are polynomial functions is also continuous for all x for which $g(x) \neq 0$. The exponential functions are continuous for all x. Specifically, the following functions are continuous for all x, and the respective limits are as indicated.

a. $f(x) = 3x^3 - 2x + 1$ is continuous for all x and

$$\lim_{x \to -1} (3x^3 - 2x + 1) = 3(-1)^3 - 2(-1) + 1 = f(-1) = 0$$

b. $g(x) = (x^3 - 1)/(x - 1)$ is continuous for all $x \neq 1$ and

$$\lim_{x \to 1} \frac{x^3 - 1}{x - 1} = \lim_{x \to 1} (x^2 + x + 1) = 1^2 + 1 + 1 = 3$$

At $x = 1$ the function assumes the meaningless (or indeterminant) form $0/0$, and the value of $g(1)$ is not defined (Figure 9.18). However, since $\lim (x^3 - 1)/(x - 1) = 3$, the function can be made continuous by defining its value at $x = 1$ to be its limit, three, as x approaches one.

Figure 9–18

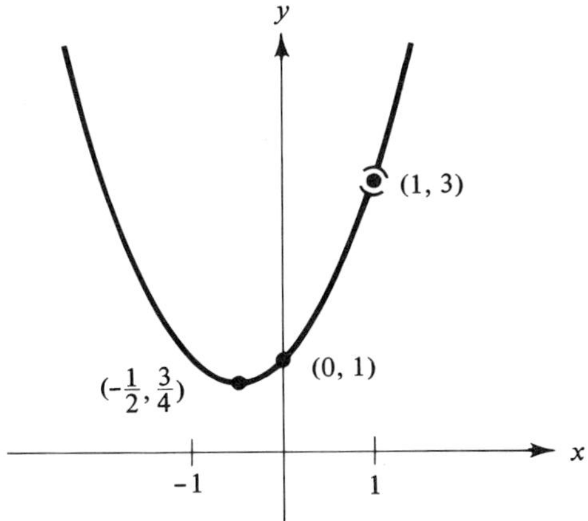

In other words the function

$$g(x) = \begin{cases} \dfrac{x^3 - 1}{x - 1} & \text{for} \quad x \neq 1 \\ 3 & \text{for} \quad x = 1 \end{cases}$$

is continuous for all x. This, of course, is the case because the function is defined for all x and $\lim_{x \to a} g(x) = g(a)$ for all a. Geometrically, $g(x) = 3$ at $x = 1$ yields the point $(1, 3)$ which plugs the hole in the graph of $g(x) = (x^3 - 1)/(x - 1)$ (Figure 9.18).

c. $f(x) = 2^{x^2 - x}$ is continuous for all x and

$$\lim_{x \to 4} f(x) = \lim_{x \to 4} (2^{x^2 - x}) = 2^{16-4} = 2^{12}$$

On the other hand the following functions are discontinuous at the indicated points.

a. $f(x) = \log x$

is discontinuous for all $x < 0$ since it is not even defined for such x's.

b. $$g(x) = \begin{cases} x^2 & \text{for} \quad x < 2 \\ 1 & \text{for} \quad x > 2 \end{cases}$$

is discontinuous at $x = 2$ because it is not defined for this value of x (Figure 9.19). In fact, the graph of the function shows a jump at $x = 2$. This jump cannot be repaired and still retain a function (why?).

c. $$h(x) = \begin{cases} \dfrac{x^2 - 4}{x + 2} & \text{if} \quad x \neq -2 \\ -5 & \text{if} \quad x = -2 \end{cases}$$

Figure 9–19

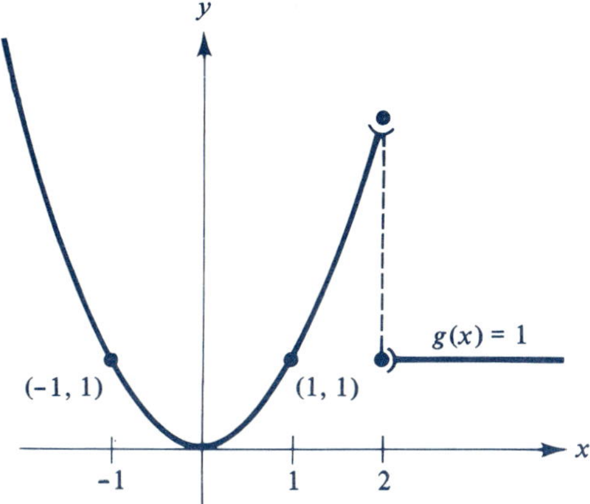

is discontinuous at $x = -2$ although it is defined as -5 at $x = -2$. Of course, this is the case because

$$\lim_{x \to -2} \frac{x^2 - 4}{x + 2} = \lim_{x \to -2} (x - 2) = -4 \neq f(-2) = -5$$

that is, the limit of the function as $x \to -2$ is not equal to the value of the function at $x = -2$.

Thus part (b) of Definition 9.3 is not satisfied. Geometrically, the point $(-2, -5)$ does not plug the hole at $(-2, -4)$ shown on the graph of the function (Figure 9.20).

Figure 9–20

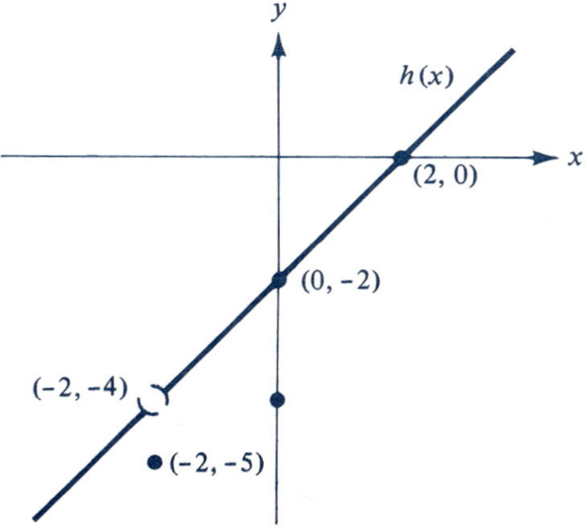

As might be expected, the extended definition of the limit and continuity concepts concerning multivariable functions parallels closely those of functions of one variable. The definitions for functions of two variables follow.

DEFINITION 9.4 If $f(x, y)$ is a function of two variables, and if the values of $f(x, y)$ approach the number L as both x and y approach the values a and b, respectively, and independent of any path, we say that L is the limit of $f(x, y)$ as x and y approach a and b, and we write it as

$$\lim_{x \to a;\, y \to b} f(x, y) = L$$

DEFINITION 9.5 We say that the function $f(x, y)$ is continuous at the point (a, b) if and only if the following are true.

a. The value $f(a, b)$ of the function exists at $x = a$ and $y = b$.
b. $\lim_{x \to a;\, y \to b} f(x, y) = f(a, b)$.

Example 9.8 Discuss the behavior of the function

$$f(x, y) = \frac{x}{x + y}$$

as x and y approach zero.

Discussion From Definition 9.4 it is clear that the function will possess a limit at a point (a, b) only if the limit of the function exists and is the same for all modes (or paths) of approach of x and y to a and b, respectively. However, if the functional values approach different numbers as x and y approach a and b along different paths, then the limit of $f(x, y)$ does not exist. Thus if $x \to 0$ and $y \to 0$ along the lines $y = 2x$ and $y = 3x$, the respective limits are

a. $$\lim_{x \to 0;\, y \to 0} \frac{x}{x + y} = \lim_{x \to 0} \frac{x}{x + 2x} = \lim_{x \to 0} \frac{x}{3x}$$

$$= \frac{1}{3} \qquad \text{along the line } y = 2x.$$

b. $$\lim_{x \to 0;\, y \to 0} \frac{x}{x + y} = \lim_{x \to 0} \frac{x}{x + 3x} = \lim_{x \to 0} \frac{x}{4x}$$

$$= \frac{1}{4} \qquad \text{along the line } y = 3x.$$

Since the values of the two limits are different along these two paths, $\lim_{x\to 0; y\to 0} x/(x+y)$ does not exist, and the function

$$f(x, y) = \frac{x}{x+y}$$

is discontinuous at the point $(0, 0)$.

Example 9.9

Discuss the limit of the function

$$f(x, y) = \frac{x^2 - y^2}{x^2 + y^2}$$

as x and y approach zero.

Discussion The function

$$f(x, y) = \frac{x^2 - y^2}{x^2 + y^2}$$

is defined except at the origin $(x = 0, y = 0)$. The limit of the function does not exist since along the x axis we have

$$\lim_{x\to 0;\, y\to 0} \left(\frac{x^2 - y^2}{x^2 + y^2}\right) = \lim_{x\to 0} \left(\frac{x^2 - 0}{x^2 + 0}\right) = 1$$

and along the y axis we obtain

$$\lim_{x\to 0;\, y\to 0} \left(\frac{x^2 - y^2}{x^2 + y^2}\right) = \lim_{y\to 0} \left(\frac{0 - y^2}{0 + y^2}\right) = -1$$

Thus the limit of the given function changes as we change the path of approach to the origin, the limit of the function does not exist as x and y approach the origin, and the function is discontinuous at the origin.

In the next chapter we shall consider differentiation of both functions of one and more than one variable. This we shall do by first establishing the concept of the ordinary derivative or simply the derivative of a function of one variable, and then extending it to that of partial derivatives of functions in two or more variables. Then we shall explore the formal properties of both ordinary and partial derivatives. Fortunately no new rules of differentiation are involved in computing partial derivatives, and in general we shall see that the problems of partial differentiation are quite similar to those of ordinary differentiation.

Exercises

1. Determine the following limits of sequences, and draw corresponding graphs:

a. $\lim_{n\to\infty} \frac{(3n + 1)}{n}$. b. $\lim_{n\to\infty} \frac{(2 - 5n)}{n}$.

c. $\lim_{n\to\infty} \frac{(3n - 1)}{n + 1}$. d. $\lim_{n\to\infty} (2n + 3)$.

Ans. (a) 3; (c) 3

2. By evaluating the lim $(a - ar^n)/(1 - r)$ where a is the first term, r the common ratio, and n the number of terms of any geometric progression, determine the following infinite sums:

a. $1 + \frac{1}{2} + \frac{1}{4} + \frac{1}{8} + \cdots$. b. $1 - \frac{1}{2} + \frac{1}{4} - \frac{1}{8} + \cdots$.
c. $3/10 + 3/100 + 3/1000 + \cdots$.
d. $1 + 3/2 + 9/4 + 27/8 + \cdots$.

Ans. (a) $\frac{1}{2}$; (b) $\frac{2}{3}$

3. By applying the technique of Exercise 2, determine the following rational numbers:

a. 0.6666 . . . . b. 0.131313 . . . .
c. 2.444 . . . . d. 1.212212212 . . . .

4. Using the method of exhaustion of Archimedes find the area bounded by each of the following parabolas, the x axis, y axis, and the line $x = 2$:

a. $y = x^2$. b. $y = x^2 + 1$. c. $y = 2x^2$.

Ans. (b) 14/3; (c) 16/3

5. Evaluate the following limits, discuss the continuity of the corresponding function, and draw appropriate graphs:

a. $\lim_{x\to 1} (x^2 - 1)/(x - 1)$. b. $\lim_{x\to 6} (x^2 - 36)/(x - 6)$.

c. $\lim_{x\to 1} 1/(x - 1)$.

Ans. (a) 2, discontinuous at $x = 1$;
(c) ∞, discontinuous at $x = 1$

6. Discuss the limits of the following functions as x and y approach zero:

   a. $f(x, y) = x/(x + y)$. b. $f(x, y) = (x^2 + y^2)/(x^2 - y^2)$.

References for Supplementary Readings

1. A. E. Taylor, *Advanced Calculus*, Chapter 5 (Boston: Ginn, 1955).
2. G. B. Thomas, *Calculus and Analytic Geometry*, 4th ed., Chapter 10 (Reading, Mass.: Addison-Wesley, 1968).
3. R. E. Johnson and F. L. Kiokemeister, *Calculus with Analytic Geometry*, 3d ed., Chapter 4 (Boston, Mass.: Allyn and Bacon, 1964).

CHAPTER 10

Ordinary and Partial Differentiation

10.1 Rates of Change and Slope

In this chapter we shall primarily be concerned with the application of the limit concept to what is commonly known in calculus as the "rate of change" in functions. This application facilitates the development of the very important concepts of the "derivative," "partial derivative," and their utilization in the areas of optimization and stability theory. The relevancy of rates of change in functions becomes evident as we examine the following real-life situations. If a certain manufacturer, for instance, increases his production by 2000 items a month, then he expects a yield of corresponding change in his costs, revenue, and certainly his profits. He is particularly interested in the relative change of these quantities (functions); that is, he will want to know the extent to which his profits increase or decrease per unit production. Similarly, a psychologist might be interested in determining a patient's attitude change per unit time of treatment, or he might be interested in evaluating the change in an individual's habit strength per repetition. To the student of calculus all these problems constitute a major part of the general study of "rates of change" in a function $f(x)$ and the related concept of the limit. Thus we might be concerned with the rate of change in the growth of a certain bacterial culture or the rate of change in the velocity of an automobile or perhaps the rate of change in an individual's blood pressure.

Specifically, if the rate R at which a chemical substance is absorbed into a bacterium and then distributed throughout its whole volume is proportional to S/V where S is the surface area and V the volume of a spherical bacterium, then we usually wish to know the rate of change of R relative to a change in the radius of the bacterial shape. Similarly, in a normal distribution

$$Y = \frac{N}{s\sqrt{2\pi}} e^{-x^2/2s^2}$$

where Y is the frequency of the distribution, N the number of observations, s the standard deviation of the distribution, and x the deviation of the measurement from the mean, we would be interested in knowing the rate of change in the frequency Y per unit change in the deviation x. These and other similar problems shall be considered, in some depth, in Chapter 11.

The problems on rates of change and slopes of tangent lines to certain curves concerned the Greeks at a time when Archimedes wrote a complete work on the tangent and area problem of a curve called the *spiral of Archimedes.* On the other hand Galileo is famous for his work dealing with rates of change of falling bodies and particularly for his experiments from the leaning tower of Pisa. The genius of this man can perhaps be appreciated even more by knowing that as early as 1620 he had become aware of almost all the concepts involved in both "differential" and "integral" calculus. Newton and Leibniz, credited with the invention of calculus, completely understood and developed general principles and techniques for solving problems related to the original area and tangent problems.

Geometrically the rate of change of a function $f(x)$ may be described as follows. In the graph of $y = f(x)$ (see Figure 10.1) the slope of the line joining the points $P(a, f(a))$ and $Q(b, f(b))$ is calculated by dividing $b - a$ into $f(b) - f(a)$; that is, the slope of the line PQ is obtained by dividing the difference in the heights of the curve $y = f(x)$ at the points P and Q by the horizontal distance $b - a$. Thus the slope m of the line PQ is

$$m = \frac{f(b) - f(a)}{b - a}$$

or

$$m = \frac{f(b) - f(a)}{h}, \qquad h = b - a$$

Figure 10-1

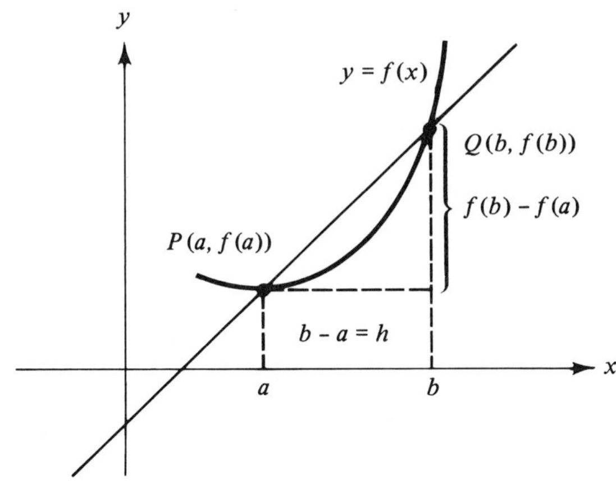

The slope of the line PQ is the average rate of change in $f(x)$ as x changes from a to b. Generally, if $y = f(x)$ is a function defined over the closed interval

$$I = [a, b] = \{x \mid a \leq x \leq b\}$$

and if x_1 and x_2 are any two values of x in I such that $y_1 = f(x_1)$ and $y_2 = f(x_2)$, then the changes in x and y are defined as follows.

DEFINITION 10.1 By the increment of x, denoted by $\Delta x = x_2 - x_1$, we shall mean the change of x from x_1 to x_2. Similarly, by the increment of $y = f(x)$, denoted by either Δy or $\Delta f(x)$, we shall mean

$$\Delta y = \Delta f(x) = y_2 - y_1 = f(x_2) - f(x_1)$$

whenever x changes from x_1 to x_2 within the domain of $f(x)$.

Clearly, for any increment Δx of x the corresponding increment Δy in the function $y = f(x)$ is

$$\Delta y = f(x + \Delta x) - f(x) \qquad \text{(Why?)}$$

The following examples further demonstrate the above definition.

Example 10.1 If $f(x) = x^2 + 1$, find $\Delta f(x)$ when x changes from $x = 2$ to $x = 2.5$.

Solution Since

$$\Delta x = x_2 - x_1 = 2.5 - 2 = 0.5$$

and since

$$\Delta f(x) = f(x + \Delta x) - f(x)$$

where $x = 2$, we obtain

$$\Delta f(x) = f(2.5) - f(2)$$

$$\Delta f(x) = (2.5^2 + 1) - (2^2 + 1)$$

or

$$\Delta f(x) = 2.25$$

that is, a 0.5 change in x results in a 2.25 change in $f(x)$.

Example 10.2 Find the slope of the line determined by the points $P(1, 1)$ and $Q(3, 9)$ of the curve $y = x^2$.

Solution Since the coordinates of the points P and Q satisfy the equation $y = x^2$ ($1 = 1^2$ and $9 = 3^2$), the points P and Q lie on the curve of $y = x^2$ (Figure 10.2). Thus the slope of the line PQ is

$$m = \frac{f(3) - f(1)}{3 - 1} \qquad \text{or} \qquad m = \frac{9 - 1}{2} = 4$$

Example 10.3 Let $H(x) = 100(1 - e^{-kx})$ be the function of habit strength in terms of the number of repetitions x and a positive constant k. Find the increment $\Delta H(x)$ of $H(x)$ if x changes from 2 to 10.

Solution Since $\Delta x = 10 - 2 = 8$ and

$$\Delta H(x) = H(x + \Delta x) - H(x)$$

Figure 10-2

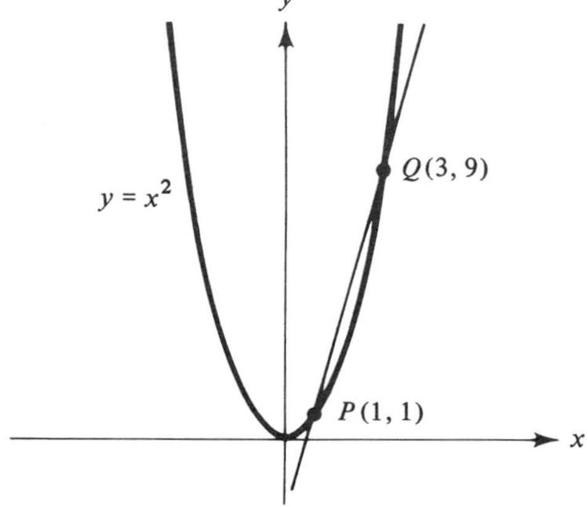

where $x = 2$, we find

$$\Delta H(x) = H(2 + 8) - H(2) = H(10) - H(2)$$

$$\Delta H(x) = 100(1 - e^{-10k}) - 100(1 - e^{-2k}) = 100(e^{-2k} - e^{-10k})$$

or

$$\Delta H(x) = 100 \frac{e^{8k} - 1}{e^{10k}}$$

Thus a change of 8 units in x results in a $100(e^{8k} - 1)/e^{10k}$ change in habit strength $H(x)$.

DEFINITION 10.2 By the average rate of change of a function $y = f(x)$, as x changes from x to $x + \Delta x$ within the domain of $f(x)$, we shall mean

$$\frac{\Delta y}{\Delta x} = \frac{f(x + \Delta x) - f(x)}{\Delta x}$$

Once again the average rate of change of the function $y = f(x)$ over the interval $[x_1, x_2]$ is clearly the slope of the line determined by the points $(x_1, f(x_1))$ and $(x_2, f(x_2))$ which lie on the curve $y = f(x)$.

Example 10.4 Evaluate the average rate of change of the function $f(x) = x^3$ as x changes from 3 to 5.

Solution Since

$$\Delta x = 5 - 3 = 2 \qquad \text{and} \qquad f(x) = f(x + \Delta x) - f(x)$$

where $x = 3$, the average rate of change is given by

$$\frac{\Delta y}{\Delta x} = \frac{f(x + \Delta x) - f(x)}{\Delta x} = \frac{f(3 + 2) - f(3)}{5 - 3}$$

$$\frac{\Delta y}{\Delta x} = \frac{5^3 - 3^3}{2}$$

or

$$\frac{\Delta y}{\Delta x} = 49$$

Example 10.5 If the total profit function of a certain commodity is

$$P(x) = -4x^2 + \frac{44}{5} x$$

find the average rate of change in $P(x)$ as x changes from 1 to 1.1.

Solution Since

$$\Delta x = 1.1 - 1 = 0.1 \quad \text{and} \quad \Delta P(x) = P(x + \Delta x) - P(x)$$

where $x = 1$, we find that

$$\frac{\Delta P(x)}{\Delta x} = \frac{P(1 + 0.1) - P(1)}{1.1 - 1}$$

$$\frac{\Delta P(x)}{\Delta x} = \frac{[-4(1.1)^2 + (44/5)1.1] - [-4(1)^2 + (44/5)1]}{0.1}$$

or

$$\frac{\Delta P(x)}{\Delta x} = 0.4$$

Thus an 0.1 change in x produces an 0.4 average rate of change in the profit function $P(x)$.

10.2 The Derivative

The question is often raised as to what values of x maximize or minimize a function $f(x)$? For example, if $P(x)$ is the total profit function of a certain commodity, then what are the values of x for which $P(x)$ is a maximum or a minimum? Such problems are exceedingly important, and their consideration constitutes the branch of calculus known as "optimization theory." The study of these problems requires the development and understanding of the derivative concept which we now define.

DEFINITION 10.3 If $f(x)$ is a function for which

$$\lim_{h \to 0} \frac{f(x + h) - f(x)}{h}$$

exists, we say the function is differentiable at x, and its derivative $f'(x)$ with respect to x is the value of this limit. Moreover, we say that a function is differentiable if it has a derivative at each point of its domain.

Many times, the derivative $f'(x)$ of a function $f(x)$ is referred to as the instantaneous rate of change or simply the rate of change of $f(x)$ at x. One of the most significant consequences of differentiability is the following theorem.[2]

Theorem 10.1 If the function $f(x)$ is differentiable, then the function is continuous.

All polynomial functions $f(x) = a_0x^n + a_1x^{n-1} + \cdots + a_n$, all rational functions $p(x)/q(x)$ where $p(x)$ and $q(x) \neq 0$ are polynomials, all exponential functions, and all logarithmic functions are typical examples of differentiable and therefore continuous functions.[2]

FOUR-STEP RULE

Since the derivative of a differentiable function is given by

$$f'(x) = \lim_{h \to 0} \frac{f(x+h) - f(x)}{h}$$

finding the derivative of $f(x)$ at $x = x_0$ requires the following four steps:

a. Evaluate $f(x_0)$.
b. Evaluate $f(x_0 + h)$.
c. Divide $f(x_0 + h) - f(x_0)$ by h.
d. Evaluate $\lim_{h \to 0} \dfrac{f(x_0 + h) - f(x_0)}{h}$.

These steps constitute what is commonly known as the "four-step rule" of differentiation, and the process comprises one of two major branches of calculus known as "differential calculus." The other branch is commonly known as "integral calculus." Invariably, the derivative of a function $y = f(x)$ at x and relative to x is denoted by either $f'(x)$ or $D_xf(x)$ or dy/dx. The last notation is sometimes called the "Leibniz notation" for the derivative of $f(x)$ in honor of the great mathematician Leibniz, who is credited as one of the inventors of calculus. Whenever the variable of differentiation is clearly understood, $Df(x)$ is used rather than $D_xf(x)$.

Example 10.6 Use the four-step rule to find the derivative of $f(x) = x^2$ at $x = 2$.

Solution Application of the four-step rule yields

$$f(2) = 2^2 = 4 \tag{10.1}$$

$$f(2+h) = (2+h)^2 = 4 + 4h + h^2 \tag{10.2}$$

$$\frac{f(2+h) - f(2)}{h} = \frac{(4 + 4h + h^2) - 4}{h} = \frac{4h + h^2}{h} = 4 + h \tag{10.3}$$

$$\lim_{h \to 0} (4 + h) = 4 \tag{10.4}$$

Figure 10-3

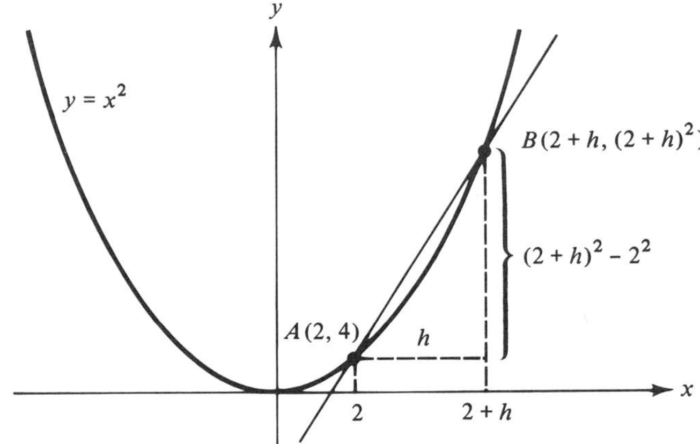

Thus the required derivative is 4 and the geometric features of the problem are shown in Figure 10.3.

From Figure 10.3 it follows that the slope of the line AB is

$$m = \frac{f(2 + h) - f(2)}{h} \qquad \text{or} \qquad m = 4 + h$$

Now, the application of the four-step rule and the limit properties of functions establish the truth of the following important theorem.

Theorem 10.2 Given the functions $f(x)$ and $g(x)$ and the constants a and b, prove that

$$D_x f(x) = 0 \qquad \text{if and only if} \quad f(x) = c \qquad \text{(constant)} \tag{10.5}$$

$$D_x[af(x) + bg(x)] = aD_x f(x) + bD_x g(x) \tag{10.6}$$

$$D_x[af(x) - bg(x)] = aD_x f(x) - bD_x g(x) \tag{10.7}$$

Proof (a) If $f(x) = c$, then

$$f'(x) = \lim_{h \to 0} \frac{f(x + h) - f(x)}{h}$$

or

$$f'(x) = \lim_{h \to 0} \frac{c - c}{h} = \lim_{h \to 0} \frac{0}{h} = 0$$

that is, the derivative of a constant is always zero. On the other hand if the derivative of a function is zero, the function can easily be shown to be a constant.

(b) If we let $F(x) = af(x) + bg(x)$, then

$$F'(x) = \lim_{h\to 0} \frac{F(x+h) - F(x)}{h}$$

$$F'(x) = \lim_{h\to 0} \frac{[af(x+h) + bg(x+h)] - [af(x) + bg(x)]}{h}$$

$$F'(x) = \lim_{h\to 0} \left[a\frac{f(x+h) - f(x)}{h} + b\frac{g(x+h) - g(x)}{h} \right]$$

$$F'(x) = a\lim_{h\to 0} \frac{f(x+h) - f(x)}{h} + b\lim_{h\to 0} \frac{g(x+h) - g(x)}{h}$$

or

$$F'(x) = D[af(x) + bg(x)] = a\,Df(x) + b\,Dg(x)$$

as asserted. Likewise it can be shown, as an exercise by the student, that

$$D[af(x) - bg(x)] = a\,Df(x) - b\,Dg(x)$$

Therefore the derivative of any constant function is always zero; the derivative of a constant times a function is equal to the constant times the derivative of the function; and the derivative of the sum (or difference) of functions is the sum (or difference) of their derivatives (why?).

Example 10.7 If a and b are constants, find the derivative of the cost function $C(x) = ax - bx^2$ where x is the output level of a certain commodity.

Solution Since

$$D(x) = \lim_{h\to 0} \frac{(x+h) - x}{h} = \lim_{h\to 0} (1) = 1$$

and since

$$D(x^2) = \lim_{h\to 0} \frac{(x+h)^2 - x^2}{h} = \lim_{h\to 0} \frac{2xh + h^2}{h}$$

$$= \lim_{h\to 0} (2x + h) = 2x$$

then these and application of Theorem 10.2 on $C(x)$ imply that

$$DC(x) = D(ax - bx^2) = a\,D(x) - b\,D(x^2) = a\cdot 1 - b(2x)$$

or

$$DC(x) = a - 2bx$$

A more extensive application of calculus in the areas of business and economics is suggested by the following definition.

DEFINITION 10.4 Let $C(x)$ be the total cost function associated with the output level x of a certain commodity. Let $f(z)$ be the function associated with the amount z of input of labor (or raw material or capital). Let $p(x)$ be the demand function whose value at output level x specifies the price leading to the sale of x units of output. Let $r(x) = x \cdot p(x)$ be the revenue function. Finally, let $u(x)$ be the utility function at output level x. Then

a. $DC(x)$ is called marginal cost at output level x.
b. $Df(z)$ is the marginal product.
c. $Dr(x) = D(xp(x))$ is known as marginal revenue.
d. $Du(x)$ represents marginal utility at the point x.

The marginal cost of the cost function

$$C(x) = ax - bx^2$$

was found to be $a - 2bx$ (see Example 10.7).

Geometrically, the derivative of a function $y = f(x)$ may be interpreted as follows. In the graph of $y = f(x)$ (Figure 10.4) the slope of the line joining the points $P(x, f(x))$ and $Q(x + h, f(x + h))$

Figure 10-4

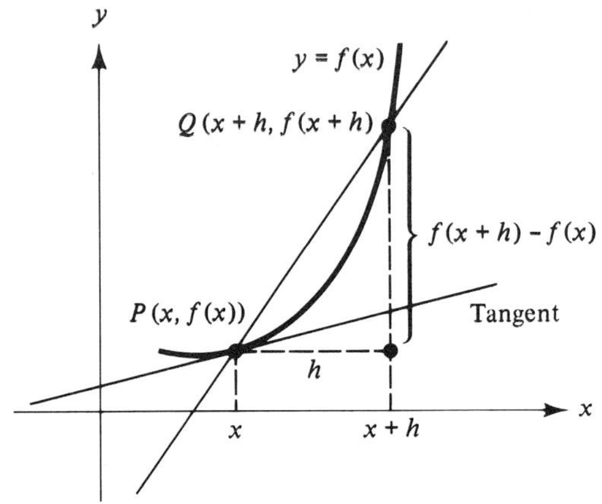

is calculated by dividing the difference in the heights of the curve $y = f(x)$ at the points P and Q by the horizontal distance h. That is, the slope of the line PQ is

$$m = \frac{f(x + h) - f(x)}{h}$$

However, as h approaches zero, the point Q approaches the point P, and the line PQ becomes tangent to the curve $y = f(x)$ at P. Thus the slope m of the line PQ becomes the slope of the tangent line to the curve at P as $h \to 0$; that is, the slope of the tangent line to a curve $y = f(x)$ at any point $P(x, f(x))$ is given by

$$m = \lim_{h \to 0} \frac{f(x + h) - f(x)}{h}$$

or

$$m = f'(x)$$

Consequently the derivative $f'(x)$ of the function $f(x)$ represents the slope of the curve $y = f(x)$ at the point $P(x, f(x))$ or equivalently the slope of the tangent line to $y = f(x)$ at P.

Example 10.8 Find the derivative of the function $f(x) = x^2 + 1$ at the point $(1, 2)$. What is the equation of the tangent line to $f(x) = x^2 + 1$ at $(1, 2)$?

Solution The derivative of $f(x)$ at any point (x, y) is given by

$$f'(x) = \lim_{h \to 0} \frac{f(x + h) - f(x)}{h}$$

Hence the derivative of $f(x) = x^2 + 1$ at $(1, 2)$ must be

$$f'(1) = \lim_{h \to 0} \frac{f(1 + h) - f(1)}{h} = \lim_{h \to 0} \frac{[(1 + h)^2 + 1] - (1^2 + 1)}{h}$$

or

$$f'(1) = \lim_{h \to 0} \frac{2h + h^2}{h} = \lim_{h \to 0} (2 + h) = 2$$

Thus the required derivative is 2, and the slope of the tangent line to $f(x) = x^2 + 1$ at $(1, 2)$ must also be 2. Since $y - y_1 = m(x - x_1)$ is the equation of the line on the point (x_1, y_1) with slope m, the equation of the required tangent must be $y - 2 = 2(x - 1)$ or $y = 2x$ (Figure 10.5).

Figure 10-5

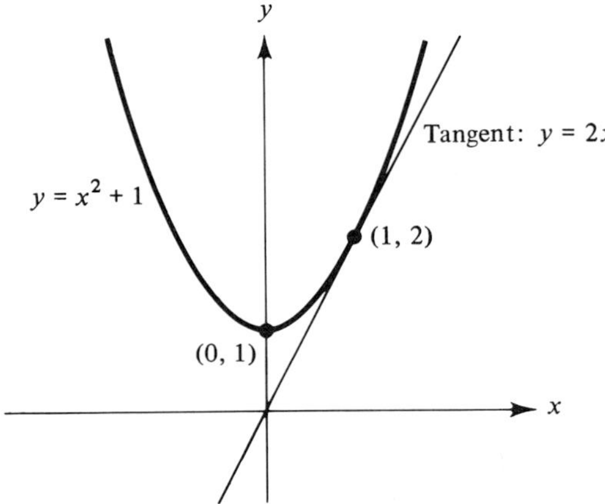

Example 10.9 Suppose that the total cost for a firm is given by the function

$$C(x) = 6 + 4x + 3x^2$$

where x represents the units of output. Show that the marginal cost is

$$M(x) = 4 + 6x$$

Solution Since the marginal cost was defined as the derivative of the cost function, we need show only that the derivative of $C(x)$ is indeed $4 + 6x$. Hence, differentiating $C(x)$, we obtain

$$DC(x) = D(x + 4x + 3x^2)$$

$$= \lim_{h \to 0} \frac{[6 + 4(x + h) + 3(x + h)^2] - (6 + 4x + 3x^2)}{h}$$

$$DC(x) = \lim_{h \to 0} \frac{4h + 6xh + h^2}{h} = \lim_{h \to 0} (4 + 6x + h)$$

or

$$DC(x) = 4 + 6x$$

as asserted.

Example 10.10 If the learning curve is

$$f(x) = \frac{Lx + Lc}{x + c + a}$$

where L is the limit of practice, x the amount of formal practice, C the equivalent previous practice, and a the rate of learning, and if $a = 4$, $c = 1$, and $L = 16$, find the equation of the tangent to $f(x)$ at $x = 0$.

Solution Since the slope of the tangent is the derivative and since $a = 4$, $c = 1$, and $L = 16$, we need only determine the derivative of

$$f(x) = \frac{16x + 16}{x + 5}$$

at $x = 0$. Therefore

$$f'(0) = \lim_{h \to 0} \frac{f(0 + h) - f(0)}{h} = \lim_{h \to 0} \frac{(16h + 16)/(h + 5) - 16/5}{h}$$

$$f'(0) = \lim_{h \to 0} \frac{64h}{(5h + 25)h} = \lim_{h \to 0} \frac{64}{5h + 25}$$

or

$$f'(0) = \frac{64}{25}$$

is the slope of the tangent line

Since the slope of the tangent is $64/25$ and since the point $(0, 16/5)$ is on the tangent, the required equation must be

$$y - \frac{16}{5} = \frac{64}{25}(x - 0) \qquad \text{or} \qquad 64x - 25y = -80$$

Exercises

1. Find the increment of each function for the indicated change in x, and draw appropriate graphs.

   a. $f(x) = 3x - 4$; $x = 2$, $\Delta x = 0.1$.
   b. $f(x) = x^2 - 1$; $x = 3$, $\Delta x = 2.5$.
   c. $g(t) = 1 - 3t^2$; $t = 0$, $\Delta t = 1.1$.
   d. $y = x^2 + 2x$; $x = 0$, $\Delta x = 2.1$.
   e. $f(x) = 2x - 3x^2$; $x = a$, $\Delta x = h$.

Ans. (a) 0.3; (c) -3.63

2. Determine the average rate of change of each function for the indicated change in x.

   a. $y = 2x + 3$; $x = 2$, $\Delta x = 0.1$.
   b. $f(x) = 3x^2$; $x = -2$, $\Delta x = 0.5$.
   c. $g(t) = 16t^2$; $t = 0$, $\Delta t = 2.2$.
   d. $f(x) = 2/x$; $x = 3$, $\Delta x = 1.2$.
   e. $f(x) = x^2 - x$; $x = a$, $\Delta x = h$.

   Ans. (a) 2; (c) 35.2

3. Using the four-step rule, find the derivative of the following functions at the indicated point.

   a. $f(x) = x^2 - 1$, $(2, 3)$.
   b. $f(x) = x - x^2$, $(\frac{1}{2}, \frac{1}{4})$.
   c. $f(x) = x^2 - 5x + 6$, $(0, 6)$.
   d. $f(x) = x^3$, $(1, 1)$.

   Ans. (a) 4; (c) -5

4. For each of the following functions, determine the derivative and the equation of the tangent line at the indicated point. Draw appropriate graphs.

   a. $y = x^2 - 2$, $(2, 2)$.
   b. $f(t) = 1/(1 + t)$, $(1, \frac{1}{2})$.
   c. $g(x) = (x + 1)/(x - 1)$, $(2, 3)$.
   d. $f(x) = \sqrt{x}$, $(4, 2)$.

   Ans. (a) 4, $4x - y = 6$; (c) -2, $2x + y = 7$

5. If $C(x) = 100 + 2x + 1/x$ is the total cost of producing x items of a certain commodity, determine the marginal cost as a function of x.

   Ans. $2 - 1/x^2$

6. If $p(x) = 200 + x + 2/x$ is the demand function whose value at output level x specifies the price that leads to sale of x units of output, find the revenue function $r(x)$ and the marginal revenue.

7. Knowing that the derivative $f'(x)$ is the rate of change of $f(x)$, translate each of the following statements into equivalent mathematical equations.

   a. The rate of increase of a population relative to time is always proportional to the population size at any time t.
   b. The rate of adoption of any educational innovation is proportional to the number of people who adopted it at time t multiplied by the number of people who have not adopted it.
   c. The level of learning difficulty decreases proportionally to

the product of the level of difficulty at any time and the total amount of practice from the beginning of formal practice.

d. Under certain conditions it is observed that the rate of change of atmospheric pressure relative to altitude is proportional to the pressure at any altitude h.

e. Radium disintegrates at a rate proportional to the amount of radium instantaneously present.

f. The rate at which an organization expends energy over time is proportional to the difference between the progress $p(t)$ of the organization at any time t and the anticipated progress $p_a(t)$.

g. A tank is initially filled with 100 gallons of salt solution containing 2 pounds of salt per gallon. Fresh brine containing 3 pounds of salt per gallon runs into the tank at a rate of 5 gallons per minute, and the mixture, which is assumed to be kept uniform by stirring, runs out at the same rate. What is the rate of change of the salt in the tank at any time t?

Ans. (a) $dP/dt = kP$; (g) $dQ/dt = 15 - 5Q/100$

10.3 Rules of Differentiation

Thus far we have illustrated by means of examples the manner in which the definition of the derivative and the four-step rule can be applied to determine the derivative of a function. However, this process often requires considerably more time and effort than necessary. Thus in the subsequent sections we shall attempt to establish some basic rules that will facilitate differentiation of certain functions. These rules are stated as theorems whose proofs we recommend as exercises for the more energetic and ambitious student. We will illustrate such theorems via special examples.

Theorem 10.3 *Power Rule* If n is any real number and $f(x) = x^n$, then

$$Dx^n = nx^{n-1}$$

For example,

$$Dx^5 = 5x^4$$

Theorem 10.4 *Sum and Difference Rules* If $f(x)$ and $g(x)$ are differentiable functions, then

$$D[f(x) + g(x)] = Df(x) + Dg(x) \tag{10.8}$$

$$D[f(x) - g(x)] = Df(x) - Dg(x) \tag{10.9}$$

For example,

$$D(x^4 + x^3) = Dx^4 + Dx^3 = 4x^3 + 3x^2$$

Theorem 10.5 *Constant Times a Function Rule* For any differentiable function $f(x)$ and any constant c we have

$$D[cf(x)] = c\,Df(x)$$

For example,

$$D(3x^8) = 3D(x^8) = 24x^7$$

Theorem 10.6 If $f_i(x)$ is a differentiable function for each $i = 1, 2, \ldots, n$, and if a_i is a constant for each i, then

$$D\sum_{i=1}^{n} a_i f_i(x) = \sum_{i=1}^{n} a_i\,Df_i(x)$$

For example,

$$D(3x^4 + 5x^3 + 6x^2) = 3Dx^4 + 5Dx^3 + 6Dx^2$$
$$= 12x^3 + 15x^2 + 12x$$

Example 10.11 Differentiate the function

$$f(x) = 5x^{12} - 6x^2 + 5x - 8$$

Solution An application of the rules on differentiation yields

$$Df(x) = D(5x^{12} - 6x^2 + 5x - 8) = D(5x^{12}) - D(6x^2) + D(5x) - D(8) \quad \text{(Theorem 10.6)}$$

$$Df(x) = 5D(x^{12}) - 6D(x^2) + 5D(x) - 0 \quad \text{(Theorem 10.5)}$$

$$Df(x) = 5(12x^{11}) - 6(2x) + 5(1) \quad \text{(Theorem 10.3)}$$

or

$$Df(x) = 60x^{11} - 12x + 5$$

Although the derivative of the sum of two functions is the sum of their derivatives, and the derivative of their difference is the difference of their derivatives, in general it is not true that the derivative of the product is the product of their derivatives; neither is it true that the derivative of their quotient is the quotient of their derivatives. For example, if $f(x) = x^6$ and $g(x) = x^3$, then

$$f(x)g(x) = x^9 \qquad \text{and} \qquad \frac{f(x)}{g(x)} = x^3$$

However,

$$D[f(x)g(x)] = 9x^8 \neq [Df(x)][Dg(x)] = (6x^5)(3x^2) = 18x^7$$

and

$$D\left[\frac{f(x)}{g(x)}\right] = 3x^2 \neq \frac{Df(x)}{Dg(x)} = \frac{6x^5}{3x^2} = 2x^3$$

Theorem 10.7 *Product and Quotient Rules* If $f(x)$ and $g(x)$ are differentiable functions, then

$$D[f(x)g(x)] = f(x)\,Dg(x) + g(x)\,Df(x) \qquad \textbf{(10.10)}$$

$$D\left[\frac{f(x)}{g(x)}\right] = \frac{g(x)\,Df(x) - f(x)\,Dg(x)}{g^2(x)} \quad \text{for} \quad g(x) \neq 0 \qquad \textbf{(10.11)}$$

Example 10.12 Find the derivative of

$$F(x) = 2x^4\sqrt{x}$$

Solution Let $f(x) = 2x^4$ and $g(x) = \sqrt{x}$. Then using the product rule, we obtain

$$DF(x) = D[f(x)g(x)] = D[(2x^4)\cdot\sqrt{x}]$$

$$DF(x) = 2x^4\,D(x^{1/2}) + x^{1/2}\,D(2x^4)$$

$$DF(x) = 2x^4(\tfrac{1}{2}x^{1/2-1}) + x^{1/2}(8x^3)$$

or

$$DF(x) = 9x^{7/2}$$

Another approach to finding the derivative of $F(x)$ is to perform the indicated operations first and then to differentiate; that is,

$$F(x) = 2x^4x^{1/2} = 2x^{9/2}$$

and

$$DF(x) = 2\left(\frac{9}{2}\right)x^{9/2-1} \qquad \text{(power rule)}$$

or

$$DF(x) = 9x^{7/2}$$

Example 10.13 Differentiate

$$F(x) = \frac{3x^3 - 2x + 5}{x^{5/2}}$$

Solution Let $f(x) = 3x^3 - 2x + 5$ and $g(x) = x^{5/2}$. Then, applying the quotient rule we obtain

$$D\left(\frac{3x^3 - 2x + 5}{x^{5/2}}\right)$$

$$= \frac{x^{5/2}\, D(3x^3 - 2x + 5) - (3x^3 - 2x + 5)\, D(x^{5/2})}{(x^{5/2})^2}$$

$$DF(x) = \frac{x^{5/2}(9x^2 - 2) - (3x^3 - 2x + 5)5/2x^{3/2}}{x^5}$$

$$DF(x) = \frac{9x^{9/2} - 2x^{5/2} - 15/2x^{9/2} + 5x^{5/2} - (25/2)x^{3/2}}{x^5}$$

or

$$DF(x) = 1.5x^{-1/2} + 3x^{-5/2} - 12.5x^{-7/2}$$

Example 10.14 If x is the number of migrants competing for opportunities in city B, if k and a are constants, and if the number y of migrants from city A to B during a fixed time interval is given by $y = k/x^a$, find the rate of change of y relative to x.

Solution Since the required rate of change is the derivative of y with respect to x, we need only find dy/dx. Therefore

$$\frac{dy}{dx} = \frac{d}{dx}\left(\frac{k}{x^a}\right) = \frac{d}{dx}(kx^{-a}) = k(-a)x^{-a-1} \qquad \text{(power rule)}$$

or

$$\frac{dy}{dx} = \frac{-ak}{x^{a+1}}$$

Exercises

Differentiate each of the following functions:

1. $y = 3x^4 - 2x^3 - 6x^2 + 10.$
2. $y = 3x^{5/3} - 3x^{1/3}.$
3. $f(x) = (2x^4 - 3)/(x^2 - 3x).$
4. $g(t) = t^2/(t^2 + 8).$
5. $y = (3x^4 - 5x^2 + 6)\sqrt[3]{x}.$
6. $y = (2x + 7x^3 - 6)(x^2 + 4).$

Ans. (1) $y' = 12x^3 - 6x^2 - 12x$; (4) $g' = 16t/(t^2 + 8)^2$

Find the value of the derivative at the indicated point for each of the following functions:

7. $y = (2x + 3)(x^2 + 4x - 5),\ x = 2.$

8. $g(t) = (t^3 + 1)(t^4 + 2),\ t = 1.$

9. $f(x) = 1/(x^2 + 1),\ x = 0.$

10. $y = 5/(x^3 - 2x + 6).$

Ans. (8) 17; (10) −1

Find the point(s) on each of the following functions at which the tangent line is horizontal (that is, $y' = 0$), and draw appropriate graphs!

11. $y = 2x^2 - 3x + 1.$

12. $y = x^4 - 2x^2 + 1.$

13. $y = -x^2 + 5x - 6.$

14. $y = x^3 - 3x^2 + 2.$

Ans. (11) $(3/4, -1/8)$; (12) $(1, 0)$, $(0, 1)$, $(-1, 0)$

15. If the number of votes a candidate receives is related to the amount of money spent, and if for x thousand dollars of expenditure the popularity $P(x)$ of votes he will receive is given by

$$P(x) = \tfrac{1}{3}x^3 - 4x^2 + 12x$$

where $x \geq 0$, determine:

a. The rate of change of his popularity.
b. The values of x for which the rate of change in his popularity is zero.
c. Draw appropriate graphs.

Ans. (b) 2, 6

16. If $p(x) = 100 + x + 4/x$ is the demand function, then determine

a. The revenue function. b. The marginal revenue.
c. The value(s) of x for which the marginal revenue is zero.

17. The profit function $P(x)$, and the functions $R(x)$ and $C(x)$ of revenue and total cost, respectively, are related by the equation

$$P(x) = R(x) - C(x)$$

Thus for each of the following pairs of revenue and cost functions, determine the marginal profit and the value(s) of x that makes it zero:

a. $R(x) = 5x,\ C(x) = -10x^3/3 + 4x^2 + 2x + 8.$
b. $R(x) = 6x,\ C(x) = 14x^3/3 - 6x^2 + 4x + 6.$
c. $R(x) = 7x,\ C(x) = 56x^3/3 - 5x^2 + 4x + 9.$

Ans. (a) $P' = -10x^2 + 8x + 2,\ x = 1, -\frac{1}{5}$

18. If u is the utility function that attaches utility $u(x) = 20 + 2x - 2/x$ to output level x, find the marginal utility at output level $x = 500$.

19. Given the learning curve $f(x) = (Lx + Lc)/x + c + a$ where L is the limit of practice, x the amount of formal practice, c the equivalent previous practice, and a the rate of learning, find the equation of the tangent line at $x = 0$ if $L = 32$, $a = 8$, and $c = 2$.

20. If the distance S required to stop an automobile under normal conditions is proportional to the square of its velocity v, and if $S = 25$ when $v = 50$, find the rate of change of S when $v = 60$ and also when $v = 80$.

Ans. 1.2, 1.6

10.4 Rolle's Theorem and the Law of the Mean

Throughout both theory and applications of calculus a very significant and useful theorem concerning the derivative of functions is that of the mean value. The simplest form of this theorem is commonly known as Rolle's theorem, and an extension of it constitutes the extended theorem of the mean.

Theorem 10.8 *Rolle's Theorem* If $f(x)$ is continuous over the closed interval $a \leq x \leq b$, if $f(a) = f(b) = 0$ and if $f'(x)$ exists at least over the open interval $a < x < b$, then the derivative of $f(x)$ is zero for at least one value of x between a and b. In other words there is at least one value x_1 of x such that

$$f'(x_1) = 0, \qquad a < x_1 < b$$

An analytic proof of Rolle's theorem is possible. However, for our practical purposes its truth is sufficiently clear from the geometric evidence of Figure 10.6(a). The geometry of Figure 10.6(b) substantiates the truth of the following corollary.

Corollary 10.1 If $f(x)$ is continuous over the closed interval $a \leq x \leq b$ if $f(a) = f(b)$, and if $f'(x)$ exists at least over the open interval $a < x < b$, then $f'(x)$ becomes zero for at least one value of x between a and b.

There certainly is nothing in Rolle's theorem that suggests the existence of only one point between a and b at which $f'(x)$ is zero, that is, at which the tangent to the graph of $f(x)$ is horizontal. In

Figure 10-6

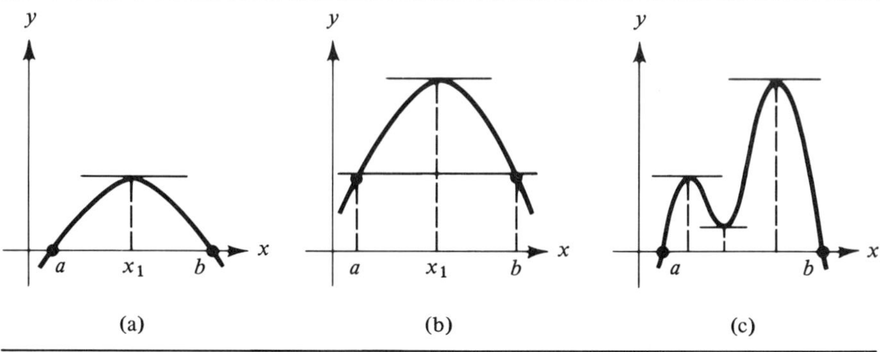

fact, there may be several such points as is clearly indicated by Figure 10.6(c).

Closely related to Rolle's theorem, of course, is another important theorem of calculus known as the "law of the mean for derivatives" or "mean value theorem."

Theorem 10.9 *Mean Value Theorem* If $f(x)$ is continuous over the closed interval $a \leq x \leq b$ and if $f'(x)$ exists at least over the open interval $a < x < b$, then there is one (or more) value x_1 of x between a and b such that

$$f(b) - f(a) = (b - a)f'(x_1), \qquad a < x_1 < b$$

The validity of the mean value theorem can be geometrically demonstrated, for it is clear from Figure 10.7 that $(f(b) - f(a))/(b - a)$

Figure 10-7

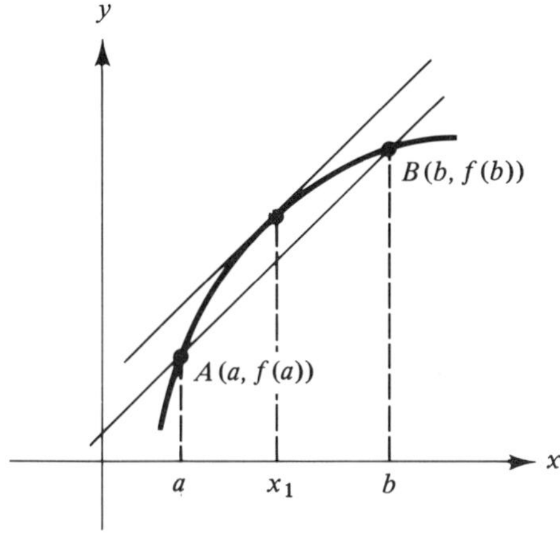

is simply the slope of the secant of the curve $y = f(x)$ determined by the points A and B. Moreover, it is obvious that somewhere on the arc between A and B there is a point at which the tangent must be parallel to the secant. Thus if x_1 is the x coordinate of such a point, then since the slope of the tangent at this point is $f'(x_1)$, it follows that

$$\frac{f(b) - f(a)}{b - a} = f'(x_1)$$

or

$$f(b) - f(a) = (b - a)f'(x_1), \qquad a < x_1 < b \tag{10.12}$$

Equation (10.12) is in fact the mean value theorem.

Finally, an extension of the mean value theorem constitutes the so-called "extended theorem of the mean."

Theorem 10.10 If $f(x)$ and $g(x)$ are both continuous over the closed interval $a \leq x \leq b$, if the derivatives $f'(x)$ and $g'(x)$ exist at least in the open interval $a < x < b$, and if $g'(x) \neq 0$ for $a < x < b$, then there is at least one value x_1 of x between a and b, such that

$$\frac{f(b) - f(a)}{g(b) - g(a)} = \frac{f'(x_1)}{g'(x_1)}, \qquad a < x_1 < b$$

Example 10.15 Verify Rolle's theorem for $f(x) = x^2 + x - 2$ over the interval $-2 \leq x \leq 1$.

Solution Since $f(x)$ is continuous and differentiable over $-2 \leq x \leq 1$ and since $f(-2) = f(1) = 0$, Rolle's theorem is applicable; and therefore

$$f'(x_1) = 0, \qquad -2 < x_1 < 1$$

or

$$2x_1 + 1 = 0 \qquad \text{or} \qquad x_1 = -\tfrac{1}{2}$$

that is, the tangent to $f(x)$ at $x = -\frac{1}{2}$ is horizontal as Figure 10.8 indicates.

Example 10.16 If $R(x)$ is the revenue function obtained by a firm from the sale of x units of output, then the average revenue is $R_a(x) = R(x)/x$. Furthermore, if $R(0) = 0$, then

$$R_a(x) = \frac{R(x) - R(0)}{x - 0}$$

Figure 10-8

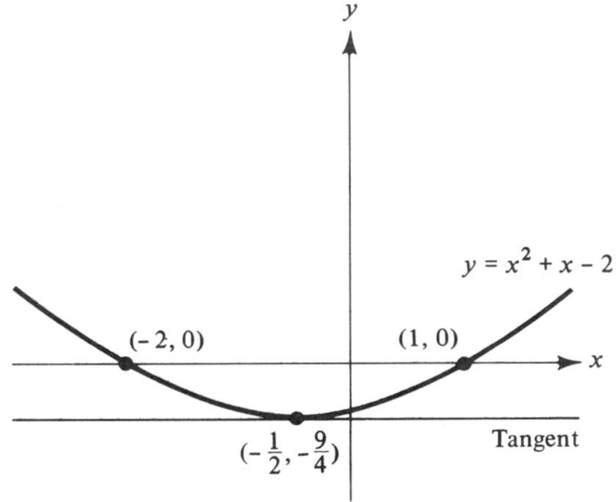

However, the mean value theorem implies that there is an x_1 such that $0 < x_1 < x$ and

$$R'(x_1) = \frac{R(x) - R(0)}{x - 0} = \frac{R(x)}{x}$$

or

$$R'(x_1) = R_a(x)$$

that is, there exists a production level $x_1 < x$ such that the average revenue $R_a(x)$ and the marginal revenue $R'(x_1)$ are the same.

Exercises

1. Determine which of the following functions satisfy the conditions of Rolle's theorem on the given interval. When the required conditions are fulfilled, find at least one value of x for which $f'(x) = 0$.

   a. $y = x^5 - 5x^4 + 6x^3 + 3x^2 - 9x, \; -1 \leq x \leq 3.$
   b. $y = 1 - x^{2/3}, \; -1 \leq x \leq 1.$
   c. $y = \begin{cases} x + 1 & \text{if } x < 0 \\ 1 - x & \text{if } x > 0, \end{cases} \quad -1 \leq x \leq 1.$
   d. $y = x^2 - x, \; 0 \leq x \leq 1.$

   Ans. (c) Conditions not fulfilled; (d) $x = \frac{1}{2}$

2. Verify that the following functions satisfy the conditions of the mean value theorem on the given interval, and exhibit at least one value of x in the indicated interval such that

$$f(b) - f(a) = (b - a)f'(x_1)$$

a. $y = 9x - 2x^3, -2 \leq x \leq 0.$
b. $y = 4x - 5x^2 + x^3/(1 - x), -2 \leq x \leq 0.$
c. $y = x^2 + 3x + 2, 1 \leq x \leq 2.$

Ans. (a) $x = -(2\sqrt{3})/3$; (c) $x_1 = 3/2$

10.5 The Chain Rule

In our discussion of functions and their interpretations (Chapter 2), we thought of a function such as $f(x) = (2x - 1)^3$ as a system or machine that accepts an input x and produces a corresponding output $(2x - 1)^3$ as indicated in Figure 10.9. It is often much more convenient to think of a function such as $f(x) = (2x - 1)^3$ as a two-stage system in which the input x is first transformed by means of a function g into $2x - 1$, and then $2x - 1$ is transformed by means of another function h into $(2x - 1)^3$ as indicated in Figure 10.10,

In terms of equations the function $f(x) = (2x - 1)^3$ may then be expressed as

$$f(x) = h(g(x)) = g^3(x)$$

where $g(x) = 2x - 1$. When this is the case, we say that we have a function h of another function g and that f is the composition of h with g. Similarly,

$$f(u) = u^2 - 3u + 5$$

where u is the function $u(x) = x^2 - 1$, is another example of composition of functions in which f is a function of the function u. Theorem 10.11 provides the most convenient way of differentiating composite functions.

Figure 10-9

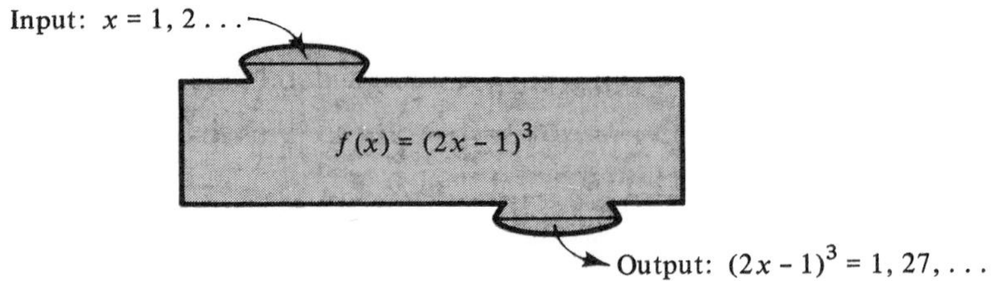

Figure 10-10

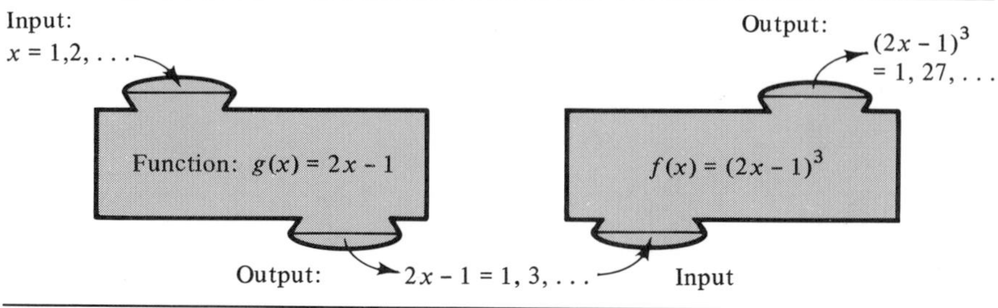

Theorem 10.11 *Chain Rule* If $y = f(u)$ and $u = g(x)$, that is, if $y = f(g(x))$ is a function of another function $g(x)$, then

$$\frac{dy}{dx} = \frac{dy}{du} \cdot \frac{du}{dx}$$

It is important to observe that the chain rule operates as though the derivatives were ratios of the quantities dx, dy, and du whenever these quantities are treated as ordinary numbers.

Corollary 10.2 *Power Rule* For any real number n and for any differentiable function $g(x)$, the derivative of $f(x) = [g(x)]^n$ is

$$f'(x) = n[g(x)]^{n-1}g'(x)$$

Theorem 10.12 If $y = f(t)$ and $x = g(t)$, that is, if y and x are related parametrically through the parameter t, then

$$\frac{dy}{dx} = \frac{dy}{dt} \bigg/ \frac{dx}{dt}, \qquad \frac{dx}{dt} \neq 0$$

The following examples further demonstrate the chain rule and its consequences in differentiation.

Example 10.17 Differentiate the following

a. $y = (x^4 - 3x + 2)^{14}$.

b. $y = \left(x^2 + \dfrac{1}{x^3}\right)^{-3}$.

c. $y = \left(\dfrac{x^2 - 1}{2x + 6}\right)^7$.

d. $y = (x^3 - 1)^{15}(2x^2 + 3)^{17}$.

Solution a. If we let $u(x) = x^4 - 3x + 2$, then we have $y = u^{14}$. Thus the application of the chain rule, together with the power rule, yields

$$\frac{dy}{dx} = \frac{dy}{du} \cdot \frac{du}{dx}$$

$$\frac{dy}{dx} = 14u^{13} \cdot (4x^3 - 3)$$

or

$$\frac{dy}{dx} = 14(x^4 - 3x + 2)^{13} \cdot (4x^3 - 3)$$

b. If we let $u(x) = x^2 + 1/x^3$, then $y = u^{-3}$. Then applying the chain rule and the power rule once again, we obtain

$$\frac{dy}{dx} = \frac{dy}{du} \cdot \frac{du}{dx}$$

$$\frac{dy}{dx} = -3u^{-4}(2x - 3x^{-4})$$

or

$$\frac{dy}{dx} = -3\left(x^2 + \frac{1}{x^3}\right)^{-4}\left(2x - \frac{3}{x^4}\right)$$

c. If we let $u(x) = (x^2 - 1)/(2x + 6)$, then $y = u^7$. Thus application of the chain and quotient rules results in

$$\frac{dy}{dx} = \frac{dy}{du} \cdot \frac{du}{dx}$$

$$\frac{dy}{dx} = 7u^6 \frac{du}{dx}$$

$$\frac{dy}{dx} = 7\left(\frac{x^2 - 1}{2x + 6}\right)^6 D\left(\frac{x^2 - 1}{2x + 6}\right)$$

$$= 7\left(\frac{x^2 - 1}{2x + 6}\right)^6 \frac{2x(2x + 6) - 2(x^2 - 1)}{(2x + 6)^2}$$

or

$$\frac{dy}{dx} = 7\frac{(x^2 - 1)^6(2x^2 + 12x + 2)}{(2x + 6)^8}$$

d. Initial application of the product rule reveals that

$$D_y = (x^3 - 1)^{15} D(2x^2 + 3)^{17} + (2x^2 + 3)^{17} D(x^3 - 1)^{15}$$

Next, applying the chain rule to $D(2x^2 + 3)^{17}$ and $D(x^3 - 1)^{15}$, we obtain

$$D_y = (x^3 - 1)^{15} \cdot 17(2x^2 + 3)^{16}(4x) + (2x^2 + 3)^{17} \cdot 15(x^3 - 1)^{14} \cdot (3x^2)$$

or

$$D_y = 68x(x^3 - 1)^{15}(2x^2 + 3)^{16} + 45x^2(x^3 - 1)^{14}(2x^2 + 3)^{17}$$

Example 10.18 If $y = t^2 - 1$ and $x = 2t^3 + 3t$, find dy/dx.

Solution By Theorem 10.12

$$\frac{dy}{dx} = \frac{dy/dt}{dx/dt}$$

Thus

$$\frac{dy}{dx} = \frac{2t}{6t^2 + 3}$$

Exercises

Find dy/dx for each of the following functions:

1. $y = (2x - x^2)^6$.
2. $y = 2/(1 - x^2)$.
3. $y = (t + 1)/t,\ x = t^2 - 2t$.
4. $y = 1 + t^2,\ x = 1 + (t^2 - 2)^{10}$.
5. $y = (1 + x^3)^5$.
6. $y = (x^2 - 2x + 3)^2$.
7. $y = \sqrt[3]{(1 + x^2)(1 - 4x)}$.
8. $y = x/(\sqrt{4 + x^2})$.
9. $y = x(4 - x^2)^{-1/2}$.

Ans. (1) $12(2x - x^2)^5(1 - x)$; (3) $-1/[2t^2(t - 1)]$; (5) $15x^2(1 + x^3)^4$.

10. At what point(s) does the following curves possess the indicated slope?

    a. $y = x/(3x - 4),\ m = -1$.
    b. $y = 2t - t^3,\ x = -5t + t^3, \quad m = 2$.

    Ans. (a) $(2, 1)$, $(\frac{2}{3}, -\frac{1}{3})$

11. At what point(s) of the curve $y = 3x^3 + 14x^2 + 3x + 8$ does the tangent pass through the origin?

    Ans. $x = -2, -1, \frac{2}{3}$

10.6 Differentiation of Logarithmic and Exponential Functions

The logarithm of a number A to a base b has previously been defined as the power L of b that yields the number A. In other words when we say that the logarithm of A to the base b is L, we mean $b^L = A$, and we express this by

$$\log_b A = L$$

In general, we say that $y = \log_b x$ if and only if $b^y = x$. Thus $y = \log_b x$ and $y = b^x$ are inverses of each other.

Clearly, any positive number other than one can be used as the base of the system of logarithms (why?). Of the infinitely many possible systems of logarithms, however, only two are of importance. The first is the so-called system of common logarithms, or logarithms to the base 10. The other is the system known as natural logarithms, logarithms to the base $e = 2.71828 \ldots$. Common logarithms are almost always denoted by $\log_{10} x$ and are particularly applicable whenever numerical calculations must be carried out. On the other hand natural logarithms are denoted by $\log_e x$ or $\ln x$ and are particularly useful in theoretical work. To change logarithms to the base a into logarithms to the base b, as we proved in Chapter 2, we utilize the equation

$$\log_b N = \log_a N \log_b a$$

The general behavior of the two logarithmic functions $y = \ln x$ and $y = \log_{10} x$ and their inverses $y = e^x$ and $y = 10^x$ is illustrated by their graphs shown in Figure 10.11.

The application of the four-step rule, together with some special limits, establishes the following two important theorems on the derivative of logarithmic and exponential functions.

Theorem 10.13 If u is a function of x, and if $y = \ln u$, then

$$\frac{dy}{dx} = \frac{1}{u}\frac{du}{dx}$$

Since the derivative of $u(x) = x$ is one, the following corollary is true.

Corollary 10.3 If $y = \ln x$, then

$$\frac{dy}{dx} = \frac{1}{x}$$

Figure 10-11

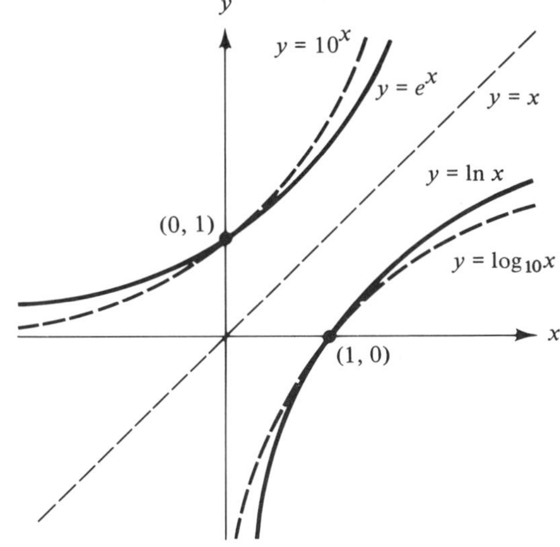

Theorem 10.14 If u is a function of x, and if $y = e^u$, then

$$\frac{dy}{dx} = e^u \frac{du}{dx}$$

Once again, since the derivative of $u(x) = x$ is one, the following corollary can be established.

Corollary 10.4 If $y = e^x$, then

$$\frac{dy}{dx} = e^x$$

To illustrate the mechanics of the above theorems and corollaries, we consider the following examples.

Example 10.19 What is the derivative of

$$y = \ln (x^2 - 3x)$$

Solution If we let $u = x^2 - 3x$, then $y = \ln u$. Thus Theorem 10.13 implies

$$\frac{dy}{dx} = \frac{1}{u}\frac{du}{dx} \qquad \text{or} \qquad \frac{dy}{dx} = \frac{1}{x^2 - 3x} \cdot (2x - 3)$$

or

$$\frac{dy}{dx} = \frac{2x - 3}{x^2 - 3x}$$

Example 10.20 What is the derivative of

$$y = \ln \left[\frac{(x - 1)^2 (x + 1)^3}{x^2 + 2x + 2} \right]$$

Solution It is possible to differentiate this function by a straightforward application of Theorem 10.13 taking

$$u = \frac{(x - 1)^2 (x + 1)^3}{x^2 + 2x + 2}$$

and then computing y' by the quotient rule. In many problems like this, however, it is more convenient to apply the various laws of logarithms before proceeding to differentiate. In this case we rewrite y as

$$y = 2 \ln (x - 1) + 3 \ln (x^2 + 1) - \ln (x^2 + 2x + 2)$$

and then apply Theorem 10.13 to each one of its three terms. Thus

$$\frac{dy}{dx} = \frac{2}{x - 1} + \frac{6x}{x^2 + 1} - \frac{2x + 2}{x^2 + 2x + 2}$$

Example 10.21 What is the derivative of

$$y = \sqrt{\frac{x^3 - 1}{x^2 + 1}}$$

Solution Method (a). If we let

$$u = \frac{x^3 - 1}{x^2 + 1}$$

then $y = u^{1/2}$, and

$$\frac{dy}{dx} = \frac{1}{2} u^{-1/2} \frac{du}{dx}$$

$$\frac{dy}{dx} = \frac{1}{2} \left(\frac{x^3 - 1}{x^2 + 1} \right)^{-1/2} \cdot \frac{3x^2(x^2 + 1) - 2x(x^3 - 1)}{(x^2 + 1)^2}$$

or

$$\frac{dy}{dx} = \frac{(x^3 - 1)^{-1/2}(x^4 + 3x^2 + 2x)}{2(x^2 + 1)^{3/2}}$$

Thus the combined application of the chain rule and the quotient rule yields

$$\frac{dy}{dx} = \frac{x^4 + 3x^2 + 2x}{2\sqrt{(x^2 + 1)^3(x^3 - 1)}}$$

Method (b). This approach is again based on the properties of the logarithmic function and is often the most convenient way of obtaining the derivative of certain functions. Thus taking the natural logarithm of both sides of the given equation, we obtain

$$\ln y = \frac{1}{2} \ln \left(\frac{x^3 - 1}{x^2 + 1}\right)$$

or

$$\ln y = \tfrac{1}{2}[\ln (x^3 - 1) - \ln (x^2 + 1)]$$

Furthermore, differentiating both sides of the last equation, in accordance with Theorem 10.13, we have

$$\frac{1}{y}\frac{dy}{dx} = \frac{1}{2}\frac{3x^2}{x^3 - 1} - \frac{1}{2}\frac{2x}{x^2 + 1}$$

$$\frac{dy}{dx} = \frac{x^4 + 3x^2 + 2x}{2(x^3 - 1)(x^2 + 1)} \cdot \sqrt{\frac{x^3 - 1}{x^2 + 1}}$$

or

$$\frac{dy}{dx} = \frac{x^4 + 3x^2 + 2x}{2\sqrt{(x^2 + 1)^3(x^3 - 1)}}$$

Example 10.22 What is the derivative of $y = e^{x/(x^2-1)}$?

Solution If we let $u = x/(x^2 - 1)$, then $y = e^u$. Thus application of Theorem 10.14 and the quotient rule results in the following equation:

$$\frac{dy}{dx} = e^u \frac{du}{dx} = e^{x/(x^2-1)} \left[\frac{1(x^2 - 1) - 2x \cdot (x)}{(x^2 - 1)^2}\right]$$

or

$$\frac{dy}{dx} = e^{x/(x^2-1)} \frac{-x^2 - 1}{(x^2 - 1)^2}$$

To obtain the derivatives of exponential and logarithmic functions to a base b different from e we now employ the following two theorems.

Theorem 10.15 If $u(x)$ is a function of x and if $y = \log_b u$,

then

$$\frac{dy}{dx} = \frac{1}{u} \log_b e \left(\frac{du}{dx}\right) \quad \text{and} \quad \frac{dy}{dx} = \frac{1}{x} \log_b e$$

whenever $u(x) = x$.

Theorem 10.16 If $u(x)$ is a function of x and if $y = b^u$, then

$$\frac{dy}{dx} = b^u \ln b \left(\frac{du}{dx}\right) \quad \text{and} \quad \frac{dy}{dx} = b^x \ln b$$

whenever $u(x) = x$.

The application and the mechanics of Theorems 10.15 and 10.16 are illustrated in the following examples.

Example 10.23 What is the derivative of $y = \log_{10} (x^2 - 1)$?

Solution If we let $u = x^2 - 1$, then $y = \log_{10} u$, and by Theorem 10.15

$$\frac{dy}{dx} = \frac{1}{u} \log_{10} e \left(\frac{du}{dx}\right) \quad \text{or} \quad \frac{dy}{dx} = \frac{1}{x^2 - 1} 2x \log_{10} e$$

Thus

$$\frac{dy}{dx} = \frac{2x \log_{10} e}{x^2 - 1}$$

Example 10.24 What is the derivative of $y = 10^{x^2}$?

Solution If we let $u(x) = x^2$, then $y = 10^u$, and by Theorem 10.16

$$\frac{dy}{dx} = 10^u \ln 10 \left(\frac{du}{dx}\right) = 10^{x^2} (\ln 10)(2x)$$

Thus

$$\frac{dy}{dx} = 2x10^{x^2} \ln 10$$

Example 10.25 If $y = Ae^{kt}$, where A and k are constants, show that dy/dt is proportional to y, and conversely if $dy/dt = ky$, then $y = Ae^{kt}$.

Solution Since

$$\frac{dy}{dt} = Ake^{kt} = k(Ae^{kt})$$

if we substitute y for Ae^{kt} in the derivative, we find

$$\frac{dy}{dt} = ky$$

as asserted. Conversely, let us assume that $dy/dt = ky$, and let us differentiate the function y/e^{kt} with respect to t. Then

$$\frac{d}{dt}\left(\frac{y}{e^{kt}}\right) = \frac{e^{kt}(dy/dt) - y(de^{kt}/dt)}{(e^{kt})^2} = \frac{e^{kt}(dy/dt) - yke^{kt}}{(e^{kt})^2}$$

or substituting ky for dy/dt, we find

$$\frac{d}{dt}\left(\frac{y}{e^{kt}}\right) = \frac{e^{kt}ky - kye^{kt}}{(e^{kt})^2} = \frac{0}{(e^{kt})^2} = 0$$

Since the derivative of a function is zero if and only if the function is a constant, the last equation implies

$$\frac{y}{e^{kt}} = A \text{ (constant)} \qquad \text{or} \qquad y = Ae^{kt}$$

Example 10.25 results in the commonly known compound interest law or law of organic growth. Many quantities in the physical, social, and management sciences vary proportionally to their instantaneous magnitudes. Therefore we find the application of this result quite useful in many practical situations. In the subsequent examples and exercises some of its simpler applications are demonstrated.

Example 10.26 If x is the numer of repetitions, then the habit strength $H(x)$ may be expressed as

$$H(x) = A(1 - e^{-kx})$$

where A and k are positive constants. Show that the rate of habit strength acquisition is a multiple of k.

Solution The rate of change of habit strength is measured by the derivative of $H(x)$. Thus

$$\frac{dH(x)}{dx} = A[0 - (-k)e^{-kx}]$$

of

$$\frac{dH(x)}{dx} = k(Ae^{-kx})$$

as asserted.

Example 10.27 In the growth of a bacterial culture, let 100 be the number of bacteria at time $t = 0$ and let the number of reproductions taking place at any time be proportional to the number of bacteria present at that time. Find the number of bacteria at $t = 1$ and $t = 5$.

Solution If we let $N(t)$ be the number of bacteria at any time t, then the rate of change is

$$\frac{dN(t)}{dt} = kN(t)$$

where k is a constant of proportionality. Thus Example 10.25 implies that

$$N(t) = Ae^{kt}$$

for some constant A. Since at $t = 0$, $N(0) = 100$, we have $100 = Ae^{k0}$ or $A = 100$. Thus the formula for the number of bacteria at any time t becomes

$$N(t) = 100e^{kt}$$

Hence, at $t = 1$ and $t = 5$ we have $N(1) = 100e^{k}$ and $N(5) = 100e^{5k}$ bacteria, respectively. Clearly, if the constant of proportionality k were either experimentally or otherwise calculated, we would be able to determine the population of the bacteria at any given time.

Example 10.28 The concentration of a drug in the body fluids depends on the time elapsed after administration. In general, the velocity of elimination is proportional to the concentration $c(t)$. Find the time required for the concentration to be $1/100$ of its initial value.

Solution The velocity of elimination or the rate of change of the concentration $c(t)$ is $dc(t)/dt$. Since the velocity of elimination is proportional to the concentration with constant of proportionality k, we have

$$\frac{dc(t)}{dt} = kc(t)$$

Again, Example 10.25 yields

$$c(t) = Ae^{kt}$$

for the concentration at any time t where A and k are constants. The value of the function $c(t)$ at $t = 0$ is the initial concentration $c(0) = A$. Thus to determine the time required for the concentration to become $1/100$ of the initial $c(0) = A$ we need only solve for t the equation

$$\frac{1}{100}A = Ae^{kt} \qquad \text{or} \qquad \frac{1}{100} = e^{kt}$$

Taking the natural logarithm of both sides in the last equation, we now find

$$-\ln 100 = kt \qquad \text{or} \qquad t = -\frac{k}{\ln 100}$$

Once again if the constant of proportionality (negative in this case) were experimentally or otherwise determined, the last equation would provide the required time.

Exercises

Find the derivative of the following functions:

1. $y = \ln(2x^2 - 1)$.
2. $y = x^x$.
3. $y = \ln(x + e^{-x})$.
4. $y = \ln(\ln x)$.
5. $y = \ln(1 - x^2)/(1 + x^2)$.
6. $y = 10^{x^2}$.
7. $y = e^x - e^{-x}$.
8. $y = \sqrt{e^x + e^{2x}}$.
9. $y = [\ln(2 - x)]^5$.
10. $y = x \ln x - x$.
11. $y = \ln^3(1 + x)$.
12. $y = \ln\sqrt{\dfrac{(1+x)(1+2x)(1+3x)}{(1-x)(1-2x)(1-3x)}}$.

Ans. (2) $x^x[1 + \ln x]$; (4) $1/(x \ln x)$; (5) $-4x/(1 - x^4)$; (10) $\ln x$; (12) $1/(1 - x^2) + 2/(1 - 4x^2) + 3/(1 - 9x^2)$

13. In a certain chemical reaction it is observed that the rate at which a substance is transformed (used up) is proportional to the quantity of the substance instantaneously present. If initially 10 grams of the substance is present and if after 30 minutes only 2 grams remains, find the formula giving the quantity present at any time t.

14. Under certain conditions, the rate at which atmospheric pressure changes with altitude is proportional to the pressure. If the pressure is 14.7 pound per square inch at sea level and if it falls to one half this value at 18,000 feet, find the formula for the pressure at any altitude.

Ans. $14.7e^{-0.000035h}$

15. Radium disintegrates at a rate proportional to the amount of radium instantaneously present. If one half of any given amount of radium will disappear in 1590 years, what fraction will disintegrate during the first century? During the tenth century?
Ans. 4.27%, 2.91%

10.7 Implicit and Successive Differentiation

Thus far the functions differentiated have been of the form $y = f(x)$, that is, they have been explicit functions or equivalently functions in which the variable y was explicitly given in terms of the variable x. However, this is not always the case, and we shall often find it necessary to concern ourselves with the differentiation of implicit functions $f(x, y) = 0$. For example, we may need to find y' from the implicit functions

$$x^2y - 3xy - x = 0 \tag{10.13}$$

$$x^2\sqrt{y} - 3xy^4 + 4^4y^4 - 6x = 0 \tag{10.14}$$

Now, given an implicit function $f(x, y) = 0$, we may be able to solve it for one or the other of the variables and proceed explicitly. On the other hand it may be inconvenient or even impossible to solve the equation $f(x, y) = 0$ for either y or x. The function described in Equation (10.13), for instance, can be written explicitly as

$$y = \frac{x}{x^2 - 3x}$$

However, it is not at all clear whether we can say the same thing about the function described in Equation (10.14). In such cases we can find the derivative dy/dx by the following simple procedure. Differentiate every term in the equation $f(x, y) = 0$ with respect to x, remembering that y is indeed a function of x. This will introduce the factor dy/dx every time y or any combination involving y is differentiated. Once the differentiation has been performed, it is easy to solve the resulting equation for dy/dx. The mechanics of implicit differentiation are demonstrated in the following examples.

Example 10.29 Find dy/dx if $x^2 + y^2 = 9$.

Solution First we write $x^2 + y^2 = 9$ as $x^2 + y^2 - 9 = 0$, and then differentiate every term in both sides of this equation with respect to x. This yields the equation

$$2x + 2y\frac{dy}{dx} - 0 = 0$$

or

$$2x + 2y\frac{dy}{dx} = 0$$

Thus solving the resulting equation for the derivative dy/dx we obtain

$$\frac{dy}{dx} = -\frac{x}{y}$$

In implicit differentiation we caution the student to make proper use of the rules of differentiation, particularly that of the chain rule which in differentiating the term y yielded $2y\,dy/dx$. In fact, it is this rule that allows dy/dx to appear in the resulting equation.

Example 10.30 What is the slope of the parabola $x^2 + 4xy + 4y^2 + 3x - 2y = 2$ at the points where $x = -1$?

Solution The given function is implicit and although we can solve for y in terms of x, we find it inconvenient to do so. Thus we differentiate implicitly to obtain

$$\frac{d(x^2)}{dx} + \frac{d(4xy)}{dx} + \frac{d(4y^2)}{dx} + \frac{d(3x)}{dx} - \frac{d(2y)}{dx} = \frac{d(2)}{dx}$$

$$2x + 4y + 4x\frac{dy}{dx} + 8y\frac{dy}{dx} + 3 - 2\frac{dy}{dx} = 0$$

or

$$(4x + 8y - 2)\frac{dy}{dx} = -2x - 4y - 3$$

or

$$\frac{dy}{dx} = -\frac{2x + 4y + 3}{4x + 8y - 2}$$

To find the required slopes we must now know not only x but the corresponding values for y as well. Thus if in the original equation of the parabola we substitute $x = -1$, we obtain

$$1 - 4y + 4y^2 - 3 - 2y = 2$$

$$2y^2 - 3y - 2 = 0$$

or

$$y = -\tfrac{1}{2}, 2.$$

Substituting these values, together with $x = -1$, in the formula for

dx/dy, we obtain the required slopes to be

$$\frac{dy}{dx} = -\frac{-2-2+3}{-4-4-2} = -\frac{1}{10} \quad \text{at the point } (-1, -\tfrac{1}{2})$$

and

$$\frac{dy}{dx} = -\frac{-2+8+3}{-4+16-2} = -\frac{9}{10} \quad \text{at the other point } (-1, 2)$$

Clearly, the derivative $f'(x)$ of a function $f(x)$ is itself a function of x, and in general it will be differentiable. The derivative of $f'(x)$, commonly referred to as the second derivative, is denoted by

$$y'', f''(x), \qquad D_x{}^2y = D^2y, \qquad \text{or} \qquad \frac{d^2y}{dx^2}$$

Similarly, the second derivative $f''(x)$ is in general differentiable, and it yields the third derivative of the function $f(x)$, and so on indefinitely, unless some one of the derivatives is a nondifferentiable function. The third derivatives are usually denoted by one of the symbols y''', $f'''(x)$, $D_x{}^3y$, or d^3y/dx^3.

The nth derivative is customarily designated by any one of the symbols

$$y^{(n)}, f^{(n)}_{(x)}, D_x{}^ny, \qquad \text{or} \qquad \frac{d^ny}{dx^n}$$

Whenever the symbol $y^{(n)}$ or $f^{(n)}_{(x)}$ is used, the parenthesis around the index n must always be included in order to prevent the index from being mistaken for an ordinary exponent.

Examples of higher-order derivatives are encountered in all applications of differentiation. For instance, the acceleration of a moving body is the first derivative of its velocity or the second derivative of the distance traveled with respect to time. To illustrate higher order differentiation, we consider the following examples.

Example 10.31 Compute all possible derivatives of $y = 3x^4 - 2x^3 + x - 5$.

Solution

$$y' = 12x^3 - 6x^2 + 1$$

$$y'' = 36x^2 - 12x$$

$$y''' = 72x - 12$$

$$y^{(4)} = 72$$

$$y^{(5)} = 0$$

and

$$y^{(n)} = 0 \qquad \text{for } n \geq 5$$

Example 10.32 If $y = x^2 - 9x + 20$, for what values of x is y' zero, positive, and negative? Graph the function.

Solution $y' = 2x - 9$. Thus $y' = 0$ if $x = 9/2$, $y' > 0$ if $x > 9/2$, and $y' < 0$ if $x < 9/12$ (Figure 10.12).

Examining the graph of the function $y = x^2 - 9x + 20$ we now find

a. At $x = 9/2$, the function is minimum $-\frac{1}{4}$.
b. For $x > 9/2$, the function is increasing.
c. For $x < 9/2$, the function is decreasing.
d. At $x = 9/2$, y'' is positive (in fact $y'' = 2$).
e. The tangent to the curve at $x = 9/2$ is horizontal.

Example 10.33 If $y = \frac{1}{3}x^3 - 4x^2 + 12x$, for what values of x is y' zero, positive, and negative? What is the second derivative y'', and what does the graph of the function look like?

Solution

$$y' = \tfrac{1}{3}\cdot 3x^2 - 4\cdot 2x + 12 \qquad \text{or} \qquad y' = x^2 - 8x + 12$$

Thus $y' = 0$ or $x^2 - 8x + 12 = 0$ whenever $x = 2$ or 6. To find the values of x for which $y' > 0$ (or $y' < 0$), we must solve the inequality $x^2 - 8x + 12 > 0$ (or $x^2 - 8x + 12 < 0$). Thus

$$x^2 - 8x + 12 > 0 \qquad \text{or} \qquad (x - 2)(x - 6) > 0$$

Figure 10-12

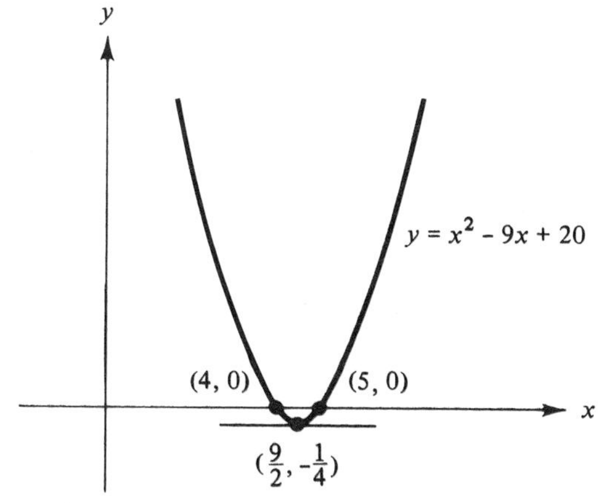

Table 10.1

x axis	$x - 2$	$x - 6$	$(x - 2)\ (x - 6) = y'$
	+	+	+
6		0	
	+	−	−
2	0		
	−	−	+

and using Table 10.1, we find that $y' > 0$ if $x > 6$ or $x < 2$, and $y' < 0$ if $2 < x < 6$.

Now, the second derivative y'' is the derivative of y'. Hence $y'' = 2x - 8$. Clearly, $y'' < 0$ if $x = 2$, and $y'' > 0$ if $x = 6$, that is, one of the values of x for which $y' = 0$ makes the second derivative negative

Figure 10–13

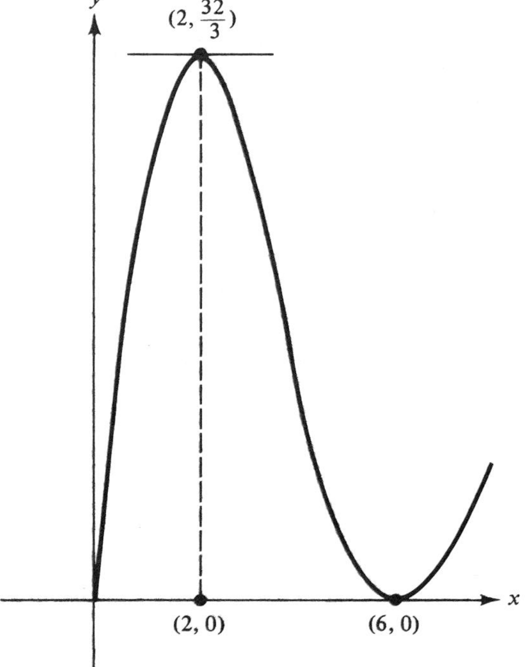

and the other makes it positive. The graph of the function is presented in the Figure 10.13.

Examining the graph of this function, we find the following:

a. At the value $x = 2$ (for which $y' = 0$) the function has a "relative" maximum, 32/3, and $y'' < 0$.
b. At the value $x = 6$ (for which $y' = 0$) the function has a "relative" minimum, 0, and $y'' > 0$.
c. For $x < 2$ or $x > 6$ (for which $y' > 0$) the function is increasing, and for $2 < x < 6$ (for which $y' < 0$) the function is decreasing.
d. At either $x = 2$ or $x = 6$ the tangent to the curve is horizontal.

Generalizations of observations made in the preceding two examples constitute a major part of optimization theory, stability, and curve analysis as we shall demonstrate in the subsequent sections on applications of the derivative.

Exercises

Differentiate the following both implicitly and explicitly, and verify that the same answer is obtained by each method:

1. $2xy + 3y + 4x - 9 = 0.$ 2. $x^2 - x^2y = 1 + y.$
3. $y^2 = x - xy^2.$ 4. $x^{1/2} + y^{1/2} = 4.$

Differentiate each of the following functions:

5. $x^3y - 3y^2 + x^4 = 0.$ 6. $\ln(x^2 - y^2) = x + y.$
7. $x^3 + y^3 + 3xy = 8.$ 8. $x \ln(x + y) - y^2 = 4.$
9. $x^2e^y + 3y^2 = 2e^y.$ 10. $x \ln y = e^{3y^4}.$

Find the second derivative of each of the following functions:

1. $y = x\sqrt{4 - x^2}.$ 2. $y = x^2/(4 + x^2).$
3. $y = (3 - x^2)/3 + x^2.$ 4. $y = (\ln x^2)^2.$
5. $y = 5e^{-x} + 6xe^{-x}.$

Find the nth derivative of the following functions:

6. $y = xe^{5x}.$ 7. $y = 1/(1 + 2x).$
8. $y = \ln(1 - x).$

10.8 Partial Differentiation

Since in the crime rate function (Exercise 6 of Chapter 9, Section 9.1)

$$z = x^3 + 2y^3 - 6x + 6y - 6xy$$

either the amount spent on welfare x or the amount spent on prisons y or both may be fixed (or constant), the function may very well be a function of one variable (the other being a constant). Thus in any function of two variables $z = f(x, y)$, although each of the variables x and y can be varied independently of the other, one variable, say y, can often be fixed and the other can be varied in order to determine the exclusive effect of changes in x on the dependent variable z; that is, the function may frequently be reduced to a function of only one independent variable, $z = g(x)$, by treating y as a constant. For example, the function

$$z = f(x, y) = x^3 + 2y^3 - 6x + 6y - 6xy$$

becomes

$$z = g(x) = x^3 - 18x + 28 \qquad \text{for } y = 2,$$

and

$$z = h(y) = 2y^3 - 5 \qquad \text{for } x = 1$$

Consequently, we may speak of the derivative of $z = f(x, y)$ with respect to x (y is kept constant) and mean essentially what was defined previously as the ordinary derivative of a function of x. Similarly, we may speak of the derivative of $z = f(x, y)$ with respect to y (x is kept constant).

The derivatives alluded to in the previous paragraph are commonly known as *partial derivatives* or first-order partials of the function $z = f(x, y)$ and are symbolized as either

$$\frac{\partial z}{\partial x} \quad \text{or} \quad \frac{\partial f}{\partial x} \quad \text{or} \quad z_x \quad \text{or} \quad f_x$$

for those with respect to x, and as either

$$\frac{\partial z}{\partial y} \quad \text{or} \quad \frac{\partial f}{\partial y} \quad \text{or} \quad z_y \text{ or } f_y$$

for those with respect to y. The process of obtaining these derivatives is called *partial differentiation*. Here, the symbol ∂ replaces the previous d in order to distinguish partial from ordinary derivatives. Formally, the partial derivatives of $z = f(x, y)$ are defined as follows.

DEFINITION 10.5 If $z = f(x, y)$ is a function of two variables, then

$$f_x = \lim_{\Delta x \to 0} \frac{f(x + \Delta x, y) - f(x, y)}{\Delta x}$$

and

$$f_y = \lim_{\Delta y \to 0} \frac{f(x, y + \Delta y) - f(x, y)}{\Delta y}$$

are the partial derivatives of $z = f(x, y)$ with respect to x and y, respectively.

Since the concept of the limit remains the same and since "partial differentiation" is identical to ordinary differentiation once all but one of the variables in a function are fixed, the results (or rules) of ordinary differentiation are applicable in partial differentiation. For instance, the partial derivatives of the function

$$g(x, y) = x^3 + 3xy + 5y^2$$

are

$$g_x = 3x^2 + 3y \qquad \text{and} \qquad g_y = 3x + 10y$$

Similarly, the partial derivatives of the following functions are as indicated.

a. $w(s, t) = s^2 + st$; $w_s = 2s + t$ and $w_t = s$.
b. $R(x, y) = 3x^2 + 2y\sqrt{x} - 4$; $R_x = 6x + yx^{-1/2}$ and $R_y = 2\sqrt{x}$.
c. $f(x, y) = (x - y)/(x + y)$; $f_x = 2y/(x + y)^2$ and $f_y = -2x/(x + y)^2$.
d. $g(x, y, z) = x^2 + y^2 + z^2$; $g_x = 2x$, $g_y = 2y$, and $g_z = 2z$.

As in the case of ordinary derivatives, each partial derivative might in turn have partial derivatives. Thus a function $z = f(x, y)$ may have four "second partial derivatives" or "second order partials" as shown in the treelike diagram below.

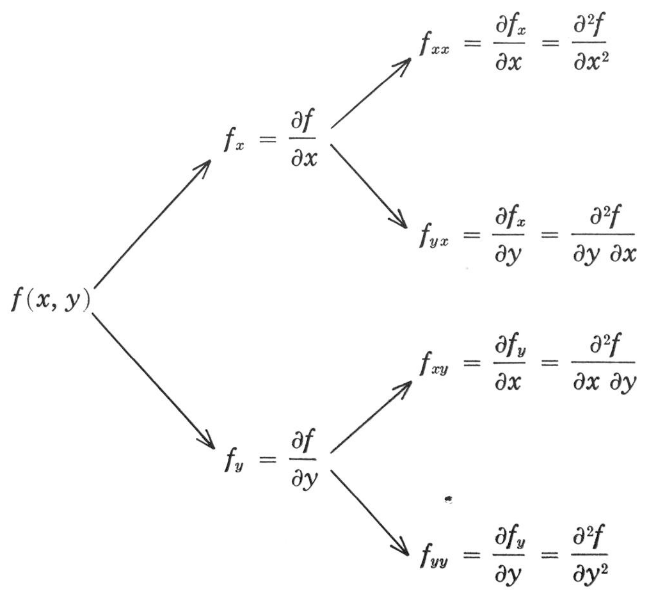

The second-order partials

$$f_{xy} = \frac{\partial^2 f}{\partial x\, \partial y} \qquad \text{and} \qquad f_{yx} = \frac{\partial^2 f}{\partial y\, \partial x}$$

are often referred to as the cross partial derivatives of $f(x, y)$. The cross partial derivatives of a function $f(x, y)$ are "almost always" equal. To demonstrate these concepts further we consider the following example.

Example 10.34 Find the four second partial derivatives of the function

$$z = x^3 + 2y^3 - 6x + 6y - 6xy$$

Solution The partial derivatives of the function are

$$z_x = 3x^2 - 6y - 6 \qquad \text{and} \qquad z_y = 6y^2 - 6x + 6$$

The first partial was obtained by treating y as a constant and then differentiating z with respect to x. Similarly, z_y was the result of differentiating z with respect to y, whereas x was considered to be a constant.

Thus the second partial derivatives can be obtained by taking the partial derivatives of both z_x and z_y with respect to x and y, while treating x and y as constants, accordingly. Therefore

$$\frac{\partial^2 z}{\partial x^2} = \frac{\partial}{\partial x}(3x^2 - 6y - 6) = 6x$$

$$\frac{\partial^2 z}{\partial x\, \partial y} = \frac{\partial}{\partial x}(6y^2 - 6x + 6) = -6$$

$$\frac{\partial^2 z}{\partial y\, \partial x} = \frac{\partial}{\partial y}(3x^2 - 6y - 6) = -6$$

and

$$\frac{\partial z}{\partial y^2} = \frac{\partial}{\partial y}(6y^2 - 6x + 6) = 12y$$

Once again, the cross partial derivatives $\partial^2 z/\partial x\, \partial y$ and $\partial^2 z/\partial y\, \partial x$ are equal to each other, and their value is -6. Generally, the following theorem, about cross partial derivatives, holds true.[3]

Theorem 10.17 If the cross partial derivatives of a function $f(x, y)$ are continuous in a region R, then they are equal throughout that region; that is, $f_{xy} = f_{yx}$ in R.

Theorem 10.17 implies that the order of partial differentiation is unimportant when the continuity condition is satisfied. This theorem, as well as the notation, can be extended to higher-order partial derivatives. Therefore when the conditions of Theorem 10.17 are satisfied for the function $u = f(x, y, z)$, then

$$u_{xyz} = u_{zyx} = u_{xzy}$$

Clearly, a function may possess an increasing number of higher-order partials. Thus for different functions the corresponding number of partials is shown in the Table 10.2. Verification of this table is left as an exercise for the student.

Theorem 10.17 and its generalizations, however, indicate that not all higher-order cross partials are distinct. Generally, the function $Z = f(x, y)$ possesses only three distinct second-order partials and four distinct third-order partials (why?). Specifically, the function

$$f(x, y) = 5x^2 + 2xy + 4y^2$$

has the following types of partial derivatives.

a. Two first-order partials:

$$\frac{\partial f}{\partial x} = 10x + 2y \qquad \text{and} \qquad \frac{\partial f}{\partial y} = 2x + 8y$$

b. Three second-order partials:

$$\frac{\partial^2 f}{\partial x^2} = 10, \qquad \frac{\partial^2 f}{\partial y^2} = 8, \qquad \text{and} \qquad \frac{\partial^2 f}{\partial x\, \partial y} = \frac{\partial^2 f}{\partial y\, \partial x} = 2$$

c. All third-order partials are equal to zero (why?).

Table 10.2

Function	Partials				
	First Order	Second Order	Third Order	...	nth Order
$f(x)$	1	1	1		1
$f(x, y)$	2	4	8		2^n
$f(x, y, z)$	3	9	27		3^n
$f(x, y, z, t)$	4	16	64		4^n
$f(x, y, z, t, u)$	5	25	125		5^n
.	.	.	.		.
.	.	.	.		.
.	.	.	.		.

In concluding our brief discussion of what is otherwise a complex and extensive mathematical study of functions in several variables,[3] we introduce the analogue to the chain rule for differentiating composite functions in one variable. For the function $y = f(u)$ of one variable, it was stated that if $u = g(x)$, then

$$\frac{dy}{dx} = \frac{df(u)}{du}\left(\frac{du}{dx}\right)$$

Analogously, the following theorem establishes a chain rule for a composite function of two variables.

Theorem 10.18 If $z = F(x, y)$ is a composite function of two variables, that is, if in $z = F(x, y)$, $x = f(u, v)$ and $y = g(u, v)$, then

$$\frac{\partial z}{\partial u} = \frac{\partial z}{\partial x}\frac{\partial x}{\partial u} + \frac{\partial z}{\partial y}\frac{\partial y}{\partial u}$$

and

$$\frac{\partial z}{\partial v} = \frac{\partial z}{\partial x}\frac{\partial x}{\partial v} + \frac{\partial z}{\partial y}\frac{\partial y}{\partial v}$$

Since many interesting applications of the derivatives of composite functions involve the special case in which $x = g(t)$ and $y = f(t)$, that is, both x and y are functions of only one variable in the function $z = F(x, y)$, the following corollary of Theorem 10.18 is very important.

Corollary 10.5 If in the function $Z = F(x, y)$, x and y are functions of a single variable t, then

$$\frac{\partial F}{\partial t} = \frac{\partial F}{\partial x}\frac{dx}{dt} + \frac{\partial F}{\partial y}\frac{dy}{dt}$$

To illustrate further the preceding theorems, corollary, and definitions, we consider the following examples.

Example 10.35 If $F(x, y) = x^2y^2$ where $x = 1/t$ and $y = t^2$, find $\partial F/\partial t$.

Solution Since Corollary 10.5 is applicable, we have

$$\frac{\partial F}{\partial t} = \frac{\partial F}{\partial x}\frac{dx}{dt} + \frac{\partial F}{\partial y}\frac{dy}{dt}$$

$$\frac{\partial F}{\partial t} = 2xy^2\left(\frac{-1}{t^2}\right) + 2x^2y(2t) = \frac{-2t^4}{t^3} + \frac{2t^2}{t^2}(2t)$$

or

$$\frac{\partial F}{\partial t} = -2t + 4t = 2t$$

Example 10.36 If $z = F(x, y) = x^2 + y^2 - 2$ where $x = u^2 + v^2$ and $y = u^2 - v^2$, find $\partial z/\partial u$, $\partial z/\partial v$, $\partial^2 z/\partial u^2$, and $\partial^2 z/\partial v^2$.

Solution Since $z = F(x, y) = x^2 + y^2 - 2$ is a composite function of two variables, the chain rule of Theorem 10.18 is applicable, and

$$\frac{\partial z}{\partial u} = \frac{\partial z}{\partial x}\frac{\partial x}{\partial u} + \frac{\partial z}{\partial y}\frac{\partial y}{\partial u} = (2x)(2u) + (2y)(2u) = 4xu + 4yu$$

or

$$\frac{\partial z}{\partial u} = 4(u^2 + v^2)u + 4(u^2 - v^2)u \qquad \text{or} \qquad \frac{\partial z}{\partial u} = 8u^3$$

Similarly,

$$\frac{\partial z}{\partial v} = \frac{\partial z}{\partial x}\frac{\partial x}{\partial v} + \frac{\partial z}{\partial y}\frac{\partial y}{\partial v} = (2x)(2v) + (2y)(-2v) = 4xv - 4yv$$

or

$$\frac{\partial z}{\partial v} = 4(u^2 + v^2)v - 4(u^2 - v^2)v \qquad \text{or} \qquad \frac{\partial z}{\partial v} = 8v^3$$

Finally,

$$\frac{\partial^2 z}{\partial u^2} = \frac{\partial}{\partial u}\left(\frac{\partial z}{\partial u}\right) = \frac{\partial}{\partial u}(8u^3) = 24u^2$$

and

$$\frac{\partial^2 z}{\partial v^2} = \frac{\partial}{\partial v}\left(\frac{\partial z}{\partial v}\right) = \frac{\partial}{\partial v}(8v^3) = 24v^2$$

Example 10.37 If a department store has two departments and if the earnings of each department are given by

$$E_1(x_1, y_1) = 5x_1 + 6y_1 + 2x_1y_1 - 2x_1^2 - y_1^2$$

and

$$E_2(x_2, y_2) = 3x_2 + 4y_2 + 3x_2y_2 - 3x_2^2 - 2y_2^2 - 5$$

where x_1 and x_2 represent investments in inventory for the respective departments in millions of dollars and y_1 and y_2 represent the floor space utilized by each department in units of 100,000 square feet, then find all first-order and second-order partial derivatives of the total earnings E.

Solution Since the total earnings E are the sum of E_1 and E_2, we must find the required partials of

$$E = E_1 + E_2 = 5x_1 + 6y_1 + 2x_1y_1 - 2x_1^2 - y_1^2 + 3x_2 + 4y_2 + 3x_2y_2 - 3x_2^2 - 2y_2^2 - 5$$

Thus

$$\frac{\partial E}{\partial x_1} = 5 + 2y_1 - 4x_1$$

$$\frac{\partial E}{\partial y_1} = 6 + 2x_1 - 2y_1$$

$$\frac{\partial E}{\partial x_2} = 3 + 3y_2 - 6x_2$$

and

$$\frac{\partial E}{\partial y_2} = 4 + 3x_2 - 4y_2$$

are the first-order partials of E. To obtain the second-order partials we take the partial derivatives of the first-order partials with respect to x_1, x_2, y_1, and y_2, respectively. Consequently,

$$\frac{\partial^2 E}{\partial x_1^2} = -4, \quad \frac{\partial^2 E}{\partial x_1\, \partial y_1} = 2, \quad \frac{\partial^2 E}{\partial x_1\, \partial x_2} = 0,$$

$$\frac{\partial^2 E}{\partial x_1\, \partial y_2} = 0, \quad \frac{\partial^2 E}{\partial y_1\, \partial x_1} = 2$$

$$\frac{\partial^2 E}{\partial y_1^2} = -2, \quad \frac{\partial^2 E}{\partial y_1\, \partial x_2} = 0, \quad \frac{\partial^2 E}{\partial y_1\, \partial y_2} = 0,$$

$$\frac{\partial^2 E}{\partial x_2\, \partial y_1} = 0, \quad \frac{\partial^2 E}{\partial x_2^2} = -6$$

$$\frac{\partial^2 E}{\partial x_2\, \partial y_2} = 3, \quad \frac{\partial^2 E}{\partial y_2\, \partial x_1} = 0, \quad \frac{\partial^2 E}{\partial y_2\, \partial y_1} = 0,$$

$$\frac{\partial^2 E}{\partial y_2\, \partial x_2} = 3, \quad \frac{\partial^2 E}{\partial y_2^2} = -4$$

and

$$\frac{\partial^2 E}{\partial x_2\, \partial x_1} = 0$$

Exercises

1. Find the first-order partials of the following functions:

   a. $f(x, y) = (x - y)/(x + y)$. b. $g(s, t) = \sqrt{s^2 + t^2}$.

   c. $h(u, v) = \ln (u^2 + v^2)$. d. $H(x, y) = e^{x^2+y^2}$.

   Ans. (d) $H_x = 2xe^{x^2+y^2}$, $H_y = 2ye^{x^2+y^2}$

2. Find the partials f_x, f_y, f_{xx}, f_{xy}, and f_{yy} for the following functions:

   a. $f(x, y) = x^2 + xy + y^2 + 24$.
   b. $f(x, y) = 2x^3 - 2xy + y^2 + 2$.
   c. $f(x, y) = x^2 - 12y^2 - 4y^3 + 3y^4 - 6$.
   d. $f(x, y) = (x^2 - y^2)/x^2 + y^2$.
   e. $f(x, y) = 1/(x - y) + 1/(x + y)$.
   f. $f(x, y) = e^2xy/(x^2 + y^2)$.

3. For each function of Exercise 2, (a), (b), and (c), determine

   a. The values of x and y for which $f_x = f_y = 0$.
   b. The values of the variables x and y such that $f_{xy}{}^2 - f_{xx}f_{yy}$ is (i) positive, (ii) negative, and (iii) zero.

4. If in each of the following functions $F(x, y)$, $x = 1/t^2$, and $y = t^2 - 1$, then find $\partial F/\partial t$.

   a. $F(x, y) = 3x^2 - 4y^2 + 6xy - 4$.
   b. $F(x, y) = (x + 2y)/(x - 2y)$.
   c. $F(x, y) = x^3y^2 - 2$.

5. If in each of the following functions $G(x, y)$, $x = u^2 - 3v^2$, and $y = u^2 + 3v^2$, find $\partial G/\partial u$ and $\partial G/\partial v$.

   a. $G(x, y) = 2x^2 - 3y^2 + xy - 7$. b. $G(x, y) = e^{xy}/(x + y)$.
   c. $G(x, y) = 2x^2 + 5xy + y^2$.

6. If x represents the dollars spent on labor and y the dollars spent on soil improvement per acre of a wheat farm, then it has been

found that the function

$$P(x, y) = 50x + 60y + 10xy - 10x^2 - 5y^2$$

represents the total profit per acre of the wheat farm. Determine the values of x and y for which both $\partial P/\partial x$ and $\partial P/\partial y$ are zero.

Ans. $x = 11$, $y = 17$

7. If x, y, and z represent the amounts of a certain commodity produced at three separate plants of some firm and if

$$c_1(x) = 120 + \frac{3}{110}x^2$$

$$c_2(x) = 100 + 3y + \frac{y^3}{200}$$

and

$$c_3(x) = 200 + 6z$$

are the corresponding cost functions, find the first partial derivatives of the total cost function of producing x, y, and z units at the three plants.

8. The rate R at which a chemical substance is absorbed into a bacterium and then distributed throughout its whole volume is proportional to s/v, that is, $R = K(S/V)$ where S is the surface area, v the volume of the bacterium, and K a constant of proportionality. Thus if a certain bacterium is in the shape of a cylinder of radius r and length l, compute $\partial R/\partial l$ and $\partial R/\partial r$ to see how an increase in the length or in the radius affects the rate of metabolism of this bacterium.

10.9 Summary and Table of Derivatives

This chapter was devoted to illustrating the basic laws of formal differentiation, partial differentiation, and some of their fundamental applications. Using the four-step rule, we derived some of the formulas of differentiation and we suggested that the others could be similarly established. Moreover, the important theorems of Rolle and the mean value, related to the derivative of a function, were discussed. Finally, we introduced implicit, successive, and partial differentiation.

The established methods of differentiation are summarized in Table 10.3.

Table 10.3

1. $Dc = 0$		c = constant
2. $Dx = 1$		
3. $Dx^n = nx^{n-1}$ for all n		Power rule
4. $Du^n = nu^{n-1}\,Du$ for all n		Power and chain rule
5. $D[f(x) + g(x)] = Df(x) + Dg(x)$		Sum rule
6. $D[f(x) - g(x)] = Df(x) - Dg(x)$		Difference rule
7. $D[cf(x)] = c\,Df(x)$		Constant times a function rule
8. $D[f(x)g(x)] = f(x)\,Dg(x) + g(x)\,Df(x)$		Product rule
9. $D\left(\frac{f(x)}{g(x)}\right) = \frac{g(x)\,Df(x) - f(x)\,Dg(x)}{g^2(x)}$	if $g(x) \neq 0$	Quotient rule
10. $D[f(x)]^n = nf(x)^{n-1}\,Df(x)$ for all n		Chain rule
11. $D[\ln x] = 1/x$		
12. $D[\ln u(x)] = 1/u(x)\,Du(x)$		Chain rule on $\ln u(x)$
13. $D[e^x] = e^x$		
14. $D[e^{u(x)}] = e^{u(x)}\,Du(x)$		Chain rule on $e^{u(x)}$
15. $D(\log_b x) = 1/x \log_b e$		
16. $D(b^x) = b^x \ln b$		
17. Implicit differentiation		
18. Multiple differentiation		
19. Partial differentiation		

References for Supplementary Readings

1. Joan Welkowitz, R. B. Ewen, and J. Cohen, Chapter 8, *Introductory Statistics for the Behavioral Sciences* (New York: Academic Press, 1972).
2. E. E. Moise, *Calculus* (Reading, Mass.: Addison-Wesley, 1967).
3. A. E. Taylor, *Advanced Calculus*, Chapters V and VI (Boston: Ginn, 1955).

CHAPTER 11

Applications of Differentiation

11.1 Introduction

People at the decision making level, particularly in business and economics, are frequently concerned with optimizing functions of one or more variables;[1] that is, they are involved in manipulating one or more variables in given data until they arrive at some "better" or "more favorable" outcome. Differentiation is generally used to determine the optimum value or values of a function. For example, the head of a manufacturing firm whose cost of production is given by

$$C(x) = 10 + 2x^2 - 8x$$

where x is the number of commodity units produced (in millions), knows that the price per million units is given by

$$p(x) = \frac{30}{x} + 24 - 22x + 8x^2 - x^3$$

and therefore the revenue received from the sales of x units is

$$R(x) = xp(x) \qquad \text{or} \qquad R(x) = 30 + 24x - 22x^2 + 8x^2 - x^4$$

Consequently, if he wishes to maximize the profits of his firm, he must maximize the function

$$P(x) = R(x) - C(x) \qquad \text{(revenue minus cost)}$$

that is, he must find the values of x for which $P(x)$ is maximum. Subsequently we shall find that, in general, $P(x)$ is maximum (or minimum) at those values of x for which the derivative of $P(x)$ is zero. Thus $P(x)$ shall be maximum at all x for which

$$P'(x) = (R(x) - C(x))' = R'(x) - C'(x) = 0$$

$$P'(x) = (30 + 24x - 22x^2 + 8x^3 - x^4)' - (10 + 2x^2 - 8x)'$$

$$= (24 - 44x + 24x^2 - 4x^2) - (4x - 8) = 0$$

$$P'(x) = -4(x - 1)(x - 2)(x - 3) - 4(x - 2) = 0$$

$$P'(x) = -4(x - 2)(x^2 - 4x + 4) = 0$$

or

$$P'(x) = -4(x - 2)^2 = 0$$

or

$$x = 2.$$

Therefore the head of the firm will maximize the profits at $x = 2$ (million) units of production.

Often the derivative of a function is not zero for any finite value(s) of the independent variable. In such cases the existing maxima or minima must be determined by means other than the vanishing of the derivative. For instance, differentiation of the habit strength function

$$H(n) = 100(1 - e^{-kn})$$

yields

$$H'(n) = 100ke^{-kn} \qquad \text{or} \qquad H'(n) \neq 0 \qquad \text{for any} \quad n$$

However,

$$\lim_{n\to\infty} H'(n) = 0$$

Since the derivative of the function is not zero for any finite values of n, its maxima and minima cannot be determined from the vanishing of the function's derivative. In this case the graph of the function (Figure 11.1) shows that at $n = 0$ the function is minimum, namely, $H(0) = 0$. Furthermore, the function approaches a maximum

Figure 11-1

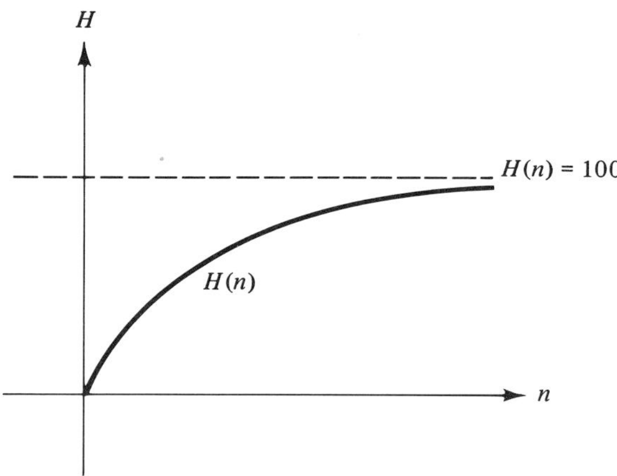

of 100 as n becomes larger and larger; that is,

$$\lim_{n\to\infty} H(n) = \lim_{n\to\infty} 100(1 - e^{-kn}) = 100$$

In the remainder of this chapter we shall be concerned with the applications of the derivative in (a) various geometric problems, (b) the solution of algebraic equations using Newton's method, and (c) problems of optimization and related mathematical techniques.

11.2 Geometric Applications

In this section we shall investigate some of the more important geometric applications of the derivative concept. We shall begin with the problem of finding the tangent and the normal lines to a curve at a given point. Then we shall establish criteria that determine whether a function increases or decreases, whether its graph concaves up or concaves down, and if and where the function possesses maximum and minimum value(s).

TANGENTS AND NORMALS

Since the derivative of a function $y = f(x)$ at $x = a$ represents the slope of the tangent line at $x = a$ and since the normal to a curve at a point is nothing but the line perpendicular to the tangent at P, it is a simple matter to obtain the equation of both the tangent and the normal lines to a curve, $y = f(x)$, at one of its points (x_0, y_0). All we have to do is find the derivative of $y = f(x)$ at (x_0, y_0) and then substitute this for the value of the slope m into the point-slope equation $y - y_0 = m(x - x_0)$ of a straight line. Thus the slope of the tangent to $y = f(x)$ at (x_0, y_0) is $m = f'(x_0)$ and the equations of the tangent and the normal lines at (x_0, y_0) are

$$y - y_0 = f'(x_0)(x - x_0)$$

and

$$y - y_0 = -\frac{1}{f'(x_0)}(x - x_0)$$

respectively. Specifically, the derivative of

$$x^2 + 3xy + 2y^2 - x + 2y = -2 \tag{11.1}$$

at any point is given by

$$2x + 3(xy' + y) + 4yy' - 1 + 2y' = 0$$

or

$$y' = \frac{1 - 2x - 3y}{2 + 3x + 4y}$$

Thus the derivative at the point $(1, -2)$ on the curve specified by Equation (11.1) is given by

$$y' = \frac{1 - 2(1) - 3(-2)}{2 + 3(1) + 4(-2)} \qquad \text{or} \qquad y' = -\frac{5}{3}$$

Therefore the required equations of the tangent and the normal at $(1, -2)$ are

$$y + 2 = -\frac{5}{3}(x - 1) \qquad \text{or} \qquad 5x + 3y = -1$$

and

$$y + 2 = \frac{3}{5}(x - 1) \qquad \text{or} \qquad 3x - 5y = 13$$

respectively.

CURVE ANALYSIS AND SKETCHING

In studying continuous functions, we, intuitively, considered continuity of a function as meaning that the function shows no holes, breaks, or jumps on its graph; that is, we interpreted continuity to imply that the graph of a continuous function can be drawn without removing the pencil from the paper. In discussing the function

$$y = A(1 + i)^x$$

where A is the amount deposited in a bank paying compounded interest at the rate $i\%$ per year and y is the amount on deposit at the end of x years, we concluded, intuitively or by approximate graphing, that it was an increasing function; similarly, we concluded that the function

$$N = N_0(2.718)^{kt}$$

describing the multiplication of a bacterial culture of N_0 bacteria is an increasing one. However, the behavior of a function is not always so easy to determine, neither can we afford to depend on intuition and guesswork. Fortunately, much information about the behavior and graph of a function $y = f(x)$ can be obtained from its first and second derivatives. For functions of more than one variable, similar information is obtained from the function's first- and second-order partials. For example, if

$$P(x) = \tfrac{1}{3}x^3 - 4x^2 + 12x$$

is the function denoting the popularity of a political candidate—that is, the number of votes he receives per x thousands of dollars spent in political campaigning—then

$$P'(x) = x^2 - 8x + 12 = (x - 2)(x - 6)$$

and

$$P''(x) = 2x - 8$$

Thus

$$P'(x) = 0 \qquad \text{for} \quad x = 2 \qquad \text{and} \qquad x = 6$$

and for these values of x we obtain

$$P''(2) < 0 \qquad \text{and} \qquad P''(6) > 0$$

that is, the second derivative of the function $P(x)$ is negative at $x = 2$ and positive at $x = 6$.

Now, an analysis of the graph of $P(x)$ (Figure 11.2) reveals the following:

a. The function is increasing for all x for which $0 < x < 2$ or $x > 6$, that is, those values of x for which $P'(x) > 0$.
b. The function is decreasing for all x such that $2 < x < 6$, that is, the values of x satisfying $P'(x) < 0$.

Figure 11-2

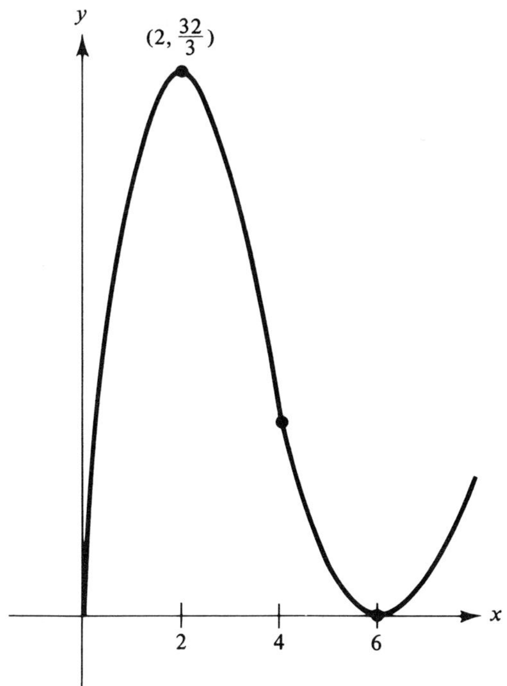

c. The function concaves up for $x > 4$, that is, all x such that $P''(x) > 0$. On the other hand the function concaves down for all $x < 4$ or those x for which $P''(x) < 0$.
d. The function has a maximum, $32/3$, at $x = 2$ where $P''(2) < 0$; the function possesses a minimum, zero, at $x = 6$ at which $P''(6) > 0$.
e. For $x = 4$ the second derivative is zero or $P''(4) = 0$. Since points at which the second derivative of a function is zero and the concaveness of the function changes are called points of inflection, the function $P(x)$ has an inflection at $x = 4$.

Thus the popularity of the politician becomes maximum at $x =$ \$2000 campaign expenditures. However, at $x =$ \$6000 campaign expenditures, his popularity becomes zero.

As a general rule in curve analysis and sketching one must be guided by the following important theorem. This theorem includes some of the observations previously made about functions and their derivatives, and its proof may be found in many books on calculus.[2]

Theorem 11.1 If $y = f(x)$ is a function defined over the interval $[a, b]$, then the following are true:

a. If $f(x)$ is continuous in (a, b) and if $f(a)f(b) < 0$ or if the signs of $f(a)$ and $f(b)$ are not the same, then $f(x) = 0$ for at least one x in (a, b).
b. The function is continuous if it is differentiable.
c. The function possesses a tangent line at all points of continuity.
d. The function is increasing for all x such that $f'(x) > 0$ and decreasing for those x satisfying $f'(x) < 0$.
e. The function concaves up (or down) for all x for which $f''(x) > 0$ (or $f''(x) < 0$).
f. The function has a "relative" maximum or minimum at all x for which $f'(x) = 0$. In fact, if $f'(x_0) = 0$ and $f''(x_0) < 0$, then $f(x_0)$ is a maximum. On the other hand if $f'(x_0) = 0$ and $f''(x_0) > 0$, then $f(x_0)$ is a minimum for $f(x)$.

A good illustration of this theorem is provided by the popularity function of the politician described in the previous paragraph. However, to demonstrate further some of the concepts involved in the theorem we consider the following additional examples.

Example 11.1 Discuss the function $y = x^4 - 10x^2$ and plot its graph.

Discussion Since $y = x^4 - 18x^2$, we have

$$y' = 4x^3 - 36x = 4x(x^2 - 9) \qquad \text{and} \qquad y'' = 12x^2 - 36$$

Now the information about the function can easily be placed in Table 11.1. Following the information contained in Table 11.1 we can easily present the graph of the given function (Figure 11.3).

Table 11.1

$f'(x) = 4x(x^2 - 9)$	Values of x	$f''(x) = 12x^2 - 36$	$f(x) = x^4 - 18x^2$
0	$0, 3, -3$	$f''(0) < 0$	Maximum, 0
		$f''(3) > 0$	Minimum, -81
		$f''(-3) > 0$	Minimum, -81
>0	$x > 3$ and $0 > x > -3$		Increasing
<0	$0 < x < 3$ and $x < -3$		Decreasing
	$x > \sqrt{3}$ and $x < -\sqrt{3}$	>0	Concaves up
	$-\sqrt{3} < x < \sqrt{3}$	<0	Concaves down
	$-\sqrt{3}, \sqrt{3}$	0	Inflexions

Figure 11-3

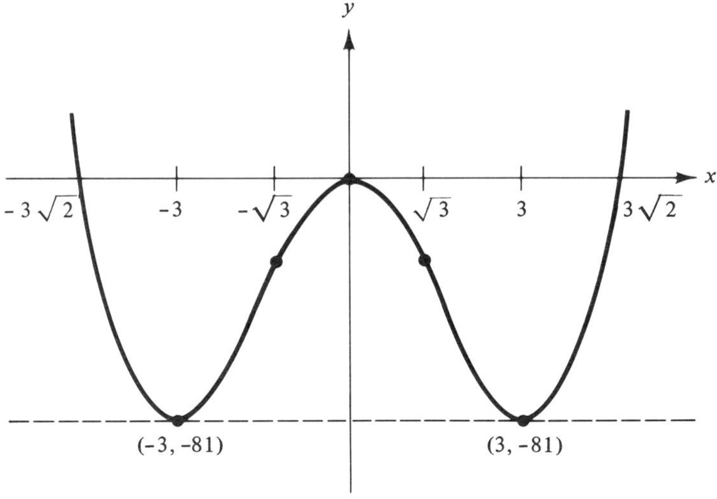

Example 11.2 Discuss the graph of the function $y = 3xe^{-x}$

Discussion Since $y = 3xe^{-x}$, we have

$$y' = 3e^{-x} - 3xe^{-x} \quad \text{and} \quad y'' = 3xe^{-x} - 6e^{-x}$$

or

$$y' = 3e^{-x}(1 - x) \quad \text{and} \quad y'' = 3e^{-x}(x - 2)$$

Having found the function's first and second derivative, we may place our information about the function in Table 11.2. From Table 11.2 we obtain the graph of the function $y = 3xe^{-x}$ as shown in Figure 11.4.

Table 11.2

$f'(x) = 3e^{-x} - 3xe^{-x}$	Values of x	$f''(x) = 3e^{-x}(x - 2)$	$f(x) = 3xe^{-x}$
0	1	$f''(1) < 0$	Maximum, $3e^{-1}$
>0	$x < 1$		Increasing
<0	$x > 1$		Decreasing
	$x > 2$	>0	Concaves up
	$x < 2$	<0	Concaves down
	2	0	Inflexion

Figure 11-4

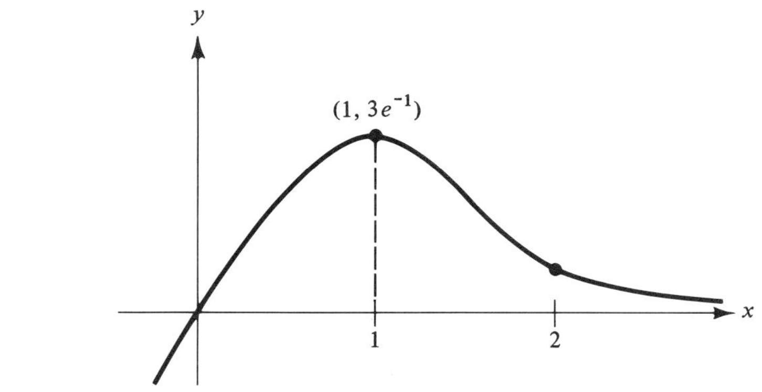

NEWTON'S METHOD OF SOLVING EQUATIONS

It frequently becomes necessary to obtain the numerical value for one or more solutions of a given equation, particularly when such a solution (s) is an irrational number. Many methods exist for such a purpose. However, among the best is that of Newton which is based on the assumption that the tangent at a point P of a certain curve is a relatively good approximation of the curve in the neighborhood of P.

To demonstrate Newton's method, let $f(x) = 0$ be the equation to be solved. Let $x = x_1$ be a "reasonable" approximation of a solution obtained by some process, such as graphing the function of $y = f(x)$ (Figure 11.5). Then the ordinate $y = f(x_1)$ and the slope $y = f'(x_1)$ of the tangent line at $x = x_1$ can be found by substituting $x = x_1$ in $y = f(x)$ and $y' = f'(x)$, respectively. Thus the equation of the tangent line at $P_1(x_1, f(x_1)$ is

$$y - f(x_1) = f'(x_1)(x - x_1) \tag{11.2}$$

This tangent crosses the x axis whenever $y = 0$. Thus Equation (11.2) yields

$$0 - f(x_1) = f'(x_1)(x - x_1) \qquad \text{or} \qquad x = x_1 - \frac{f(x_1)}{f'(x_1)} \tag{11.3}$$

If x_2 denotes the value of x obtained in Equation (11.3), then x_2 is the point of intersection of the tangent at P_1 and the x axis. Figure 11.5 suggests that, in general, the point x_2 is closer to the point where

Figure 11-5

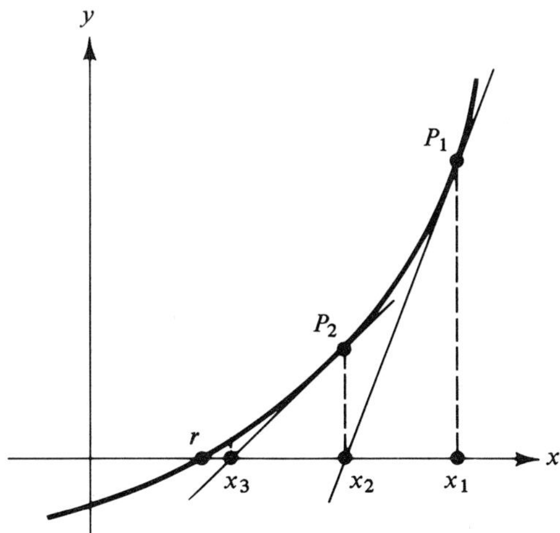

$f(x)$ actually crosses the axis than is x_1; that is, x_2 is a better approximation to the root $x = r$. Starting now with the approximation $x = x_2$ of r and repeating the process, we obtain the better estimate

$$x_3 = x_2 - \frac{f(x_2)}{f'(x_2)}$$

for the root r of $f(x) = 0$. In general, the root r can be approximated to whatever degree of accuracy we desire by simply repeating the process above a sufficient number of times.

A complete statement on the applicability of Newton's method is included in the following theorem.[2]

Theorem 11.2 If $y = f(x)$ is a function over $[a, b]$, if $f''(x)$ exists and is continuous over $[a, b]$, and if (i) $f(a)$ and $f(b)$ have opposite signs, while (ii) $f'(x)$ and $f''(x)$ are never zero in $[a, b]$, then there is exactly one root $x = c$ of $f(x) = 0$ in (a, b). Moreover, if x_1 is chosen such that

$$a \leq x_1 - \frac{f(x_1)}{f'(x_1)} \leq b$$

then the sequence

$$x_{n+1} = x_n - \frac{f(x_n)}{f'(x_n)}$$

converges to the root $x = c$.

To demonstrate Newton's method, we consider the following example.

Example 11.3 Find a decimal approximation of the solution of $x^2 - 10 = 0$.

Solution Let $f(x) = x^2 - 10$, then $f(3) < 0$ and $f(4) > 0$. Thus $f(x) = x^2 - 10 = 0$ has a solution between three and four. Moreover, $f'(x) = 2x, f''(x) = 2$, and neither is zero in the closed interval $[3, 4]$. Since the conditions of Theorem 11.2 are met, we may apply Newton's method to approximating the root of $x^2 - 10 = 0$ in $(3, 4)$. For this purpose let $x_1 = 3$ be our first approximation; then the second approximation is

$$x_2 = x_1 - \frac{f(x_1)}{f'(x_1)} = 3 - \frac{3^2 - 10}{2(3)} = \frac{19}{6} = 3.1666$$

the third is

$$x_3 = x_2 - \frac{f(x_2)}{f'(x_2)} = \frac{19}{6} - \frac{(19/6)^2 - 10}{2(19/6)} = 3.1622$$

and the fourth is

$$x_4 = x_3 - \frac{f(x_3)}{f'(x_3)} = 3.1622 - \frac{(3.1622)^2 - 10}{2(3.1622)} = 3.16227$$

Thus if we wish the solution to be correct to the fourth decimal place we may pick it to be

$$x = 3.1623$$

Evidently the calculations involved in applying Newton's method are many and tedious. Fortunately the discovery and use of high-speed calculators and computers relieve us from such cumbersome tasks. Many good computer programs on Newton's method exist, and their utilization quite easily approximates the solution of a given equation.[3]

Exercises

1. Find the equation of the tangent and the normal lines to each of the following curves at the indicated point.

   a. $y = x^3 - 4x$, $(2, 0)$.
   b. $x = t^3 - t$ and $y = t^2 + t$, $\quad t = 1$.

c. $y = 3 \ln (1 + x^2)$, $(1, 3 \ln 2)$.
d. $x^3 + y^3 - 3xy - 27 = 0$, $(3, 0)$.

Ans. (a) $8x - y = 16$, $x + 8y = 2$; (b) $3x - 2y = -4$, $2x + 3y = 6$

2. Discuss the behavior and graphs of the following functions. For what value(s) is each function increasing, decreasing, maximum, and minimum?

a. $y = 5xe^{-x}$.
b. $y = x^{2/3}(1 - x)^{1/3}$.
c. $y = 2x^3 - 3x^2 - 72x$.
d. $y = x/(x^2 - 1)$.
e. $y = (x - 2)/(x^2 + x)$.
f. $y = 1/(x^2 - 1)$.
g. $y = 1 + 1/x + 1/x^2$.

Ans. (d) $x < -\frac{1}{2}$ increasing, $x > -\frac{1}{2}$ decreasing

3. Graph each of the functions,

a. The total revenue function $R(x) = 14x - 3x^2$.
b. The total cost function $C(x) = x^2 + 5x + 2x$.
c. The profit function $P(x) = R(x) - C(x)$ when x is the units of production of a certain commodity.

4. Using Newton's method, find to three decimal places all the real roots of the following equations:

a. $x^3 + x^2 - 1 = 0$.
b. $x^3 - 3x^2 + 3 = 0$.
c. $x^2 - 11 = 0$.
d. $4x^3 - 8x + 1 = 0$.
e. $x^3 - 15 = 0$.

Ans. (a) 0.755; (b) 2.532

5. If a polynomial equation $P(x) = 0$ has a k-fold root or a root of multiplicity k, say $x = a$, then show that $x = a$ is also a root of the equations

$$P'(x) = 0, P''(x) = 0, \ldots, P^{(k-1)}(x) = 0$$

but is not a root of the equation

$$P^{(k)}(x) = 0$$

Hint: If $P(x) = 0$ has an r-fold root $x = a$, then $(x - a)^r$ is a factor of $P(x)$.

11.3 Optimization Theory

A very powerful and useful application of differentiation is found in the optimization of certain functions, that is, the process employed in locating those values of the variables in a given function that will maximize or minimize the function. In general, there exist the following two types of maxima and minima.

DEFINITION 11.1 We say that a function $f(x)$ possesses a *relative* or *local* maxima (or minimum) at $x = x_0$ if and only if $f(x_0)$ is greater (or less) than all other values of the function in a neighborhood of x_0. On the other hand we say that the function possesses an *absolute* maximum (or minimum) at $x = x_0$ if and only if $f(x_0)$ is greater (or less or equal to) than any other value of the function.

Since in most applications we are interested in the behavior of a function in different intervals (or neighborhoods), we shall be primarily concerned with relative maxima and minima, henceforth referred to simply as "maxima" or "minima." In most instances such maxima and minima will be determined by applying the last part of Theorem 11.1, which when strengthened becomes the following important theorem.

Theorem 11.3 If $f(x)$ is a function defined in the open interval (a, b) and if $f'(x)$ exists for every x in (a, b), then

a. $f(x)$ continuous in $[a, b]$ implies that $f(x)$ has a maximum and a minimum in $[a, b]$.
b. $f(x_0)$ is a maximum value of $f(x)$ in (a, b) if $f'(x_0) = 0$ and $f''(x_0) < 0$.
c. $f(x_0)$ is a minimum value of $f(x)$ in (a, b) if $f'(x_0) = 0$ and $f''(x_0) > 0$.

Although the general proof and applicability of the above theorem can be found in many books on calculus,[2] there exist some special cases in which the theorem fails, and therefore an independent examination for maxima and minima is required. These cases we enumerate as follows.

1. When the function has an extreme point x_0 (either a maximum or a minimum) for which $f'(x)$ may not exist (Figure 11.6).

Figure 11-6

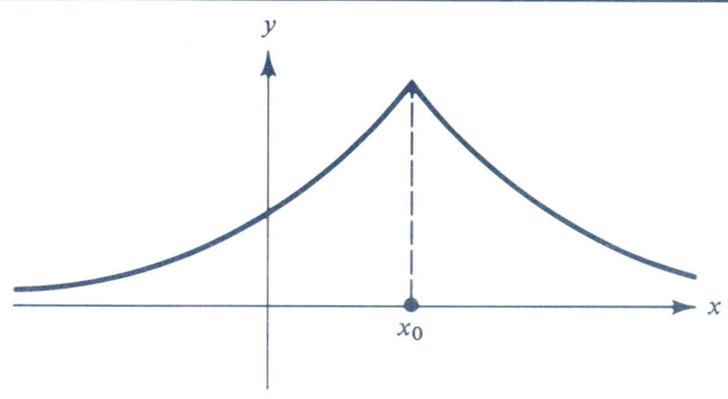

2. When the derivative of the function is not zero for any finite values of the independent variable.
3. When the endpoints of an interval or points of a discontinuity happen to be the maxima or minima of the function.
4. When both $f'(x) = 0$ and $f''(x) = 0$.

To demonstrate further the usefulness of Theorem 11.3 in problems of optimization, we consider the following examples.

Example 11.4 A jewelry store manager sells, at a constant rate, 10,000 rings a year. If it costs \$10 per 1000 to store the rings and if each reorder costs \$2 plus \$8 per 1000, then how many times a year and in what size lots should he reorder so as to minimize his yearly costs?

Solution Let x be the lot size in thousands; then $x/2$ is the average number of rings on hand at a given time. The cost we are to minimize is the carrying cost plus the reorder cost multiplied by the number of reorders. The different costs are

a. The carrying cost, namely, $10(x/2) = 5x$.
b. The reordering cost—that is, the fixed cost plus the cost per 1000, namely, $2 + 8x$.

The number of reorders is equal to the number of sales divided by the lot size, that is, $10/x$ is the number of reorders. Therefore the total cost becomes

$$C(x) = 5x + (2 + 8x)\frac{10}{x}$$

or

$$C(x) = 5x + \frac{20}{x} + 80 \qquad \textbf{(11.4)}$$

Therefore

$$C'(x) = 5 - \frac{20}{x^2} \qquad \text{and} \qquad C''(x) = \frac{40}{x^3}$$

Since

$$C'(x) = 5 - \frac{20}{x^2} = 0 \qquad \text{for} \quad x = 2 \qquad \text{and} \qquad C''(2) = 5 > 0$$

Theorem 11.3 implies that the cost function $C(x)$ is minimum, $C(2) = \$100{,}000$, at $x = 2000$ lot size, and the jeweler reorders five times a year.

This is indeed a minimum in the interval $[0, 10]$ because at the endpoints 0 and 10, $C(0)$ is not defined, and $C(10) = 132 > 100$.

Example 11.5 If the cost of running an airplane is \$500 per hour plus fuel, if the cost of fuel per hour is proportional to the square of the speed and is \$300 per hour when the speed is 600 mph, and if the plane's maximum speed is 1000 mph, then at what speed should the plane be flown in order to minimize the per mile cost?

Solution Let C be the cost of running the plane for a distance D, then

$$C = (\text{cost per hour}) \times (\text{number of hours})$$

Since the cost per hour is given by

$$C_1 = 500 + kv^2$$

where k is a constant of proportionality and v the speed; since at $v = 600$ the cost $C_1 = 300$ and therefore

$$300 = k(600)^2 \qquad \text{or} \qquad k = \frac{1}{1200}$$

and since $D = vt$ or $t = D/v$; the cost of running the plane becomes

$$C = \left(500 + \frac{1}{1200}v^2\right)\cdot\frac{D}{v} \qquad \text{or} \qquad C = \left(\frac{500}{v} + \frac{v}{1200}\right)D$$

Thus the cost per mile is given by

$$\frac{C}{D} = \frac{500}{v} + \frac{v}{1200}$$

Again, an application of Theorem 11.3 yields

$$\left(\frac{C}{D}\right)' = -\frac{500}{v^2} + \frac{1}{1200}, \qquad \left(\frac{C}{D}\right)'' = \frac{1000}{v^3}$$

$(C/D)' = 0$ for $v = 200\sqrt{15}$, and $(C/D)''$ is positive for all v. Hence the optimum speed is

$$v = 200\sqrt{15} \text{ mph}$$

Example 11.6 If in a certain manufacturing firm the cost of production is given by

$$C(x) = 10 + 2x^2 - 8x$$

in millions of dollars, where x is the number of commodity units produced (in millions), and if the price per million units is obtained from

$$S(x) = \frac{30}{x} + 24 - 22x + 8x^2 - x^3$$

then

$$R(x) = xS(x) = 30 + 24x - 22x^2 + 8x^3 - x^4$$

is the revenue received from the sale of x units. For what value(s) of x will the firm have maximum profit?

Solution If $P(x)$ is the profit of the firm, then

$$P(x) = R(x) - C(x)$$

or

$$P(x) = 20 + 32x - 24x^2 + 8x^3 - x^4$$

Thus

$$P'(x) = 32 - 48x + 24x^2 - 4x^3 = -4(x^3 - 6x^2 + 12x - 8)$$

or

$$P'(x) = -4(x - 2)^3, \qquad P''(x) = -12(x - 2)^2$$

$P'(x) = 0$ for $x = 2$, and $P''(2) = 0$. Since the second derivative is zero at $x = 2$, Theorem 11.3 fails. However, $P'(x) > 0$ for all $x < 2$ or $P(x)$ is increasing for all $x < 2$ and $P'(x) < 0$ for all $x > 2$, that is, $P(x)$ is decreasing for all $x > 2$ (Figure 11.7). Therefore the function $P(x)$ attains its maximum at $x = 2$ million units.

A more general illustration of optimization theory and Theorem 11.3 is provided by the branch of the management sciences known as break-even or marginal analysis.[1] To introduce break-even analysis, let us consider two simple cases as follows.

Assume that a firm produces a product X at x number of units, it receives a g amount of revenue and an amount h of expenditures is incurred in its production, and it therefore realizes $P = g - h$ of

Figure 11-7

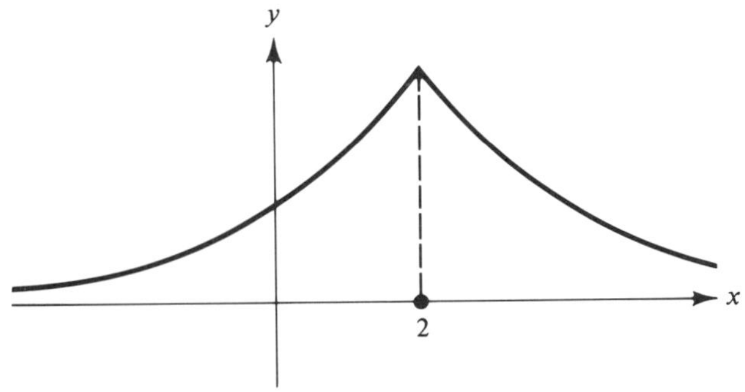

profit. Generally, g, h, and P will be functions of x, and the firm will be interested in knowing

a. For what values of x will it break even?
b. For what values of x will its profits be maximized?

To answer these questions, let the functions of revenue, cost, and profit be

$$g = g(x), \qquad h = h(x), \qquad \text{and} \qquad P = g(x) - h(x)$$

respectively. Then the following possibilities exist:

a. The functions g and h are linear.
b. Either both or one of the functions g and h is nonlinear.

Case 1. Suppose that both g and h are linear; that is, suppose that

$$g(x) = px$$

where p is the unit price considered to be constant, and

$$h(x) = a + bx$$

where a and b are constants representing the fixed cost and the unit variable cost, respectively. Then

$$P = g(x) - h(x) \qquad \text{or} \qquad P = px - a - bx$$

Consequently, the value of x for which

$$P = g(x) - h(x) = 0, \qquad g(x) = h(x), \qquad \text{or} \qquad px = a + bx$$

namely,

$$x_0 = \frac{a}{p - b}$$

is the break-even point of the activity of the firm or the production at which the firm has zero profits. Evidently,

$$x_0 = \frac{a}{p - b} = \frac{\text{fixed costs}}{\text{unit price—unit variable cost}}$$

is unique for $p \neq b$, a nontrivial positive number for $p > b$, and undefined for $p = b$. Graphically, this value, $x = x_0$, is shown in Figure 11.8.

Since the diagram shows that $P(x)$ does not have a finite number for its maximum, the answer to the second question is as one would expect, that there is no level of production x at which the firm's profits are maximized unless, of course, there was a fixed maximum level

Figure 11-8

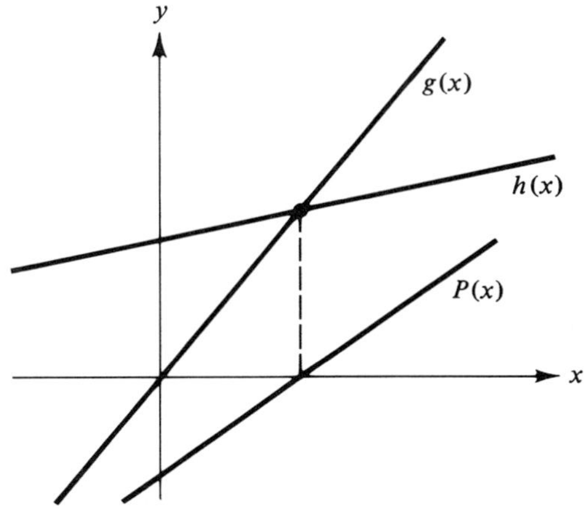

of production x_0 to begin with, in which case x_0 would be the optimum level. Theoretically the function $P(x)$ will be maximum if $P'(x) = 0$, and $P''(x) < 0$. Thus

$$P(x) = px - a - bx$$

implies

$$P'(x) = p - b$$

or $P'(x)$ is a constant $p - b$ which becomes zero when $p = b$. Moreover, $P''(x) = 0$, and consequently Theorem 11.3 fails to identify a maximum or a minimum. However, the geometric analysis confirms once again that profits cannot be maximized unless a maximum capacity level was introduced to begin with.

Case 2. Let us assume the more interesting case when either both or one of the functions $g(x)$ and $h(x)$ is nonlinear. Specifically, suppose that

$$g(x) = 7x - \frac{x^2}{1000} \qquad \text{and} \qquad h(x) = 5x$$

Then

$$P(x) = g(x) - h(x) = 2x - \frac{x^2}{1000}$$

and the break-even points occur when

$$P(x) = 2x - \frac{x^2}{1000} = 0$$

or at $x = 0$ and $x = 2000$ number of units of production (Figure 11.9).

Figure 11-9

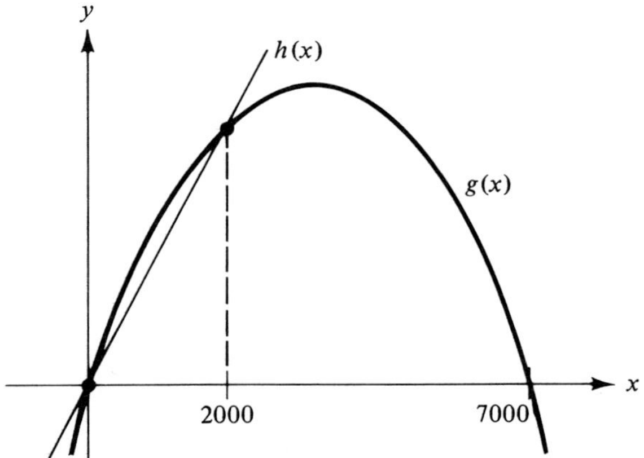

On the other hand $P(x)$ will be maximized when $P'(x) = 0$ and $P''(x) < 0$. Thus

$$P(x) = 2x - \frac{x^2}{1000}$$

implies

$$P'(x) = 2 - \frac{x}{500} \qquad \text{and} \qquad P''(x) = -\frac{1}{500}$$

Hence

$$P'(x) = 2 - \frac{x}{500} = 0 \qquad \text{for } x = 1000$$

Since $P''(x) = -1/500$ is negative for all x and therefore for $x = 1000$, the profit function $P(x)$ is maximum at production level of $x = 1000$ units (Figure 11.10).

Clearly, many other possibilities exist for the functions $g(x)$, $f(x)$, and $P(x)$. In all cases the analysis and solution of the related problems are similar to those already discussed.

The break-even analysis problem is a small but integral part of the more general economic theory of marginalism[1] which deals with optimum activity levels; that is, it concerns itself with such crucial decision problems as how many units to produce, how many people to hire, how much material to order, how much money to invest and spend on advertising, how much of each of a number of products should be produced, and so on. These and similar questions we shall leave for those who intend to pursue studies in economic theory and management.

Figure 11-10

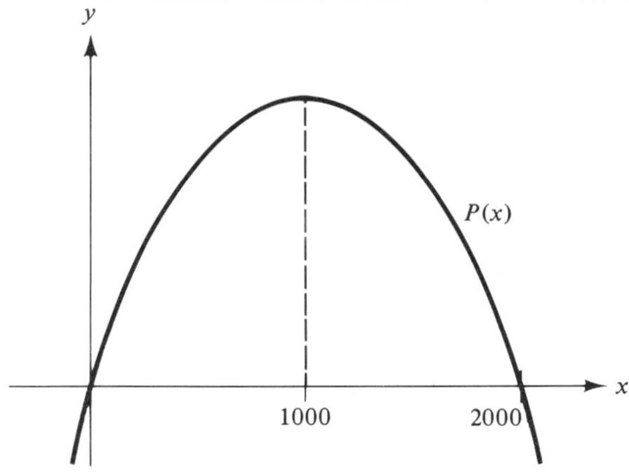

Exercises

Find the values of x for which each of the following functions is maximum or minimum, and determine its maximum or minimum value:

1. $f(x) = x^2/2 - 3x + 9.$
2. $f(x) = 2x^3 - 3x^2 - 12.$
3. $f(x) = 2/x + 8x.$
4. $f(x) = (1 - x^2)/x.$
5. $f(x) = x^3 - 5x^2 + 3x + 12.$
6. $f(x) = (4 - x)^3/(3 - 2x).$
7. $f(x) = 2 + (x - 3)^2$

Ans. (1) Minimum 9/2 at $x = 3$; (3) Relative minimum 8 at $x = 0.5$, Relative maximum -8 at $x = -0.5$; (5) Maximum $12\frac{13}{27}$ at $x = \frac{1}{3}$, Minimum 3 at $x = 3$; (7) Minimum 2 at $x = 3$

8. If the revenue and cost functions of a firm producing a product at x number of units are

$$R(x) = 12x - x^2 \quad \text{and} \quad C(x) = \frac{x^3}{3} - 3x^2 + 12x$$

respectively, then determine the break-even point and the value(s) of x that maximize its profits.

Ans. $x = 6$, $x = 4$

9. Work Problem 8, and draw the appropriate graphs, if

a. $R(x) = 5x - x^2/2000$ and $C(x) = 3x$.
b. $R(x) = 6x - x^2/1000$ and $C(x) = 5x$.
c. $R(x) = 7x - x^2/1000$ and $C(x) = 5x - 3x^2/5000$.

10. If the total cost function for a firm is

$$C(x) = \tfrac{1}{2}x^3 - 2x^2 + 2x + 75$$

where x denotes the output level, find

a. The optimum level for which the total cost is a minimum.
b. The output level x that minimizes the average cost. Verify that at this level marginal and average costs are the same.

Hint: $AC = TC/x$ and $MC = (TC)'$.

Ans. (a) $x = 2$, (b) $x = 5$, $MC = AC = 39/2$

11. If x items of a certain commodity priced at $P = 100 - 0.10x$ each are sold per week and if their total cost is $C = 50x + \$1000$, how many should be produced each week in order to maximize profits?

Ans. 250

12. Given that the sum of two numbers is 30, determine the numbers so that

a. Their product is maximum.
b. The sum of their squares is a minimum.

13. If the perimeter of a rectangular field is 2000 feet, find its dimensions so that the area of the field is maximum.

Ans. 500 × 500 feet

14. What are the dimensions of an open box, of largest volume, that can be constructed from a 12-inch square piece of cardboard by cutting equal squares from the corners and turning up the sides?

11.4 Maxima and Minima of $f(x, y)$

As in the case of functions of one variable, we often are concerned with the optimization of a function $z = f(x, y)$; that is, we are interested in the manipulation of the variables x and y in the function $f(x, y)$ to the extent that the function becomes maximum or minimum. For example, manipulation of x and y in the function

$$f(x, y) = 2x^3 - 2xy + y^2$$

yields the functional values listed in Table 11.3. Moreover, the partial derivatives

$$f_x = 6x^2 - 2y \qquad \text{and} \qquad f_y = -2x + 2y$$

of the given function become zero at the points $(0, 0)$ and $(\tfrac{1}{3}, \tfrac{1}{3})$. These points are commonly known as critical or extreme points and, in

Table 11.3

x	y	f(x, y)
0	0	0
0	1	1
1	0	2
1	1	1
2	2	12
.	.	.
.	.	.
.	.	.
1/3	1/3	—1/27
.	.	.
.	.	.
.	.	.

general, constitute the points at which the function is a maximum or a minimum. In our particular case the point (0, 0) is obviously not a maximum or minimum of the function. However, it appears that at $(\frac{1}{3}, \frac{1}{3})$ the function is a minimum.

The partial derivatives of the function

$$f(x, y) = x^2 + xy + y^2 + 24$$

are

$$f_x = 2x + y \qquad \text{and} \qquad f_y = x + 2y$$

Clearly, these partials are zero for $x = y = 0$. Thus the point (0, 0) is an extreme point, and various manipulations of x and y seem to indicate that at $x = 0$ and $y = 0$ the given function is a minimum.

In an ordinary business situation let us assume that a store has two departments whose earnings are given by

$$E_1(x_1, y_1) = x_1 + y_1 + x_1y_1 - x_1^2 - y_1^2 - 3$$

and

$$E_2(x_2, y_2) = 4x_2 + 2y_2 + 2x_2y_2 - 2x_2^2 - y_2^2 - 4$$

where x_1, x_2 are investments in inventory in millions of dollars and y_1, y_2 are floor space in units of 10,000 square feet used by each department. Clearly the management of the store would like to know the allocation of capital and floor space, to each department, which maximize the total earnings. Since the total earnings are given by

$$E = E_1 + E_2 = x_1 + y_1 + x_1y_1 - x_1^2 - y_1^2 - 3 + 4x_2 + 2y_2 + 2x_2y_2 - 2x_2^2 - y_2^2 - 4$$

since

$$E_{x1} = 1 + y_1 - 2x_1$$

$$E_{y1} = 1 + x_1 - 2y_1 \tag{11.5}$$

$$E_{x2} = 4 + 2y_2 - 4x_2$$

$$E_{y2} = 2 + 2x_2 - 2y_2 \tag{11.6}$$

and since Equations (11.5) do not involve x_2 and y_2 just as Equation (11.6) do not involve x_1 and y_1, we find that the equations

$$E_{x1} = 0, \qquad E_{y1} = 0, \qquad E_{x2} = 0, \qquad \text{and} \qquad E_{y2} = 0 \tag{11.7}$$

are satisfied by those values of x_1 and y_1 and those of x_2 and y_2 that are solutions of the independent systems

$$\begin{aligned} 1 + y_1 - 2x_1 &= 0 \\ 1 + x_1 - 2y_1 &= 0 \end{aligned} \qquad \text{and} \qquad \begin{aligned} 4 + 2y_2 - 4x_2 &= 0 \\ 2 + 2x_2 - 2y_2 &= 0 \end{aligned} \tag{11.8}$$

Thus solving the last two systems we find that Equations (11.8) are satisfied for

$$x_1 = y_1 = 1, \qquad x_2 = 3, \qquad \text{and} \qquad y_2 = 4$$

or that the points (1, 1) and (3, 4) are critical (or extreme) points of the functions $E_1(x_1, y_1)$ and $E_2(x_2, y_2)$, respectively. Furthermore, substituting different values for x_1, y_1, x_2, and y_2 and evaluating $E = E_1 + E_2$ seem to indicate that $E(x_1, y_1, x_2, y_2)$ is maximum when

$$x_1 = y_1 = 1, \qquad x_2 = 3, \qquad \text{and} \qquad y_2 = 4$$

and that this maximum is

$$E(1, 1, 3, 4) = 4 \text{ millions}$$

Thus an optimum allocation requires

$$x_1 + x_2 = 4 \text{ millions}$$

of inventory investment and

$$y_1 + y_2 = 5000 \text{ square feet}$$

of floor space.

A more efficient way to determine the maxima and minima in the previous three examples, as well as those of many others, requires the generalized theory and techniques introduced by Theorem 11.4.[2]

Theorem 11.4 If $f(x, y)$ is a function of two variables and if its second-order partial derivatives are continuous in some region, then the point (a, b) in that region is a critical or an extreme point of the function if

$$f_x(a, b) = f_y(a, b) = 0 \qquad \text{and} \qquad f_{xx}(a, b)f_{yy}(a, b) > [f_{xy}(a, b)]^2$$

The function has a maximum at (a, b) if $f_{xx}(a, b) < 0$ and a minimum if $f_{xx}(a, b) > 0$. The point (a, b) is a saddle point (neither maximum nor minimum) of $f(x, y)$ if

$$f_x(a, b)f_y(a, b) < [f_{xy}(a, b)]^2$$

and no conclusion can be drawn when

$$f_x(a, b)f_y(a, b) = [f_{xy}(a, b)]^2$$

The introductory examples clearly illustrate this theorem. However, for its further demonstration we consider the following additional examples.

Example 11.7 Determine the maximum value of the function

$$f(x, y) = 2x + 4y - x^2 - y^2 - 1$$

Solution Since

$$f_x = 2 - 2x \qquad \text{and} \qquad f_y = 4 - 2y$$

we have $f_x = f_y = 0$ for $x = 1$ and $y = 2$. Moreover, $f_{xx} = -2$, $f_{yy} = -2$, and $f_{xy} = 0$. Thus

$$f_{xx}(1, 2)f_{yy}(1, 2) = 4 > [f_{xy}(1, 2)]^2 = 0$$

Hence, the point $(1, 2)$ is an extreme or critical point of this function, and since

$$f_{xx}(1, 2) = -2 < 0$$

the function possesses the maximum value

$$f(1, 2) = 4$$

at the point $(1, 2)$.

Example 11.8 If the profit P as a function of the prices x and y of two competing items is given by

$$P(x, y) = 65 - 2(x - 7)^2 - 4(y - 7)^2 - 4x$$

determine the values of x and y that maximize profits.

Solution Since

$$P_x = -4(x - 7) - 4, \qquad P_y = -8(y - 7), \qquad P_{xx} = -4,$$

$$P_{yy} = -8, \qquad \text{and} \qquad P_{xy} = 0$$

we obtain

$$P_x = P_y = 0$$

for $x = 6$ and $y = 7$. Moreover,

$$P_{xx}(6, 7)P_{yy}(6, 7) = (-4)(-8) = 32 > [P_{xy}(6, 7)]^2 = 0$$

Thus the point $(6, 7)$ is a critical or extreme point of the function P, and since

$$P_{xx}(6, 7) = -4 < 0$$

the function possesses the maximum value

$$P(6, 7) = 39$$

at the point $(6, 7)$.

Exercises

Whenever possible, determine the critical points, maxima, and minima for each of the following functions.

1. $f(x, y) = 2xy - 5y^2 - 2x^2 + 4x + 4y - 2.$
2. $f(x, y) = x^2 - y^2 - 2x + 4y - 5.$
3. $f(x, y) = 7x^2 + y^2 - 5xy - 3x + 6y - 4.$
4. $f(x, y) = y \ln x - x + 3.$
5. $f(x, y) = x^3 - 12xy + 8y^3 + 3.$
6. $f(x, y) = x^3 - 9xy + y^3.$
7. $f(x, y) = x^2 + y^2 + xy - 3x + 6.$
8. $f(x, y) = x^3 - 12xy + 8y^3 + 5.$
9. $f(x, y) = 4 \ln x + e^y - x - y + 1.$
10. $f(x, y) = 1/x + xy - 8/y + 15.$

    Ans. (1) Maximum 2 at $(4/3, 2/3)$; (2) Saddle point $(1, 2)$; (5) Minimum -5 at $(2, 1)$, saddle points $(0, 0)$, $(-2, 1)$; (7) Minimum 3 at $(2, -1)$; (10) Maximum 9 at $(-\frac{1}{2}, 4)$

11. In a certain economic model the cost C of a given item is related

to the factors x and y by the equation

$$C = x^3 - 12xy + 8y^3 + 15$$

Determine the values of x and y that will correspond to a minimum cost.

Ans. Minimum 7 at (2, 1)

12. A sociological study indicates that the measurement R of crime rate depends on the amounts x and y spent on welfare and prisons, respectively. If

$$R = x^3 + 2y^3 - 6x + 6y - 6xy + 3$$

where x and y are measured in hundreds of millions of dollars, determine the values of x and y that will minimize the crime rate.

Ans. $x = 200$, $y = 100$ millions

13. The total profit per acre on a given farm has been found to be related to the expenditure per acre for labor and soil improvements. If the profit P is given by

$$P = 20x + 60y + 10xy - 10x^2 - 6y^2 + 10$$

where x represents the dollars per acre spent on labor and y those spent on soil improvements, determine the optimum expenditure levels.

Ans. $x = 6$, $y = 10$

14. The cost C of repairs for inspections at two points in a certain manufacturing process is given by

$$C = 4x^2 + 2y^2 - 20x - 12y + 40$$

where x and y are the number of inspections at the two points, respectively. What should the number of inspections, at each point, be in order to minimize the repair costs?

11.5 Optimization with Constraints

Decisions of optimization in certain situations are often subjected to various kinds of restrictions or constraints. For example, the linear programming problem in Chapter 7 required that the optimized linear function be subjected to a certain number of constraints.

Since we already know how to solve linear programming problems, that is, since we know how to maximize or minimize linear functions under constraints, we shall restrict ourselves here to optimization of nonlinear functions subject to certain restrictions. To demonstrate this let us suppose that the department store described at the beginning of Section 11.4 had a limited amount of money for inventory investment and floor space allocation. Specifically, let us assume that exactly \$6,000,000 and 50,000 square feet of floor space were available. Then the variables x_1, y_1, x_2, and y_2 in the total earnings

$$E = x_1 + y_1 + x_1y_1 - x_1^2 - y_1^2 + 4x_2 + 2y_2 + 2x_2y_2 - 2x_2^2 - y_2^2 - 7$$

would be restricted to satisfy

$$x_1 + x_2 = 6{,}000{,}000 \qquad y_1 + y_2 = 50{,}000$$

Thus the problem of maximizing total earnings becomes the fcllowing:

Maximize:

$$E = x_1 + y_1 + x_1y_1 - x_1^2 - y_1^2 + 4x_2 + 2y_2 + 2x_2y_2 - 2x_2^2 - y_2^2 - 7 \tag{11.9}$$

subject to the constraints

$$x_1 + x_2 = 6 \qquad \text{or} \qquad x_2 = 6 - x_1$$

and

$$y_1 + y_2 = 5 \qquad \text{or} \qquad y_2 = 5 - y_1 \tag{11.10}$$

Clearly, this problem could be solved by substituting x_2 and y_2 from Equations (11.10) into Equation (11.9) and then maximizing, in the usual manner, the function $E(x_1, y_1)$ as a function of two variables, namely, x_1 and y_1 (x_2 and y_2 are eliminated).

The elimination process just introduced frequently becomes tedious, cumbersome, and downright impossible. Fortunately there exists an alternative method that optimizes functions under constraints without eliminating any of the variables and also provides a measure of how restrictive a constraint actually is. This method was first introduced by the great eighteenth century mathematician J. L. Lagrange, and it is therefore known as "Lagrange's method of multipliers" or merely as "Lagrange's method." A simplified version of the Lagrange's method follows.

LAGRANGE'S METHOD

The optimum of a function $f(x, y)$, subject to the restriction $g(x, y) = 0$, can be obtained by forming the function

$$F(x, y, t) = f(x, y) + tg(x, y)$$

and optimizing it with respect to x, y, and t as indicated by an extended form of Theorem 11.4 whose necessary conditions are

$$F_x = 0, \qquad F_y = 0, \qquad \text{and} \qquad F_t = 0$$

In the event of more than one restriction, additional t's are introduced, one for each restricting equation.

The proof of Lagrange's method is beyond the scope of this book and so is the statement of a test for extrema[4] analogous to that of Theorem 11.4. However, the context of many practical problems usually makes clear what type of extremum (maximum, minimum, or neither) is obtained, and therefore a test for extrema is most often unnecessary.

To demonstrate Lagrange's method, let us attempt to maximize the total earnings of the previously discussed department store. The function to be maximized then becomes

$$\begin{aligned} F(x_1, y_1, x_2, y_2, t_1, t_2) &= E_1(x_1, y_1) + E_2(x_2, y_2) \\ &\quad + t_1(x_1 + x_2 - 6) + t_2(y_1 + y_2 - 5) \\ &= (x_1 + y_1 + x_1y_1 - x_1^2 - y_1^2 - 3) \\ &\quad + (4x_2 + 2y_2 + 2x_2y_2 - 2x_2^2 - y_2^2 - 4) \\ &\quad + t_1(x_1 + x_2 - 6) + t_2(y_1 + y_2 - 5) \end{aligned} \tag{11.11}$$

and the necessary conditions are the following.

$$\begin{aligned} F_{x_1} &= 1 + y_1 - 2x_1 + t_1 = 0 \\ F_{x_2} &= 4 + 2y_2 - 4x_2 + t_1 = 0 \\ F_{y_1} &= 1 + x_1 - 2y_1 + t_2 = 0 \\ F_{y_2} &= 2 + 2x_2 - 2y_2 + t_2 = 0 \\ F_{t_1} &= x_1 + x_2 - 6 = 0 \\ F_{t_2} &= y_1 + y_2 - 5 = 0 \end{aligned} \tag{11.12}$$

Since the last two equations of system (11.12) yield

$$x_2 = 6 - x \qquad \text{and} \qquad y_2 = 5 - y_1$$

this system can be reduced by eliminating x_2 and y_2 to the following system of four equations in four unknowns.

$$\begin{aligned} 1 + y_1 - 2x_1 + t_1 &= 0 \\ 1 + x_1 - 2y_1 + t_2 &= 0 \\ -10 - 2y_1 + 4x_1 + t_1 &= 0 \\ 4 - 2x_1 + 2y_1 + t_2 &= 0 \end{aligned}$$

or

$$\begin{aligned} -2x_1 + y_1 + t_1 &= -1 \\ x_1 - 2y_1 + t_2 &= -1 \\ 4x_1 - 2y_1 + t_1 &= 10 \\ -2x_1 + 2y_1 + t_2 &= -4 \end{aligned} \tag{11.13}$$

Now solving system (11.13) we obtain

$$x_1 = \frac{7}{3} \qquad x_2 = 6 - x_1 = \frac{11}{3}$$

$$y_1 = 1 \qquad y_2 = 5 - y_1 = 4$$

$$t_1 = \frac{8}{3} \qquad t_2 = -\frac{4}{3}$$

Substituting these values into the total earnings function, we find

$$E = \frac{7}{3} + 1 + \frac{7}{3}\cdot 1 - \left(\frac{7}{3}\right)^2 - 1^2 + 4\frac{11}{3} + 2\cdot 4$$

$$+ 2\frac{11}{3}4 - 2\left(\frac{11}{3}\right)^2 - 4^2 - 7 = \frac{14}{9}$$

Therefore investing \$6,000,000 instead of \$4,000,000 and using the same floor space of 50,000 square feet reduces the earnings from \$4,000,000 to about \$1,555,555.

To analyze further the process involved in the solution of the department store problem, we may interpret the function $F(x_1, y_1, x_2, y_2, t_1, t_2)$ as a generalized earnings function consisting of the following three parts:

a. $E_1 + E_2$ as the unrestricted earnings in the two departments.
b. $t_1(x_1 + x_2 - 6)$ as the change in earnings due to investment restrictions.
c. $t_2(y_1 + y_2 - 5)$ as the change in earnings due to space restriction.

Furthermore the parameter t_1 represents the change in earnings due to some small change in the amount available for inventory investment, and t_2 is the change due to change in floor space. The signs of t_1 and t_2, of course, indicate the effect of the restriction. Thus t_1 positive means the earnings will decrease as $(x_1 + x_2)$ becomes larger than six, and t_1 negative implies the earnings will increase.

Since in our example t_1 is positive, the total earnings E can be increased by decreasing the amount invested in inventory, in which case the optimum solution can be obtained by solving the same problem without the restriction on inventory; that is, optimize

$$F = E_1 + E_2 + t_2(y_1 + y_2 - 5)$$

The solution of this problem we shall leave as an exercise for the student, and instead we shall present some additional examples to demonstrate Lagrange's methods.

Example 11.9 Maximize the function $f(x, y) = x^2 - y^2 - y$ subject to the restriction

$$g(x, y) = x^2 + y^2 - 1 = 0$$

Solution If we define the function

$$F(x, y, t) = f(x, y) + tg(x, y)$$

or

$$F(x, y, t) = x^2 - y^2 - y + t(x^2 + y^2 - 1)$$

we obtain the necessary conditions

$$F_x = 2x + 2xt = 0$$

$$F_y = -2y - 1 + 2yt = 0$$

and

$$F_t = x^2 + y^2 - 1 = 0$$

Now $F_x = 0$ implies $x(1 + t) = 0$; that is, either $x = 0$ or $t = -1$. If $x = 0$, then $x^2 + y^2 - 1 = 0$ yields $y = \pm 1$, and the points $(0, 1)$ and $(0, -1)$ are critical. On the other hand $t = -1$ implies $-2y - 1 - 2y = 0$ or $y = -\frac{1}{4}$, and again $x^2 + y^2 - 1 = 0$ yields $x = \pm\sqrt{15}/4$ or the points $(\sqrt{15}/4, -1/4)$ and $(-\sqrt{15}/4, -1/4)$ constitute an additional pair of critical or extreme points. Substituting each of the critical points $(0, 1)$, $(0, -1)$, $(\sqrt{15}/4, -1/4)$, and $(-\sqrt{15}/4, -1/4)$ in the given function

$$f(x, y) = x^2 - y^2 - y$$

we obtain

$$f(0, 1) = -2, f(0, -1) = 0, f\left(\frac{\sqrt{15}}{4}, -\frac{1}{4}\right)$$

$$= f\left(-\frac{\sqrt{15}}{4}, -\frac{1}{4}\right) = \frac{9}{4}$$

Thus the function is maximum, 9/4 at the critical points $(\sqrt{15}/4, -1/4)$ and $(\sqrt{15}/4, -1/4)$.

In order to appreciate the simplicity of Lagrange's method, the student should attempt to solve the same problem using the elimination method, that is, he should solve the restriction $x^2 + y^2 - 1 = 0$ for one of the variables, substitute it in $f(x, y)$, and then maximize the resulting function in the usual manner. Even better, he should solve the problem by applying the chain rule to obtain $D_x f$ and $D_x g$, that is

$$D_x f = \frac{df}{dx} = \frac{\partial f}{\partial x} + \frac{\partial f}{\partial y}\frac{dy}{dx} = 0,$$

and

$$D_x g = \frac{dg}{dx} = \frac{\partial g}{\partial x} + \frac{\partial g}{\partial y}\frac{dy}{dx} = 0$$

where

$$\frac{dy}{dx} = -\frac{x}{y} \qquad (\text{from } x^2 + y^2 - 1 = 0)$$

and then proceed accordingly.

Example 11.10 The past record of a firm shows that newspaper and television advertising increase sales by

$$F(x, y) = \frac{10^{-10}xy^2}{5}$$

where x and y are the amounts spent on newspaper and television advertising, respectively. If the advertising budget is \$1,000,000, determine the optimum advertising allocations.

Solution Since $x + y = 1{,}000{,}000$, we must optimize the function

$$F(x, y) = \frac{10^{-10}xy^2}{5}$$

subject to the restriction $x + y = 1{,}000{,}000$. If we define the function

$$f(x, y) = \frac{10^{-10}xy^2}{5} + t(x + y - 1)$$

we obtain the necessary conditions

$$f_x = \frac{10^{-10}y^2}{5} + t = 0$$

$$f_y = \frac{10^{-10}x2y}{5} + t = 0,$$

and

$$f_t = x + y - 1 = 0$$

Thus

$$f_x - f_y = \frac{10^{-10}y^2 - 10^{-10}x2y}{5} = 0 \qquad \text{or} \qquad y^2 - 2xy = 0$$

The last equation and $x = 1 - y$ imply

$$y^2 - 2(1 - y)y = 0$$

or

$$3y^2 - 2y = y(3y - 2) = 0$$

that is,

$$y = 0 \qquad \text{and} \qquad x = 1$$

or $y = 2/3$ and $x = 1/3$ in millions of dollars. Therefore the firm realizes maximum sales increase of

$$F = \left(\frac{10^6}{3}, \frac{2.10^6}{3}\right) = \frac{10^{-10} \cdot 10^6/3 \cdot 4.10^{12}/9}{5} = \frac{4.10^8}{135}$$

or

$$F = \frac{4.10^8}{135} \text{ dollars}$$

when it allocates $10^6/3$ and $2.10^6/3$ dollars to newspaper and television advertising, respectively.

Exercises

Using Lagrange's method, optimize the following functions subject to the indicated restrictions.

1. $f(x, y) = x^2 - 2y^2 + 2xy + 4x$ if $2x - y = 0$.

   Ans. Maximum 4/3 at (2/3, 4/3)

2. $g(x, y) = x^2 - 2x + xy + 2y^2$ if $x - 2y + 1 = 0$.
3. $f(x, y) = x^2 - xy + y$ if $x + y = 2$.

   Ans. Minimum 7/8 at (3/4, 5/4)

4. $g(x, y) = xy + 6x$ if $x + y = 2$.
5. $f(x, y) = (x + y)^2$ if $x - y - 3 = 0$.

   Ans. Minimum 9/49 at (9/7, −12/7)

6. $g(x, y, z) = 2x + y + 2x + x^2 - 3z^2$ if $x + y + z = 1$ and $2x - y + z = 2$.

7. If the sales of a company are given by

$$S(x, y) = 5xy - 3y^2$$

where x and y are the material and labor cost, respectively, and if $x + y = 15$, find the maximum attainable sales.

Ans. Maximum 5625/32 at (165/16, 75/16)

8. If the profit function from the production of two different items is

$$P(x, y) = 2xy$$

where x and y are the production levels of the two products and if the total production must be 1000 units, find x and y that maximize the profit.

References for Supplementary Readings

1. R. J. Thierauf and R. A. Grosse, *Decision Making Through Operations Research*, Chapter 6 (New York: Wiley, 1970).
2. A. W. Goodman, *Analytic Geometry and Calculus*, Chapters 5 and 19 (New York: Macmillan, 1974).
3. A. J. Pettofrezzo, *Introduction to Numerical Analysis*, Chapter 2 (Lexington, Mass.: Heath, 1967).
4. A. E. Taylor, *Advanced Calculus*, Chapter 6 (Boston: Ginn, 1970).

CHAPTER 12

Integration Theory

12.1 The Indefinite Integral

In Chapter 10 we presented ordinary and partial differentiation in some detail and obtained a set of formulas by which the derivatives of most familiar functions can be computed; that is, for any function $f(x)$ we associated another function $f'(x)$ called the derivative of $f(x)$. For example, when a body weighing w pounds falls from rest under the influence of gravity and without any air resistance, the distance fallen is given by the formula

$$S = 16t^2$$

and its velocity, obtained as the derivative of S, is

$$v = S' = 32t$$

We may need to find the distance fallen when the velocity is known or, more generally, we may be interested in determining the function whose derivative is known. For instance, if $y = f(x)$ is the production function[1] that relates the output level y of some commodity with the amount x of input of labor (or raw material or capital), then the marginal product[1] can be obtained as the derivative $f'(x)$. However, if the marginal product is known, we may be interested in determining the function that specifies production in terms of input of labor.

Generalization of the last two examples leads to the following important definition.

DEFINITION 12.1 By the antiderivative or indefinite integral of a function $f(x)$ we shall mean another function, $F(x)$, whose derivative is $f(x)$; that is,

$$F'(x) = f(x)$$

The process of obtaining the antiderivative (or indefinite integral) of a function $f(x)$ is called *integration*, and the operation of computing an indefinite integral (or antiderivative) is indicated by the symbol

$$\int (\quad) \, dx$$

where x might very well be y, z, or t, or any other variable in terms of which the process is to be carried out. The function to which the operation is to be applied will appear in each case in parenthesis as shown above. Thus the indefinite integral of $f(x)$ with respect to x is indicated by

$$\int f(x) \, dx$$

and that of $g(t)$ with respect to t is

$$\int g(t) \, dt$$

Clearly, integration may be viewed as the inverse operation of differentiation and vice versa. To demonstrate this relationship let us find the antiderivatives of the functions

$$v = 32t \tag{12.1}$$

and

$$g(x) = x^2 - 2 \tag{12.2}$$

Since the antiderivative of v must be a function whose derivative is $v = 32t$, differentiation "in reverse" yields

$$S = 16t^2 + c \qquad \text{or} \qquad S = \int 32t \, dt = 16t^2 + c$$

for the antiderivative or integral of $v = 32t$, where c is, of course, an arbitrary constant. That $S = 16t^2 + c$ is the required antiderivative of $v = 32t$ can be established by differentiating S to obtain $v = 32t$.

Similarly, the antiderivative of

$$g(x) = x^2 - 2$$

is

$$G(x) = \frac{x^3}{3} - 2x + A \qquad \text{or} \qquad \int (x^2 - 2)\, dx = \frac{x^3}{3} - 2x + A$$

where A is an arbitrary constant. Again, this is so because

$$G'(x) = 3\frac{x^2}{3} - 2 + 0 \qquad \text{or} \qquad G'(x) = x^2 - 2 = g(x)$$

Example 12.1 If

$$r(x) = x^2 + 2x - 1$$

is the marginal revenue[1] and

$$c(x) = 4 + 6x$$

is the marginal cost of a certain firm, find the revenue and the cost functions.

Solution Since the marginal revenue and the marginal cost are the derivatives of the revenue and the cost function, respectively, we are in fact looking for the antiderivatives or integrals of the functions $r(x)$ and $c(x)$. Thus

$$R(x) = \int r(x)\, dx = \int (x^2 + 2x - 1)\, dx = \frac{x^3}{3} + x^2 - x + a$$

and

$$C(x) = \int c(x)\, dx = \int (4 + 6x)\, dx = 4x + 3x^2 + b$$

are the required functions because

$$R'(x) = 3\frac{x^2}{3} + 2x - 1 + 0 = x^2 + 2x - 1 = r(x)$$

and

$$C'(x) = 4 + 6x + 0 = 4 + 6x = c(x)$$

A minor complication arises at the outset of the previous examples and the definition of the antiderivative in that the antiderivative of a given function is not unique; that is, not only x^2, but $x^2 + 5$, $x^2 - 3$, and, in general, $x^2 + c$, where c is any constant, are all functions whose derivative is $2x$. Whether or not all antiderivatives can be obtained in any given case by adding an arbitrary constant c to a particular antiderivative is a question that needs further investigation. The fact is if we know one condition satisfied by the antiderivative of a function, then the antiderivative can be uniquely determined. For example, if we knew that at the beginning or at $t = 0$ the falling body of the introductory example was at $S = 100$, then the substitution of $t = 0$ and $S = 100$ in the antiderivative

$$S = 16t^2 + C \qquad \text{of} \quad v = 32t$$

would yield $100 = 16(0)^2 + C$ or

$$C = 100 \qquad \text{and} \qquad S = 16t^2 + 100$$

Similarly, since $a(t) = 32$ is the acceleration of a falling body, its velocity is given by the antiderivative

$$v(t) = 32t + c \tag{12.3}$$

of $a(t)$. Thus if at $t = 0$ the body were at rest (or $v = 0$), then substituting $t = 0$ and $v = 0$ in Equation (12.3) we would obtain

$$0 = 0 + c$$

or

$$c = 0 \qquad \text{and} \qquad v(t) = 32t$$

Clearly, indefinite integrals and antiderivatives can be found by applying the theorems and rules of differentiation presented in Chapter 10 "in reverse." To demonstrate this "in reverse" application of differentiation, we state in Table 12.1 some rules of differentiation and their counterparts in integration.

Table 12.1

Differentiation	Antidifferentiation or Integration
1. $D(c) = 0$	1. $\int 0\,dx = c$, c = constant
2. $D(x) = 1$	2. $\int 1\,dx = x + c$
3. $D(x^n) = nx^{n-1}$	3. $\int x^n\,dx = x^{n+1}/(n+1) + c$
4. $D[f(x) \pm g(x)]$ $= Df(x) \pm Dg(x)$	4. $\int [f(x) \pm g(x)]\,dx = \int f(x)\,dx \pm \int g(x)\,dx$
5. $D[cf(x)] = c\,Df(x)$	5. $\int cf(x)\,dx = c \int f(x)\,dx$
6. $D(\ln x) = 1/x$	6. $\int 1/x\,dx = \ln x + c$
7. $D(e^x) = e^x$	7. $\int e^x\,dx = e^x + c$
8. $D(a^x) = a^x \ln a$	8. $\int a^x\,dx = a^x/(\ln a) + c$
9. $Du^n = nu^{n-1}\,Du$	9. $\int f^n(x)f'(x)\,dx = f^{n+1}(x)/(n+1) + c$
10. $D \ln u = 1/u \cdot du/dx$	10. $\int f'(x)/f(x)\,dx = \ln f(x) + c$
11. $De^u = e^u\,Du$	11. $\int e^{f(x)}f'(x) = e^{f(x)} + c$

The following special cases provide further demonstration of the above rules:

Since $D(x^3/3 + C) = x^2$, $\int x^2\,dx = x^3/3 + c$.

Since $D(2x + c) = 2$, $\int 2\,dx = 2x + c$.

Since $D(x^3/3 + x^4/4 + c) = x^2 + x^3$, $\int (x^2 + x^3)\,dx =$ $x^3/3 + x^4/4 + c$.

Since $D(-x^{-3}/3 + c) = x^{-4}$, $\int 1/x^4\,dx = -x^{-3}/3 + c$.

Since $D[3 \ln (x + 1) + c] = 3/(x + 1)$, $\int 3/(x + 1)\,dx =$ $3\int 1/(x + 1)\,dx = 3 \ln (x + 1) + c$.

Since $D(e^{2x}/2 + c) = e^{2x}$, $\int e^{2x}\,dx = e^{2x}/2 + c$.

Since $D(x^4 + 3x^2 + 1)^3 = 3(x^4 + 3x^2 + 1)^2(4x^3 + 6x)$, $\int (x^4 + 3x^2 + 1)^2(4x^3 + 6x)\,dx = \frac{1}{3}(x^4 + 3x^2 + 1)^3 + c$.

Since $D[\ln (x^2 + x) + c] = (2x + 1)/(x^2 + x)$, $\int (2x + 1)/(x^2 + x)\,dx = \ln (x^2 + x) + c$.

Exercises

1. Find the antiderivatives or indefinite integrals of the following functions:

a. $f(x) = x^6$.
b. $g(t) = t^{-4}$.
c. $h(x) = e^x(e^x - 2)^2$.
d. $f(x) = x\sqrt{1 + x^2}$.
e. $y = x^4(1 - x)^2$.
f. $g(t) = 2t^3 - 3t^2$.
g. $h(x) = \sqrt{3x}$.
h. $f(t) = \sqrt{3 + 4t}$.
i. $f(x) = (1 + 3x^2)^2$.
j. $g(x) = 2x/(1 + 2x^2)^3$.
k. $y = x^2/(1 + x^3)$.
l. $y = 1/3x$.
m. $y = (x + 1)/x$.
n. $y = e^{2x}/(1 + e^{2x})$.
o. $y = xe^{2x}$.

Ans. (a) $\frac{1}{7}x^7 + c$; (d) $\frac{1}{3}(1 + x^2)^{3/2} + c$; (i) $x + 2x^3 + \frac{9}{5}x^6 + c$

2. Perform the following integrations:

a. $\int 4/x\,dx$.
b. $\int x^{-4}\,dx$.
c. $\int 1/(2 - x)\,dx$.
d. $\int (e^{3x} - 1)\,dx$.
e. $\int (x^4 + 1)/x^2\,dx$.
f. $\int (1 + 2x^2)^2 x\,dx$.
g. $\int (e^{2x} + x^4)\,dx$.
h. $\int 1/\sqrt{3x}\,dx$.
i. $\int 2x/(x^2 - 1)\,dx$.

Ans. (a) $4 \ln x + c$; (e) $x^3/3 - 1/x + c$; (i) $\ln (x^2 - 1) + c$

3. Find y in terms of x in each of the following:

a. $dy/dx = x$.
b. $dy/dx = y^3$.
c. $dy/dx = \sqrt{1 + y}$.
d. $dy/dx = (1 + y)^2$ and $y = 0$ when $x = 0$.
e. $dy/dx = 1/x$.
f. $dy/dx = 1/y$ and $y = 1$ when $x = 2$.

Ans. (a) $y = x^2/2 + c$; (c) $y = -1 + [(x + c)/2]^2$; (e) $y = \ln x + c$

4. Determine the value of each integral in Exercise 2 if the value of the integral is 10 when $x = 1$. For example,

$\int x^2\, dx = x^3/3 + c$

But for $x = 1$ the integral is 10. Thus,

$10 = \frac{1}{3} + c$

or

$$C = \frac{29}{3} \quad \text{and} \quad \int x^2\, dx = \frac{x^3}{3} + \frac{29}{3}$$

Ans. (a) $4 \ln x + 10$; (e) $x^3/3 - 1/x - 1967/3$

5. If the marginal revenue relative to the sales of a certain object is given by

$$r(x) = \frac{dR}{dx} = 3x^2 - 4x + 2$$

find the total revenue R produced by the sale of three of these items.

Ans. 15

6. If the marginal cost of producing a certain item is

$$c(x) = \frac{dC}{dx} = 4 + 2x + \frac{e^{-2x}}{6}$$

find the cost C of producing five items given that there is a fixed cost of \$3.

7. If certain marginal physical productivity is

$$p(x) = \frac{dP}{dx} = 200 - 0.16x$$

then calculate the productivity $P(x)$.

8. If the marginal physical productivity for x used car salesmen is

$$p(x) = \frac{dP}{dx} = 8 - 0.4x$$

where P is in cars per day, how many salesmen should be employed in order to sell 60 cars per day?

Ans. 30

9. If the population of the earth was 3.5 billion in 1970 and is increasing at a rate of 2% ($dP/dt = 0.02P$) per year, when will the population be 50 billion?

Ans. 133 years

10. A radioactive substance disintegrates at a rate proportional to its mass. Find the amount remaining at any time t and the time necessary for the substance to disintegrate to one half of its initial amount.

11. If the rate of change at which atmospheric pressure changes with altitude is proportional to the pressure ($dP/dh = kP$), if the pressure is 14.7[16] feet in.[2] at sea level, and if it has fallen to one half this value at 18,000 feet, find the formula for the pressure at any height.

12.2 The Definite Integral

In developing the indefinite integral and the antiderivative concepts, we encountered a minor complication in the nonuniqueness of a function's antiderivative. However, we recognized that a single condition on the antiderivative determined the integration constant uniquely. Thus to avoid unnecessary calculations of such constants and to provide additional mathematical tools applicable to many branches of the physical, behavioral, and management sciences, we now introduce the concept of the definite integral of a continuous function $f(x)$ over the closed interval $[a, b]$. For this purpose let us begin by considering the area under the curve $y = f(x)$ and between the vertical lines $x = a$ and $x = b$. Let the interval $[a, b]$ be divided into n subintervals (or parts) not necessarily of equal length, by the points $x_1, x_2, x_3, \ldots, x_{n-1}$. Finally, let

Δx_i be the general subinterval $x_i - x_{i-1}$ for each i.
Y_i be the largest value of y over the subinterval Δx_i.
y_i be the smallest value of y over the subinterval Δx_i.
R_i be the rectangle having dimensions Δx_i and Y_i.
r_i be the rectangle having dimensions Δx_i and y_i.
m_i be the x coordinate of an arbtrary point in the subinterval $\Delta x_i = x_i - x_{i-1}$.
$f(m_i)$ be the value of the function at $x = m_i$ (Figure 12.1).

Figure 12-1

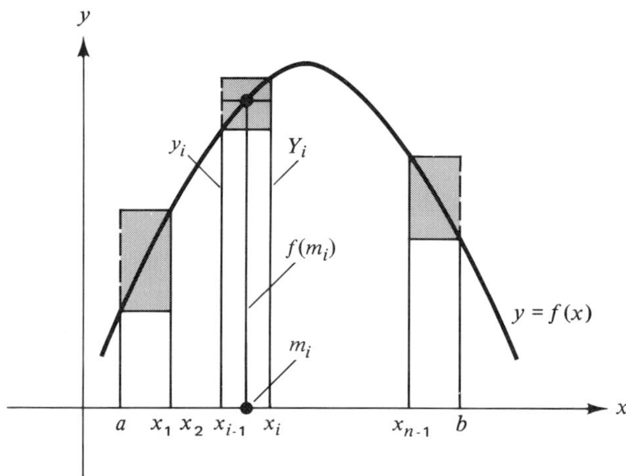

From the definitions given above and elementary geometric considerations, we obtain

$$\sum_{i=1}^{n} r_i \leq \text{area under the curve } y = f(x) \leq \sum_{i=1}^{n} R_i$$

or

$$\sum_{i=1}^{n} r_i \leq \sum_{i=1}^{n} f(m_i)\ \Delta x_i \leq \sum_{i=1}^{n} R_i$$

It can easily be shown that $\sum_{i=1}^{n} r_i$ and $\sum_{i=1}^{n} R_i$ approach the same limit as the number of subdivisions n becomes infinite in such a way that the length Δx_i approaches zero. In other words it can be shown[2] that

$$\lim_{n\to\infty} \sum_{i=1}^{n} f(m_i)\ \Delta x_i = \lim_{n\to\infty} \sum_{i=1}^{n} r_i = \lim_{n\to\infty} \sum_{i=1}^{n} R_i \tag{12.4}$$

and therefore the limit in Equation (12.4) gives the area under the curve of $y = f(x)$ and between the lines $x = a$ and $x = b$.

DEFINITION 12.2 If the function $f(x)$ is continuous over the closed interval $[a, b]$ then by the definite integral of $f(x)$ from a to b we shall mean the $\lim_{n\to\infty} \sum_{i=1}^{n} f(m_i)\ \Delta x_i$ and denote it by

$$\int_a^b f(x)\ dx = \lim_{n\to\infty} \sum_{i=1}^{n} f(m_i)\ \Delta x_i \tag{12.5}$$

The numbers a and b in Equation (12.5) are referred to as the "lower and upper limits" of integration, and the interval $[a, b]$ is called the "range" of integration.

The integral defined in Equation (12.5) represents the area under the curve $y = f(x)$ and between the lines $x = a$ and $x = b$, and is commonly known as the Riemann integral of $f(x)$ after the great mathematician Riemann (1826–1866), who initiated its application on the set of continuous functions. Some of the fundamental properties of the Riemann integral are included in the following theorems. The last of these theorems, known as the "fundamental theorem of calculus" was first formulated by Sir Isaac Newton and Gottfried Leibniz and demonstrates the important and intimate relationship between the derivative and the integral.

Theorem 12.1 If the function $f(x)$ is continuous over the closed interval $[a, b]$, then the following are true:

a. $\int_a^a f(x)\,dx = 0.$

b. $\int_a^b f(x)\,dx = -\int_b^a f(x)\,dx.$

c. $\int_a^b cf(x)\,dx = c\int_a^b f(x)\,dx.$

d. $\int_a^b [f(x) \pm g(x)]\,dx = \int_a^b f(x)\,dx \pm \int_a^b g(x)\,dx.$

e. $\int_a^c f(x)\,dx = \int_a^b f(x)\,dx + \int_b^c f(x)\,dx.$

f. $\int_a^b f(x)\,dx = \int_a^b f(t)\,dt = \int_a^b f(u)\,du = \cdots.$

Theorem 12.2 If $f(x)$ is a continuous function over the range of integration, then

$$\frac{d}{dx}\left(\int_a^x f(x)\,dx\right) = f(x)$$

For instance,

$$\frac{d}{dx}\left(\int_1^x \sqrt{1+t}\,dt\right) = \sqrt{1+x}$$

and

$$\frac{d}{dx}\left(\int_0^x e^{3t}\,dt\right) = e^{3x}$$

Theorem 12.3 *Fundamental Theorem of Calculus.* If $f(x)$ is a continuous function over the closed interval $[a, b]$, and if $F(x)$ is an antiderivative of $f(x)$, then

$$\int_a^b f(x)\,dx = F(b) - F(a)$$

In practice $F(b) - F(a)$ shall be designated by

$$F(x)\Big|_a^b$$

For example

$$\int_2^5 x^2\,dx = x^3/3\Big|_2^5 = 5^3/3 - 2^3/3 = 117/3$$

and

$$\int_1^2 1/x\,dx = \ln x\Big|_1^2 = \ln 2 - \ln 1 = \ln 2 - 0 = \ln 2$$

because $x^3/3$ and $\ln x$ are antiderivatives for x^2 and $1/x$, respectively.

To demonstrate the above theorems further, we consider the following examples.

Example 12.2 By evaluating the integrals $\int_1^4 x^2\,dx$, $\int_1^2 x^2\,dx$, and $\int_2^4 x^2\,dx$, illustrate that

$$\int_a^c f(x)\,dx = \int_a^b f(x)\,dx + \int_b^c f(x)\,dx$$

Illustration Since $a = 1$, $b = 2$, $c = 4$,

$$\int_1^4 x^2\,dx = x^3/3\Big|_1^4 = 4^3/3 - 1^3/3 = 63/3$$

$$\int_1^2 x^2\,dx = x^3/3\Big|_1^2 = 2^3/3 - 1^3/3 = 7/3$$

$$\int_2^4 x^2\,dx = x^3/3\Big|_2^4 = 4^3/3 - 2^3/3 = 56/3$$

and

$$63/3 = 7/3 + 56/3$$

we have

$$\int_1^4 x^2\,dx = \int_1^2 x^2\,dx + \int_2^4 x^2\,dx \qquad \text{or} \qquad 63/3 = 7/3 + 56/3$$

Example 12.3 By evaluating $\int_1^2 (x + 1)\,dx$ and $\int_2^1 (x + 1)\,dx$, illustrate that

$$\int_a^b f(x)\,dx = -\int_b^a f(x)\,dx$$

Illustration Since $a = 1$, $b = 2$, and

$$\int_1^2 (x + 1)\,dx = \frac{x^2}{2} + x\Big|_1^2 = \left(\frac{4}{2} + 2\right) - \left(\frac{1}{2} + 1\right) = \frac{5}{2}$$

$$\int_2^1 (x + 1)\,dx = \frac{x^2}{2} + x\Big|_2^1 \quad \left(\frac{1}{2} + 1\right) - \left(\frac{4}{2} + 2\right) = -\frac{5}{2}$$

we have

$$\int_1^2 (x + 1)\,dx = -\int_2^1 (x + 1)\,dx$$

Example 12.4 It is well known that the amount of capital stock $C(t)$ is related to the ratio of net investment $I(t)$ by

$$\frac{dC(t)}{dt} = I(t)$$

Thus if the rate of net investment at any time t is

$$I(t) = 3t^{1/2} + 2$$

then the capital stock formation in the nth year is given by

$$\begin{aligned}\int_n^{n+1} (3t^{1/2} + 2)\,dt &= (2t^{3/2} + 2t)\Big|_n^{n+1} \\ &= 2[(n + 1)^{3/2} + (n + 1)] - 2n^{3/2} - 2n \\ &= 2[(n + 1)^{3/2} - n^{3/2}] + 2\end{aligned}$$

Example 12.5 It is well known that the speed of an automobile is the rate of change of the distance traveled; that is,

$$v(t) = \frac{dS}{dt} \qquad \text{and} \qquad S = \int v(t)\,dt$$

Therefore if $v(t)$ is the velocity of a car at any time, then the distance covered between times $t = 4$ and $t = 10$ is

$$S = \int_4^{10} v(t)\,dt$$

Specifically, if $v(t) = t + 2$, then the distance covered between times $t = 4$ and $t = 10$ is

$$S = \int_4^{10} (t+2)\,dt = \left(\frac{t^2}{2} + 2t\right)\Bigg|_4^{10} = \left(\frac{100}{2} + 20\right) - \left(\frac{16}{2} + 8\right)$$

$$= 54$$

Example 12.6 If $y(x)$ is the proportion of a population scoring less than x on a given test and if

$$y(x) = \int_0^x 2te^{-t^2}\,dt$$

then find the proportion scoring between 10 and 20.

Solution Since

$$y(20) = \int_0^{20} 2te^{-t^2}\,dt$$

$$y(10) = \int_0^{10} 2te^{-t^2}\,dt = -\int_{10}^{0} 2te^{-t^2}\,dt$$

and since the proportion of the population scoring between 10 and 20 is the proportion of those below 20 minus that of those below 10, we obtain

$$y(20) - y(10) = \int_0^{20} 2te^{-t^2}\,dt + \int_{10}^{0} 2te^{-t^2}\,dt = \int_{01}^{20} 2te^{-t^2}\,dt$$

$$= -\frac{e^{-t^2}}{2}\Bigg|_{10}^{20} = -\frac{e^{-40}}{2} + \frac{e^{-20}}{2}$$

for the proportion of the population scoring between 10 and 20.

Example 12.7 Using the definition of the definite integral, as the area under the curve, evaluate the integral

$$\int_0^1 (x+1)\,dx$$

Solution Let us divide the interval [0, 1] into n equal subintervals of length

$$\Delta x_i = \frac{1-0}{n} = \frac{1}{n}$$

by the points $x_1 = 1/n$, $x_2 = 2/n$, $x_3 = 3/n, \ldots, x_{n-1} = (n-1)/n$ (Figure 12.2). Then

$$\int_0^1 (x+1)\,dx = \lim_{n\to\infty} \sum_{i=1}^{n} f(x_i)\,\Delta x_i = \lim_{n\to\infty} \sum_{i=1}^{n} f\left(\frac{i}{n}\right)\frac{1}{n}$$

or

$$\int_0^1 (x+1)\,dx = \lim_{n\to\infty} \sum_{i=1}^{n} \left(\frac{i}{n}+1\right)\frac{1}{n} = \lim_{n\to\infty} \sum_{i=1}^{n} \left(\frac{i}{n^2}+\frac{1}{n}\right)$$

$$= \lim_{n\to\infty} \left[\left(\frac{1}{n^2}+\frac{1}{n}\right)+\left(\frac{2}{n^2}+\frac{1}{n}\right)+\cdots+\left(\frac{n}{n^2}+\frac{1}{n}\right)\right]$$

$$= \lim_{n\to\infty} \left[\frac{1+2+3+\cdots+n}{n^2}+n\left(\frac{1}{n}\right)\right]$$

$$= \lim_{n\to\infty} \left(\frac{(n+1)/2\cdot n}{n^2}+1\right) = \lim_{n\to\infty} \left(\frac{n+1}{2n}+1\right)$$

Therefore

$$\int_0^1 (x+1)\,dx = \lim_{n\to\infty} \left(\frac{1}{2}+\frac{1}{2n}+1\right) = \frac{1}{2}+0+1 = \frac{3}{2}$$

Geometrically, Figure 12.2 shows that the required area is the sum of the area of the triangle CDE and that of the square $ABCD$;

Figure 12-2

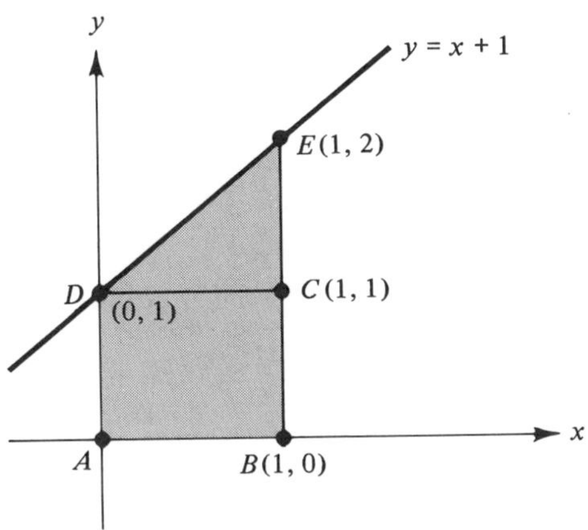

that is, the area under the curve $y = x + 1$ is

$$E = 1 + \frac{1}{2} = \frac{3}{2}$$

Exercises

1. Find the area under each of the following curves and between the indicated lines. Draw appropriate graphs.

   a. $y = x^2$, $x = 1$ and $x = 3$.
   b. $y = e^x/4$, $x = 0$ and $x = 4$.
   c. $y = 2/(x + 1)$, $x = 3$ and $x = 6$.
   d. $y = 2x + 4$, $x = 0$ and $x = 4$.

   Ans. (a) 26/3; (c) ln (49/16)

2. With respect to x, differentiate the following functions:

   a. $f(x) = \int_{10}^{x} \sqrt{1 + t^2}\, dt.$ b. $g(x) = \int_{0}^{x} 2t/(1 + t^2)\, dt.$

   c. $h(x) = \int_{x}^{4} (e^t + t)\, dt.$ d. $F(x) = \int_{x}^{0} (2t + \ln t)\, dt.$

   Ans. (a) $\sqrt{1 + x^2}$; (c) $-(e^x + x)$

3. For each of the following rate of net investment functions, find the capital stock formation in the indicated year:

   a. $I(t) = 3t^{1/2} + 2$, fourth year.
   b. $I(t) = 2t + 3$, fifth year.
   c. $I(t) = 2t^{1/2} + 3$, nth year.
   d. $I(t) = 3t - 1$, fourth year.

   Ans. (a) $10\sqrt{5} - 6$; (d) 12.5

4. If the velocity of an automobile is given by each of the following functions, find the distance traveled between the indicated times:

   a. $v(t) = t^2 + 3$, $t = 2$ and $t = 5$.
   b. $v(t) = t + t^2$, $t = 3$ and $t = 6$.
   c. $v = 3t - 1$, $t = 0$ and $t = 5$.
   d. $v = e^{-t} + t$, $t = 0$ and $t = 3$.

   Ans. (a) 48; (c) 32.5

5. If $y(x)$ is the proportion of a population scoring less than x on a

given test and if

$$y = \int_0^x 2te^{-t^2}\, dt$$

find the proportion scoring between

a. 10 and 30. b. 15 and 20. c. 10 and 15.

12.3 Special Techniques of Integration

In the last two sections we developed a number of important integration formulas by applying known rules of differentiation in reverse. In this section we shall develop several additional methods of integration of certain types of functions that often arise in applications of the integral concept. Specifically, we shall discuss "integration by parts," and "integration by partial fractions."

INTEGRATION BY PARTS

Integration by parts is merely a restatement of the formula for differentiating a product. Since

$$D[f(x)g(x)] = f'(x)g(x) + f(x)g'(x) \tag{12.6}$$

integrating both sides of Equation (12.6) yields

$$\int f'(x)g(x)\, dx + \int f(x)g'(x)\, dx = f(x)g(x)$$

or

$$\int f(x)g'(x)\, dx = f(x)g(x) - \int f'(x)g(x)\, dx \tag{12.7}$$

Thus

$$\int_a^b f(x)g'(x)\, dx = f(x)g(x)\Big|_a^b - \int_a^b f'(x)g(x)\, dx \tag{12.8}$$

The latter two formulas constitute what are commonly known as integration by parts formulas. The effectiveness of this method clearly depends on whether or not the integral on the right of Equation (12.7) is easier to find than the original.

Example 12.8 Evaluate the integral $\int_0^1 xe^x\, dx$.

Solution Since

$$\int xe^x\,dx = \int x\,d(e^x) = xe^x - \int e^x\,dx$$

we have

$$\int_0^1 xe^x\,dx = xe^x\Big|_0^1 - \int_0^1 e^x\,dx = 1e^1 - 0 - e^x\Big|_0^1$$

or

$$\int_0^1 xe^x\,dx = e - e + 1 = 1$$

Clearly, $f(x) = x$ and $g'(x) = e^x$ in Equation (12.7) led to the evaluation of the given integral. Had we written the given integral as

$$\int xe^x\,dx = \int e^x\,d\left(\frac{x^2}{2}\right) = \frac{x^2e^x}{2} - \int \frac{x^2}{2}\,d(e^x)$$

or

$$\int xe^x\,dx = \frac{x^2e^x}{2} - \int \frac{x^2}{2}\,e^x\,dx$$

which is permissible, the integral on the right would have become more complicated than the original, and therefore the substitution of of $f(x) = e^x$ and $g'(x) = x$ in Equation (12.7) would have been unwise.

Example 12.9 Evaluate the integral $\int_1^2 x \ln x\,dx$.

Solution Since

$$\int x \ln x\,dx = \int \ln x\,d\left(\frac{x^2}{2}\right) = \frac{x^2 \ln x}{2} - \int \frac{x^2}{2}\cdot d(\ln x)$$

we have

$$\int x \ln x\,dx = \frac{x^2 \ln x}{2} - \int \frac{x^2}{2}\cdot\frac{1}{x}\,dx$$

or

$$\int x \ln x\,dx = \frac{x^2 \ln x}{2} - \frac{x^2}{4}$$

and

$$\int_1^2 x \ln x \, dx = \frac{x^2 \ln x}{2}\bigg|_1^2 - \frac{x^2}{4}\bigg|_1^2 = 2 \ln 2 - \frac{5}{4}$$

INTEGRATION BY PARTIAL FRACTIONS

It can easily be verified that

$$\frac{x^3}{x^2 + 3x + 2} = x - 3 - \frac{1}{x+1} + \frac{8}{x+2} \tag{12.9}$$

and therefore

$$\int \frac{x^3}{x^2 + 3x + 2} \, dx = \int \left(x - 3 - \frac{1}{x+1} + \frac{8}{x+2} \right) dx$$

or

$$\int \frac{x^3}{x^2 + 3x + 2} \, dx = \frac{x^2}{2} - 3x - \ln (x + 1) + 8 \ln (x + 2) + c \tag{12.10}$$

Obviously Equation (12.9) facilitated the evaluation of the integral of the function

$$f(x) = \frac{x^3}{x^2 + 3x + 2} \tag{12.11}$$

The general object of the method of partial fractions is to systematize the procedure leading from the quotient of two polynomials to the simple fractions of which it is composed. As the above example demonstrates, when a quotient is decomposed into simple fractions, then its integral can readily be obtained.

Before algebraic techniques can be applied to decompose a given fraction $f(x)/g(x)$ we must always remember that the degree of $f(x)$ in $f(x)/g(x)$ ought to be less than that of $g(x)$. If this is not the case, then the denominator $g(x)$ can be divided into the numerator $f(x)$ until a remainder of degree less than the degree of $g(x)$ is obtained. Such a division yields

$$\frac{f(x)}{g(x)} = q(x) + \frac{r(x)}{g(x)} \tag{12.12}$$

where $q(x)$ is the quotient and $r(x)$ is the remainder whose degree is less than that of $g(x)$.

Once Equation (12.12) is obtained, the next step is to factor $g(x)$ into its ultimate factors. These factors must be one or the other of the following four types.

a. Linear and nonrepeated: $ax + b$.
b. Linear and repeated: $(ax + b)^n$, $n = 2, 3, \ldots$.
c. Quadratic and nonrepeated: $ax^2 + bx + c$.
d. Quadratic and repeated: $(ax^2 + bx + c)^n$, $n = 1, 2, \ldots$.

After the factors of $g(x)$ have been found, it is possible to write $f(x)/g(x)$ as the sum of simple fractions.[2] This we do according to the following rules.

Rule 1. For each factor of the form $(ax + b)$, there must be a single fraction of the form $A/(ax + b)$, where A is a constant to be determined.

Rule 2. For each factor of the form $(ax + b)^n$, there must be a sum of fractions of the form

$$\frac{A_1}{ax + b} + \frac{A_2}{(ax + b)^2} + \cdots + \frac{A_{n-1}}{(ax + b)^{n-1}} + \frac{A_n}{(ax + b)^n}$$

where $A_1, A_2, \ldots, A_n$ are constants to be determined.

Rule 3. For each "irreducible" factor of the form $ax^2 + bx + c$, there must be a single fraction of the form $(Ax + B)/(ax^2 + bx + c)$ where A and B are constants to be determined.

Rule 4. For each repeated "irreducible" factor $(ax^2 + bx + c)^n$, there must be a sum of fractions of the form

$$\frac{A_1x + B_1}{ax^2 + bx + c} + \frac{A_2x + B_2}{(ax^2 + bx + c)^2} + \cdots + \frac{A_nx + B_n}{(ax^2 + bx + c)^n}$$

where $A_1, B_1, A_2, B_2, \ldots, A_n, B_n$ are constants to be determined.

After the given fraction has been expressed as a sum of fractions specified by the above rules, these fractions are then reduced to a least common denominator and added. Thus the common denominator will be the denominator of the original fraction. The numerator of the sum will involve all the undetermined constants. These constants can in turn be calculated by making the two numerators identical either by equating coefficients of like powers of the unknowns in the two numerators or by substituting convenient numerical values for the variable of the numerators, and then solving the resulting equations for the coefficients $A, B, \ldots$. When all the coefficients are determined, the integration of each component fraction can be carried out as the following examples demonstrate.

Example 12.10 Perform the integration

$$\int \frac{x^3}{x^2 + 3x + 2}\,dx$$

Solution Since

$$\frac{x^3}{x^2 + 3x + 2} = x - 3 + \frac{7x + 6}{x^2 + 3x + 2} \qquad \text{(by division)}$$

and since

$$\frac{7x + 6}{x^2 + 3x + 2} = \frac{7x + 6}{(x + 1)(x + 2)} = \frac{A}{x + 1} + \frac{B}{x + 2} \qquad \text{(Rule 1)}$$

or

$$\frac{7x + 6}{(x + 1)(x + 2)} = \frac{A(x + 2) + B(x + 1)}{(x + 1)(x + 2)}$$

we have

$$7x + 6 = A(x + 2) + B(x + 1) = (A + B)x + (2A + B)$$

or

$$A + B = 7 \qquad \text{and} \qquad 2A + B = 6 \qquad \text{(Why?)}$$

Now, solving the last two equations for A and B we obtain $A = -1$ and $B = 8$. Therefore

$$\frac{x^3}{x^2 + 3x + 2} = x - 3 + \frac{7x + 6}{x^2 + 3x + 2}$$

$$= x - 3 - \frac{1}{x + 1} + \frac{8}{x + 2}$$

and

$$\int \frac{x^3}{x^2 + 3x + 2}\,dx = \int \left(x - 3 - \frac{1}{x + 1} + \frac{8}{x + 2}\right) dx$$

$$= \frac{x^2}{2} - 3x - \ln(x + 1) + 8\ln(x + 2) + c$$

Example 12.11 Integrate

$$\int \frac{5x + 3}{x^3 - 2x^2 - 3x}\,dx$$

Solution Since

$$\frac{5x+3}{x^3-2x^2-3x} = \frac{5x+3}{x(x+1)(x-3)}$$

$$= \frac{A}{x} + \frac{B}{x+1} + \frac{C}{x-3} \qquad \text{(Rule 1)}$$

and since

$$\frac{5x+3}{x(x+1)(x-3)}$$

$$= \frac{A(x+1)(x-3) + Bx(x-3) + Cx(x+1)}{x(x+1)(x-3)}$$

we have

$$5x+3 = A(x+1)(x-3) + Bx(x-3) + Cx(x+1)$$

for all x

Therefore the coefficients of like powers of x must be equal, and this implies

$$A + B + C = 0$$

$$-2A - 3B + C = 5$$

$$-3A = 3$$

Hence $A = -1$, $B = -1/2$, $C = 3/2$, and

$$\frac{5x+3}{x^3-2x^2-3x} = -\frac{1}{x} - \frac{1}{2}\frac{1}{x+1} + \frac{3}{2}\frac{1}{x-3}$$

or

$$\int \frac{5x+3}{x^3-2x^2-3x}\,dx = \int -\frac{1}{x}\,dx - \frac{1}{2}\int \frac{1}{x+1}\,dx + \frac{3}{2}\int \frac{1}{x-3}\,dx$$

$$= -\ln x - \frac{1}{2}\ln(x+1) + \frac{3}{2}\ln(x-3) + c$$

Example 12.12 Integrate

$$\int \frac{1}{x(x-1)^2}\,dx$$

Solution Since

$$\frac{1}{x(x-1)^2} = \frac{A}{x} + \frac{B}{x-1} + \frac{C}{(x-1)^2}$$

or

$$\frac{1}{x(x-1)^2} = \frac{A(x-1)^2 + Bx(x-1) + Cx}{x(x-1)^2}$$

we have

$$1 = A(x-1)^2 + Bx(x-1) + Cx$$

or

$$1 = (A+B)x^2 + (2A - B + C)x + A$$

Thus equating coefficients of equal powers, we obtain

$$\begin{aligned} A &= 1 \\ 2A - B + C &= 0 \\ A + B &= 0 \end{aligned}$$

Consequently, $A = 1$, $B = -1$, $C = -3$, and

$$\frac{1}{x(x-1)^2} = \frac{1}{x} - \frac{1}{x-1} - \frac{3}{(x-1)^2}$$

or

$$\begin{aligned} \int \frac{1}{x(x-1)^2}\,dx &= \int \frac{1}{x}\,dx - \int \frac{1}{x-1}\,dx - 3\int \frac{1}{(x-1)^2}\,dx \\ &= \ln x - \ln(x-1) + 3(x-1)^{-1} + c \end{aligned}$$

Exercises

Perform the following integrations:

1. $\int \ln x\,dx.$
2. $\int x^2 e^{-x}\,dx.$
3. $\int x^2 \ln x\,dx.$
4. $\int x^3 \ln x\,dx.$
5. $\int x^3 e^{-x^2}\,dx.$
6. $\int x^2 e^{-2x}\,dx.$
7. $\displaystyle\int \frac{2}{x(x+1)}\,dx.$
8. $\displaystyle\int \frac{3}{x^2(x-1)}\,dx.$
9. $\displaystyle\int \frac{2x+1}{x^3 + x^2 - 2x}\,dx.$
10. $\displaystyle\int \frac{x^2 - 3x + 2}{(x+1)^4}\,dx.$

11. $\int \frac{2x^2 + x - 3}{x^2 - 2x}\,dx.$ 12. $\int \frac{x^2 - 7}{(x + 3)^3}\,dx.$

Ans. (1) $x \ln x - x + c$; (3) $x^3/3 \ln x - x^3/9 + c$;

(9) $-\frac{1}{2} \ln \left[\frac{x(x + 2)}{(x - 1)^2}\right] + c$; (11) $2x + 3/2 \ln x +$

$7/2 \ln (x - 2) + G$

12.4 Applications of Integration

Suitable interpretations of the area under a curve concept provide many and varied applications of the theory of integration in the physical as well as in the behavioral and management sciences. For example, in probability theory and statistics the area under a curve may be viewed as the probability of the outcome of an experiment.

Since the probability of a specific outcome of an experiment is the fraction of time in which the outcome is expected to occur (or take place), whenever the experiment is repeated a very large number of times, the probability of a head is $\frac{1}{2}$ when a well-balanced coin is tossed, and $\frac{1}{6}$ is the probability that a 7 comes up in rolling a pair of dice. Further analysis of these two experiments results in Tables 12.2 and 12.3, each showing the set of mutually exclusive outcomes together with the corresponding distribution of probabilities. These probabilities were obtained on the basis of the probability principles presented in Chapter 4.

Figure 12.3 represents a probability diagram of the experiment of rolling two dice. In the figure each outcome is indicated by its total

Table 12.2

Coin Experiment	
Outcome	Probability
H	1/2
T	1/2

Table 12.3

Dice Experiment	
Outcome	Probability
2	1/36
3	2/36 = 1/18
4	3/36 = 1/12
5	4/36 = 1/9
6	5/36
7	6/36 = 1/6
8	5/36
9	4/36 = 1/9
10	3/36 = 1/12
11	2/36 = 1/18
12	1/36

number of spots represented as a number on the x axis. A series of equal-width columns are drawn with height adjusted so that the area of each column represents the same fraction of the total area of all of the columns as the probability of the outcome.

Of course, the approach applied in the experiment of rolling two dice will always provide a probability diagram as long as the number of possible outcomes of the experiment is finite and relatively small. However, when the number of outcomes becomes large, it is usually simpler to replace discrete columns with a continuous function that yields corresponding areas when integrated between the bounding values of a given column. For instance, if $y = P(x)$ is the curve shown in Figure 12.3,

$$\int_{1.5}^{2.5} P(x)\,dx = \frac{1}{36}$$

so that the area corresponding to the outcome 2 will be correct. Similarly,

$$\int_{2.5}^{3.5} P(x)\,dx = \frac{2}{36} = \frac{1}{18}$$

$$\int_{3.5}^{4.5} P(x)\,dx = \frac{3}{36} = \frac{1}{12}$$

and so on.

Figure 12-3

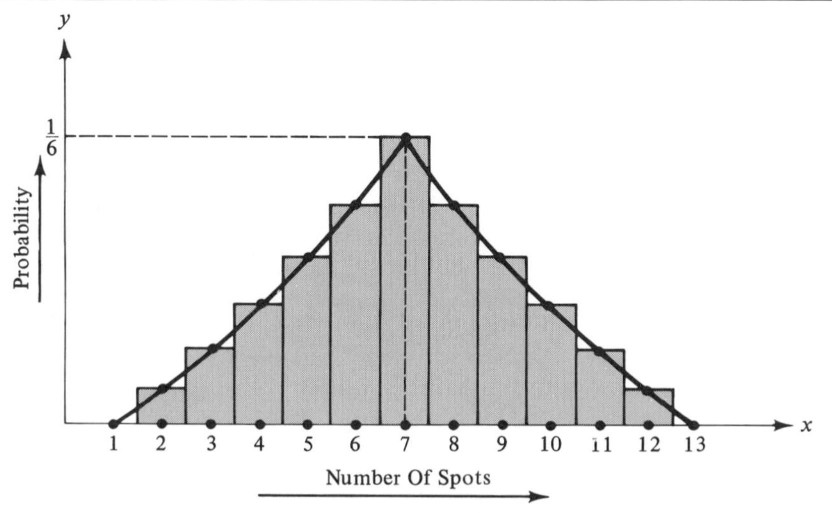

In general, the function $P(x)$ just described is called the *probability density function*[3] of the experiment. In other words a probability density function is a continuous function that can be used to replace the discrete description of probability. Integration of the related probability function of an experiment yields the required probabilities.

Although it is true that the actual nature of density functions is complex, their basic properties are relatively simple. Since we shall be studying experiments whose outcomes correspond to some set of real numbers x such that $a \leq x \leq b$, since $P(x) \geq 0$ for all $x \epsilon [a, b]$, and since

$$\int_a^b P(x)\,dx = 1 \qquad \text{(Why?)}$$

that is, the total area bounded by the probability density function and the x axis must be one, almost any nonnegative real valued function $f(x)$ whose integral over $[a, b]$ is finite might represent a probability density function. The factor

$$c = \frac{1}{\int_a^b f(x)\,dx}$$

can be used to make the area involved equal to one square unit; that is, if

$$c = \frac{1}{\int_a^b f(x)\,dx},$$

and

$$P(x) = cf(x)$$

then $P(x)$ will be a probability density function, because

$$\int_a^b P(x)\,dx = \int_a^b cf(x)\,dx = c\int_a^b f(x)\,dx$$

$$= \frac{1}{\int_a^b f(x)\,dx}\int_a^b f(x)\,dx = 1$$

For example, the function $f(x) = e^x$ on $0 \le x \le 1$ yields

$$\int_0^1 e^x\, dx = e^x \Big|_0^1 = e - 1 < 1$$

Thus $y = e^x$ does not meet the requirements of a probability density function. However, the function

$$P(x) = \frac{e^x}{e-1}$$

is nonnegative on $[0, 1]$, and

$$\int_0^1 P(x)\, dx = \int_0^1 \frac{e^x}{e-1}\, dx = \frac{e^x}{e-1}\Big|_0^1 = \frac{e}{e-1} - \frac{1}{e-1} = 1$$

Thus the function

$$P(x) = \frac{e^x}{e-1}$$

is a probability density function, and the related area is shown in Figure 12.4.

Example 12.13 If the outcomes of a certain experiment correspond to the positive real numbers and if the related probability density function is

$$P(x) = \frac{1}{(x+1)^2}$$

Figure 12-4

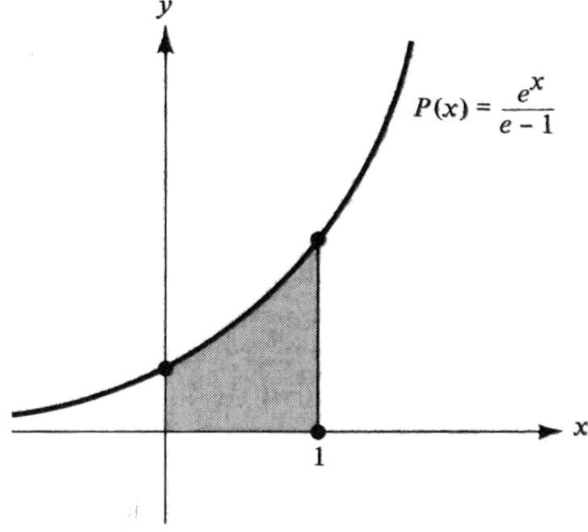

where $0 \leq x \leq \infty$, find the probability of an outcome greater than or equal to three.

Solution Since $P(x) > 0$ for all x, and since

$$\int_0^\infty \frac{1}{(x+1)^2}\,dx = -(x+1)^{-1}\Big|_0^\infty = \lim_{k\to\infty} -(x+1)^{-1}\Big|_0^k$$

$$= \lim_{k\to\infty}\left(-\frac{1}{k+1}+\frac{1}{1}\right) = 0 + 1 = 1$$

the function $P(x)$ is a probability density function. Thus the probability of an experimental outcome greater than or equal to three is given by

$$P(x \geq 3) = \int_3^\infty \frac{1}{(x+1)^2}\,dx = -(x+1)^{-1}\Big|_3^\infty$$

$$= \lim_{k\to\infty} -(x+1)^{-1}\Big|_3^k$$

$$= \lim_{k\to\infty}\left(-\frac{1}{k+1}+\frac{1}{3+1}\right) = 0 + \frac{1}{4}$$

or

$$P(x \geq 3) = \tfrac{1}{4}$$

that is, for this experiment one quarter of the time an outcome greater than or equal to three should be expected. The corresponding area is shown in Figure 12.5.

(Note: If one or both limits of integration are infinity, then we may replace them with numbers k and l, apply the fundamental theorem of calculus, and then take the limit as k and l approach infinity. The value of the limit, when it exists, is the value of the integral. Otherwise, the integral does not exist.)

Example 12.14 The probability density function

$$P(x) = \frac{1}{s\sqrt{2\pi}}\, e^{-1/2}\left(\frac{x-m}{s}\right)^2$$

where $-\infty < x < \infty$, and where m and s are the mean and standard deviation, respectively, has the widest application in problems of "normal," "Gaussian," or "bell shape," distribution. Thus if the results of an experiment are normally distributed with $m = 3$ and $s = 2$, find the probability of an outcome between three and five.

Figure 12-5

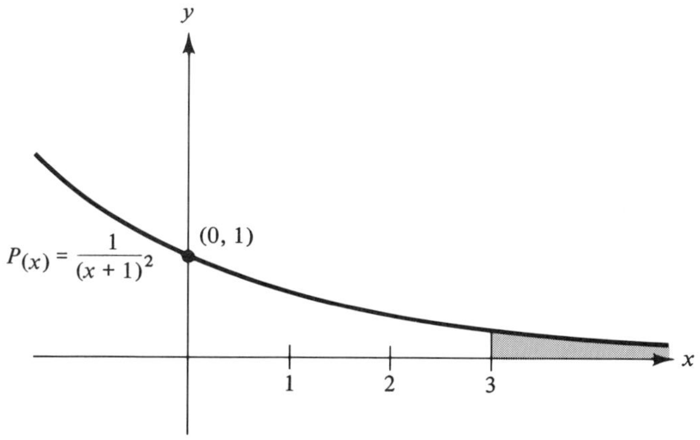

Solution Since the distribution is normal, the probability density function must be

$$P(x) = \frac{1}{2\sqrt{2\pi}} e^{-1/2}\left(\frac{x-3}{2}\right)^2$$

and the probability of an experimental outcome between three and five will be given by

$$P(3 \leq x \leq 5) = \int_3^5 P(x)\,dx = \int_3^5 \frac{1}{2\sqrt{2\pi}} e^{-1/2}\left(\frac{x-3}{2}\right)^2 dx$$

or

$$P(3 \leq x \leq 5) = \frac{1}{2\sqrt{2\pi}} \int_3^5 e^{-1/2}\left(\frac{x-3}{2}\right) dx \tag{12.13}$$

Now letting $t = (x - 3)/2$, we obtain $x = 2t + 3$ and $dx = 2dt$. Moreover $t = 0$ when $x = 3$ and $t = 1$ when $x = 5$. Thus the required probability of Equation (12.13) becomes

$$P = \frac{1}{2\sqrt{2\pi}} \int_0^1 \exp\left(-\frac{1}{2}\right)^{t2} 2\,dt$$

or

$$P = \frac{1}{2\sqrt{2\pi}} \int_0^1 \exp\,(-t^2/2)\,dt \tag{12.14}$$

Now the application of the normal curve area tables of integration on

Equation (12.14) yields

$$P = 0.3413$$

for the required probability

Example 12.15 In economic models if the function $f(t)$ specifies the flow of revenue as a function of time, the integral

$$R = \int_0^t f(t)\,dt$$

determines the total revenue. Hence suppose that the function

$$f(t) = 4000(1 + t)^{-3/2}$$

where t is given in years determines the revenue flow. Then the integral

$$R = \int_0^8 4000(1 + t)^{-3/2}\,dt = 4000(-2)(1 + t)^{-1/2}\Big|_0^8$$

$$= -8000[(1 + 8)^{-1/2} - (1 + 0)^{-1/2}]$$

or

$$R = -8000\left(\frac{1}{3} - 1\right) = \frac{16{,}000}{3}$$

provides the total revenue expected over the next eight years.

Example 12.16 If the marginal revenue relative to the sale of a certain commodity is

$$\frac{dR}{dx} = 5x^2 - 4x + 3$$

then the total revenue produced by the sale of eight of these items is given by the integral

$$R = \int (5x^2 - 4x + 3)\,dx = \frac{5}{3}x^3 - 2x^2 + 3x + c \qquad \textbf{(12.15)}$$

However, if zero items were sold (or $x = 0$), then the revenue $R = 0$, and Equation (12.15) yields

$$0 = 0 + c \qquad \text{or} \qquad c = 0$$

Thus the total revenue is given by

$$R = \frac{5}{3}x^3 - 2x^2 + 3x$$

where x is the number of units sold, and the sale of eight units would produce

$$R = \frac{5}{3}(8)^3 - 2(8)^2 + 3(8) = \frac{2560}{3} - 128 + 24 = \frac{2248}{3}$$

of total revenue.

Example 12.17 In the study of the economic concepts of supply and demand, and the related problems of producer or consumer surplus, if $y = f(x)$ is the demand function, then the consumer's surplus is given by

$$S = \int_0^a f(x)\,dx - af(a)$$

where a is the number of units sold at a price $f(a)$. Thus if $f(x) = 5 - x^2$ is the demand function and there are two units actually sold, then the consumer's surplus is

$$S = \int_0^2 (5 - x^2)\,dx - 2f(2)$$

$$S = 5x - \frac{x^3}{3}\Big|_0^2 - 2(5 - 2^2) = 5(2) - \frac{2^3}{3} - 0 - 2(5 - 4)$$

or

$$S = \frac{16}{3}$$

Exercises

1. Determine the area under each of the following curves and between the indicated lines, and draw appropriate graphs:

   a. $f(x) = (4 - x)^3$, $x = 0$ and $x = 4$.
   b. $f(x) = e^{3x}$, $x = 0$ and $x = 1$.
   c. $f(x) = 2x/(1 + x^2)$, $x = 1$ and $x = 2$.
   d. $f(x) = 2x^3 - 3x^2$, $x = 1$ and $x = 3$.
   e. $f(x) = (x - 3)(x - 2)(x + 1)$, $x = 0$ and $x = 4$.

   Ans. (a) 64; (c) ln (5/2); (e) 7 1/6

2. In each of the following, find the area bounded by the given functions. Draw appropriate graphs.

   a. $f(x) = x^2 + 2$; $y = x$, $x = 1$, and $x = 2$.

b. $f(x) = x^2$ and $g(x) = \sqrt{x}$.
c. $f(x) = x^2/4$ and $g(x) = 3x$.

Ans. (a) 17/6; (c) 72

3. If an experiment has outcomes corresponding to the positive real numbers and if its probability density function is

$$p(x) = \frac{1}{(1 + x)^2}$$

then determine the following probabilities:

a. An outcome x such that $1 \leq x \leq 5$.
b. An outcome x such that $x \geq 3$.
c. An outcome x such that $x \leq 2$.
d. An outcome x such that $0 \leq x \leq 1$.

Ans. (a) $\frac{1}{2}$; (c) $\frac{1}{3}$

4. Evaluate the following "improper" integrals.

a. $\int_1^\infty 1/x^2\,dx$. b. $\int_0^\infty e^{-x}\,dx$. c. $\int_1^\infty dx/\sqrt{1 + x}$.

Ans. (a) 1; (c) not defined

5. For the probability function

$$p(x) = \frac{1}{(1 + x)^2}, \qquad x \geq 0$$

determine the values for k such that there is (a) 90%, (b) 80%, and (c) 70% probability that $0 \leq x \leq k$.

Ans. (a) $k = 9$

6. If an experimental result is known to be normally distributed with mean 3 and standard deviation 2, find the values of k such that (a) 90%, (b) 80%, and (c) 70% of the time results between $3 - k$ and $3 + k$ can be expected.

7. If the revenue in dollars from a certain source is

$$f(t) = 2000(1 + t)^{-3/2}$$

where t is given in years, then find the total revenue over the next 15 years.

Ans. 3000

8. For each of the following revenue functions and the indicated numbers of years, find the expected total revenue:

a. $f(t) = 3000(1 + t)^{-2/3}$ for 7 years.

b. $f(t) = 2000(1 + t^2)^{1/2}t$ for 5 years.
c. $f(t) = 250/(1 + t)$ for 3.5 years.
d. $f(t) = 4000(1 + t)^{-3/5}$ for 31 years.

Ans. (a) 9000; (c) 250 ln (4.5)

9. For each of the following marginal revenues find the total revenue produced by the sales of the indicated number of items.

a. $dR/dx = 3x^2 - 6x + 7$, 4 items.
b. $dR/dx = 2x\sqrt{2 + x^2}$, 6 items.
c. $dR/dx = 3x\sqrt{4 + x^2} + 4$, 2 items.
d. $dR/dx = e^{-x}/3 + 9x$, 5 items.

Ans. (a) 44; (c) $8\sqrt{8}$

10. For each of the following demand functions and the indicated number of items find the consumer's surplus:

a. $f(x) = 2 + x^3$, 2 items.
b. $f(x) = 3 + \ln(1 + x)$, 4 items.
c. $f(x) = 3x^2 - 6x + 9$, 3 items.
d. $f(x) = 1 + x^3$, 4 items.

Ans. (a) 52; (c) 0

References for Supplementary Readings

1. J. E. Howell and D. Teichroew, *Mathematical Analysis for Business Decisions*, Chapters 6, 10, 11, and 12 (Homewood, Ill.: Irwin, 1971).
2. E. J. Purcell, *Calculus with Analytic Geometry*, Chapters 3, 5, and 13 (New York: Appleton-Century-Crofts, 1960).
3. H. D. Brunk, *Introduction to Mathematical Statistics*, Chapters 3, 5, and 13 (Boston, Mass.: Ginn, 1960).

Index